Introduction to

LINEAR ALGEBRA

Second edition

Introduction to LINEAR ALGEBRA

Second edition

Thomas A. Whitelaw, B.Sc., Ph.D.
Senior Lecturer in Mathematics
University of Glasgow

CHAPMAN & HALL/CRC

Boca Raton London New York Washington, D.C.

Library of Congress Cataloging-in-Publication Data

Whitelaw, Thomas A.
Introduction to linear algebra / Thomas A. Whitelaw -- 2nd ed.
p. cm.
Includes index.
ISBN 0-7514-0159-5
1 Algebras, Linear.
QA184.W485 1999
512′ 5—dc21 99-18118
CIP

First edition 1991
Reprinted 1994
First CRC Press reprint 1999
Originally published by Chapman & Hall

International Standard Book Number 0-7514-0159-5
Library of Congress Card Number 99-18118

Preface

The new edition of this textbook will provide students of varied backgrounds with a comprehensive and accessible introduction to the primary subject-matter of linear algebra. The contents include detailed accounts of matrices and determinants (chapters 2 to 4), finite-dimensional vector spaces (chapter 5), linear mappings (chapters 6 and 7), and (spread through chapters 2, 3, 5 and 6) the applications of these ideas to systems of linear equations, while later chapters (8 to 10) discuss eigenvalues, diagonalization of matrices, euclidean spaces, and quadratic forms. In writing the book, my constant aim has been to draw on my considerable experience of teaching this material to produce an account of it that students striving to gain an understanding of linear algebra will find helpful. In particular, therefore, I have provided suitably detailed explanations of points which students often find difficult.

The reader will see that little is taken for granted. For example, the accounts of matrices and vector spaces are self-contained and start right at the beginning. However, a basic knowledge of elementary ideas and notations concerning sets is assumed, and from chapter 5 onwards the reader must be able to understand simple arguments about sets (such as proving that one set is a subset of another). Then, from chapter 6 onwards, a knowledge of mappings becomes essential, though to help the reader an appendix on mappings provides a condensed summary of everything relevant. More generally, the level of mathematical sophistication will be found to rise as one progresses through the book, in accordance with the reasonable assumption that, when students come to tackle the material of the later chapters, they will have already acquired some degree of mathematical maturity.

Some remarks remain to be made on the so far unmentioned chapter 1, and also on chapters 4 and 10. Chapter 1 is devoted to the system of vectors arising naturally in 3-dimensional euclidean geometry. While this material is important in its own right, the main reason for its inclusion here is to prepare the way for the discussion of vector spaces in general—in the belief that the abstract concept of a vector space is much more easily assimilated when one tangible example is familiar, along with the geometrical overtones of some of the algebraic ideas.

Chapter 4 discusses determinants. Although every student should learn about determinants at an early stage, a fully watertight discussion of them would surely be inappropriately difficult for the first half of this book. So I have opted for a compromise which entails proving certain key properties of determinants for the 3×3 case only. Nevertheless, a fully general statement of properties of determinants is given, and, whenever possible, a fully general proof that property B follows from property A (even when property A has been proved only for the 3×3 case).

Lastly, chapter 10, a completely new chapter appearing for the first time in this edition of the book, is about quadratic forms. Here, without sacrificing mathematical rigour, I have deliberately avoided excessive abstraction and have sought, rather, to ensure that the reader is given a feel for quadratic forms and ready access to their applications in other parts of mathematics and in other disciplines (e.g. physics).

Following every chapter there is an ample set of exercises. In each set there is a good mixture—ranging from straightforward numerical exercises to some difficult problems which will extend the best of students. The most difficult exercises are indicated by asterisks. At the end of the book I have provided answers to the exercises—numerical answers where appropriate and hints for, or part solutions of, the less easy problems.

Inevitably, throughout the book I have drawn consciously and unconsciously on the ideas and traditions which have evolved in the teaching of linear algebra in the Mathematics Department of Glasgow University. It would be invidious to mention some individuals and not others, but it must be acknowledged that one or two of the most ingenious ideas seen in the exercises owe more than a little to Professor W. D. Munn.

Last, but certainly not least, let me record my gratitude to my friend and colleague Dr J. B. Hickey for his assistance in manuscript checking and proof correction. In all this very considerable work he has shown characteristic kindness, as well as a most helpful attention to detail.

T.A.W.

Contents

Chapter Seven MATRICES FROM LINEAR MAPPINGS

Chapter Eight EIGENVALUES, EIGENVECTORS AND DIAGONALIZATION

Chapter Nine EUCLIDEAN SPACES

Chapter Ten QUADRATIC FORMS

CHAPTER ONE

A SYSTEM OF VECTORS

1. Introduction

This textbook provides an introduction to Linear Algebra—a vast subject in its own right, and one which is useful in many parts of pure and applied mathematics. Many of the major problems discussed in linear algebra arise from systems of simultaneous linear equations and the rectangular arrays (*matrices*, as we call them) of coefficients occurring in such systems. It is also true that several ideas of importance in linear algebra can be traced to geometrical sources.

The contents of this first chapter are largely geometrical. The chapter uses geometrical notions to introduce a system of mathematical objects called *vectors*, and it shows that this system can be used to deal with various geometrical problems. In fact, however, much of the importance of the ideas of this chapter lies beyond geometry, in the essentially algebraic generalizations which these ideas admit. In particular, a later chapter (chapter 5) will introduce the concept of a *vector space*. This is a generalization of the idea of a system of vectors (as encountered in this chapter), and it is a concept which is very powerful and illuminating in the discussion of many problems, including those that arise from systems of linear equations and matrices. Some of the details and emphases of this first chapter are designed to prepare the way for the generalizations that come later.

For the time being, though, those generalizations lie over the horizon, and our first concern is with the motivation for studying vectors of the kind considered in this chapter. Such motivation is to be found in physics as well as in geometry. In physics there are many things (e.g. displacements, forces, magnetic fields) that have both a magnitude and a direction: we call such things *vectorial quantities*. Even those who claim no knowledge of physics can understand that a walk (in a straight line) from point A to point B belongs in this category because it has a magnitude (its length) and a direction (from A to B). Correspondingly, in a geometrical discussion, we may well be concerned not only with the distance between the points A and B but also with the direction of the step from A to B.

For these and other reasons it is valuable to have an algebraic system applicable *wherever* (in physics or geometry or elsewhere) vectorial quantities are encountered—just as the system of positive integers 1, 2, 3, 4, 5,... is applicable to all elementary counting problems, whatever their contexts. The set of vectors introduced in this chapter serves this purpose very well. The set will be denoted by E_3, the subscript "3" indicating that the discussion will be geared to 3-dimensional geometry. Roughly speaking, E_3 consists of a zero vector and one vector with each possible magnitude and direction in 3-dimensional space. (A more precise and formal description will be given at the beginning of §2.)

A little must be said to clarify some geometrical assumptions made in this chapter. The whole discussion will take place within the framework of 3-dimensional *euclidean* geometry, and thus we simply assume that 3-dimensional space has the straightforward properties taken for granted in the elementary study of geometry. Further, we use the word "direction" with its everyday meaning: for instance, "vertically upwards" is an example of a direction. Along, or parallel to, any given straight line there are two directions, which are regarded as different: e.g. along any vertical line there are the two opposite directions "vertically upwards" and "vertically downwards". The directions along parallel lines are the same—e.g. vertically upwards and vertically downwards in the case of all vertical lines. Finally, in giving distances or lengths, we shall assume that a unit of length has been established, so that a length can be quoted as a pure number—e.g. 6 (meaning 6 times the unit of length).

2. Description of the system E_3

After the introductory remarks of §1, we now begin the formal discussion of the system E_3 with three descriptive postulates.

(1) E_3 consists of the "zero vector" and infinitely many nonzero vectors. None of the latter is equal to the zero vector.

(2) The zero vector has magnitude 0 and does not possess a direction. Every nonzero vector in E_3 has a positive magnitude and a direction.

(3) For every positive real number α and every direction $\mathscr{D}$ in 3-dimensional space, there is in E_3 precisely one vector with magnitude α and direction $\mathscr{D}$.

Later we shall introduce addition of vectors and other similar operations. First there follows a series of remarks about notation and other matters arising from the descriptive postulates.

(*a*) When possible, we use bold-type symbols (e.g. $\mathbf{u}$, $\mathbf{v}$, $\mathbf{w}$, $\mathbf{x}$, $\mathbf{y}$, $\mathbf{z}$) to denote vectors. The zero vector is denoted by $\mathbf{0}$.

(*b*) For any vector **x**, the magnitude of **x** is denoted by $|\mathbf{x}|$. So (cf. postulate (2)) in every case $|\mathbf{x}|$ is a nonnegative number, positive unless $\mathbf{x} = \mathbf{0}$. It is essential not to confuse $|\mathbf{x}|$, which is a number, with **x**, which is a vector.

(*c*) In view of the phrase "precisely one" in postulate (3), a nonzero vector is completely specified by the giving of its magnitude and its direction. Note further that a claim that two nonzero vectors **x** and **y** are equal is essentially a twofold statement: it means that **x** and **y** have the same magnitude and that **x** and **y** have the same direction.

(*d*) A **unit vector** means a vector whose magnitude is 1. It follows from postulate (3) that there is precisely one unit vector having any prescribed direction.

(*e*) For any nonzero vector **x**, the **negative** of **x** (denoted by $-\mathbf{x}$) is defined to be the vector with the same magnitude as **x** and direction opposite to that of **x**. (E.g. if **x** has magnitude 2 and direction vertically upwards, then $-\mathbf{x}$ is the vector with magnitude 2 and direction vertically downwards.) The negative of **0** is defined to be **0** (i.e. $-\mathbf{0} = \mathbf{0}$).

(*f*) For some students it may be helpful to describe a concrete set of objects which is like the set E_3 and which could be thought of as a representation of E_3. Choose any point O as "origin" and imagine lots of arrows emanating from O—a zero arrow (of zero length) and one arrow with every possible positive length and direction in space. Since such a set of arrows obeys the descriptive postulates for E_3, it offers a satisfactory representation, or picture, of the set E_3. It would be wrong to let this picture dominate one's thinking too much, though it does suggest one important fact that we arrive at later through other considerations—the fact that there is a one-to-one correspondence between the vectors in E_3 and the points in 3-dimensional space. This is apparent from the arrow picture of E_3 when one observes that precisely one arrow ends at each point of 3-dimensional space.

3. Directed line segments and position vectors

Let A and B be any points.

The distance between A and B will be denoted by $|AB|$.

Figure I(*a*) suggests a more complex idea—the straight line segment between A and B, regarded as directed from A to B. This we denote by $\overrightarrow{AB}$: and each such object is termed a **directed line segment** (for which we use the abbreviation d.l.s.).

Every d.l.s. $\overrightarrow{AB}$ has a magnitude (its length $|AB|$) and a direction (unless A coincides with B). So it is natural to regard each d.l.s. $\overrightarrow{AB}$ as representing a certain vector in E_3—namely the vector with magnitude $|AB|$ and with (if $B \neq A$) the same direction as $\overrightarrow{AB}$ (i.e. the direction from A to B). The information that "$\overrightarrow{AB}$ represents the vector **x**" may be incorporated into diagrams as in figure I(*b*). We shall use the notation $[\overrightarrow{AB}]$ for the (one and only) vector represented by $\overrightarrow{AB}$. (Think of the square brackets as standing for "vector represented by".)

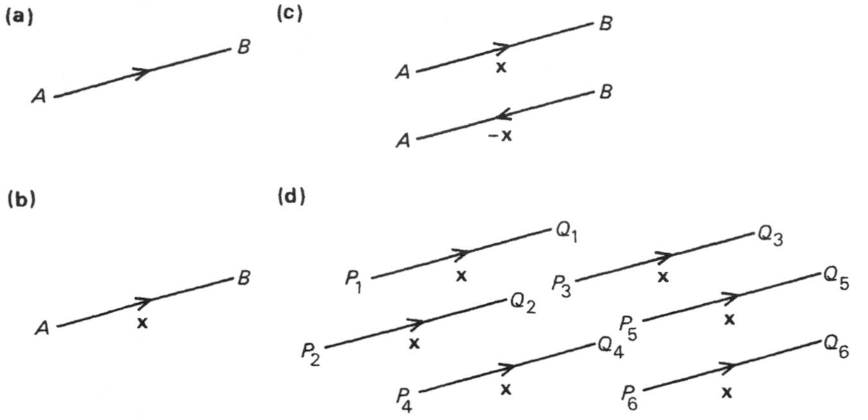

Figure I

The following simple points are worth recording explicitly.

3.1 If A and B are points,

(i) $|[\overrightarrow{AB}]| = |AB|$,

(ii) $[\overrightarrow{AB}]$ is $\mathbf{0}$ if and only if A and B coincide,

(iii) $[\overrightarrow{BA}] = -[\overrightarrow{AB}]$.

((i) is part of the explanation of what we mean by "$[\overrightarrow{AB}]$"; and (ii) and (iii) are trivialities, the latter illustrated by figure I(*c*).)

To prevent any risk of confusion, it must be appreciated that, for any given vector $\mathbf{x}$, there are *infinitely many* d.l.s.s that represent $\mathbf{x}$: indeed, for each point P, there is a d.l.s. $\overrightarrow{PQ}$ that starts at P and represents $\mathbf{x}$. Figure I(*d*) gives an indication of all this. To say that two d.l.s.s $\overrightarrow{AB}$ and $\overrightarrow{CD}$ represent the same vector means that $\overrightarrow{AB}$ and $\overrightarrow{CD}$ have the same length and (if $A \neq B$ and $C \neq D$) the same direction. This remark makes it obvious that (cf. figure II):

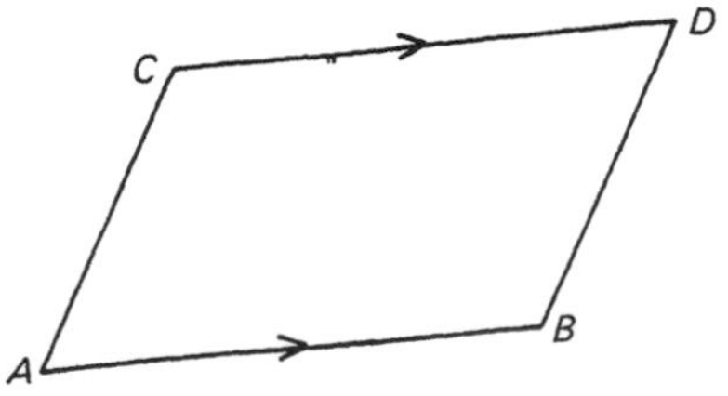

Figure II

3.2 If $ABDC$ is a quadrilateral, then $ABDC$ is a parallelogram if and only if $[\overrightarrow{AB}] = [\overrightarrow{CD}]$.

Now suppose that a point O has been chosen as origin. For any point P, the vector $[\overrightarrow{OP}]$ serves very well and naturally to specify the position of P, since this vector tells us the distance and (if $P \neq O$) the direction from O to P. We call the vector $[\overrightarrow{OP}]$ the **position vector** of P (relative to the origin O), and we denote it by $\mathbf{r}_P$ (cf. figure III).

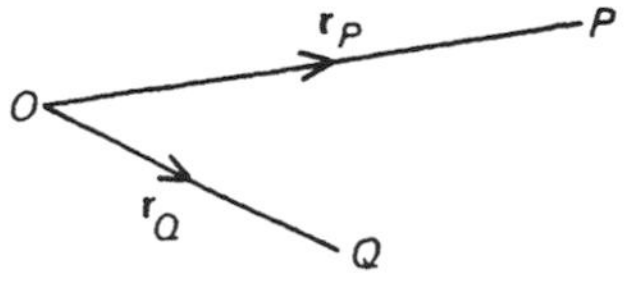

Figure III

Observe that:

(*a*) the position vector $\mathbf{r}_O$ of the origin is $\mathbf{0}$;

(*b*) for any vector $\mathbf{x}$ in E_3 there is precisely one point whose position vector is $\mathbf{x}$.

As will become apparent, position vectors are extremely useful in enabling the system E_3 to be applied to geometrical problems.

4. Addition and subtraction of vectors

Let $\mathbf{x}$ and $\mathbf{y}$ be arbitrary vectors in E_3. The sum $\mathbf{x}+\mathbf{y}$ of these vectors is defined by means of a geometrical construction as follows. Let P be any point; let Q be the point such that $[\overrightarrow{PQ}] = \mathbf{x}$; and let R be the point such that $[\overrightarrow{QR}] = \mathbf{y}$. Then $\mathbf{x}+\mathbf{y}$ is defined to be $[\overrightarrow{PR}]$ (cf. figure IV).

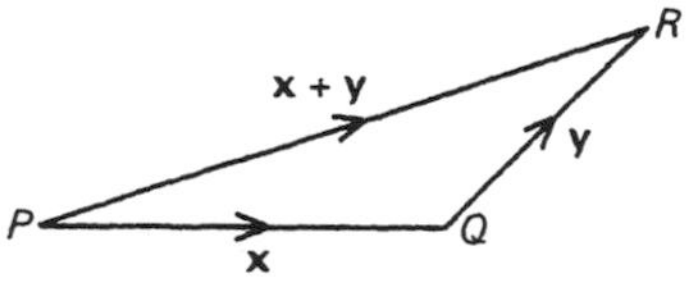

Figure IV

It should be noted that (for essentially simple geometrical reasons) variation of the choice of the point P in the above definition makes no difference to the meaning given to $\mathbf{x}+\mathbf{y}$.

This definition of addition of vectors corresponds to the way in which vectorial quantities (forces, magnetic fields) combine in physics. But, even to a non-physicist, the definition will make sense through considering "the result of a walk from P to Q followed by a walk from Q to R". Associated with this motivatory thought is the fact (very much apparent from figure IV) that:

4.1 For any points P, Q, R, $[\overrightarrow{PQ}]+[\overrightarrow{QR}] = [\overrightarrow{PR}]$.

Subtraction of vectors can now be concisely defined by the natural equation

$$\mathbf{x}-\mathbf{y} = \mathbf{x}+(-\mathbf{y}).$$

The fundamental properties of vector addition are given in the next four results (4.2, 4.3, 4.4, 4.5).

4.2 (The associative property)

$$(\mathbf{x}+\mathbf{y})+\mathbf{z} = \mathbf{x}+(\mathbf{y}+\mathbf{z}) \text{ (for all } \mathbf{x}, \mathbf{y}, \mathbf{z} \in E_3).$$

4.3 (The commutative property)

$$\mathbf{x}+\mathbf{y} = \mathbf{y}+\mathbf{x} \text{ (for all } \mathbf{x}, \mathbf{y} \in E_3).$$

4.4 $\mathbf{x}+\mathbf{0} = \mathbf{x}$ (for all $\mathbf{x} \in E_3$).

4.5 $\mathbf{x}+(-\mathbf{x}) = \mathbf{0}$ (for all $\mathbf{x} \in E_3$).

The last two (4.4 and 4.5) are immediately obvious from the definition of vector addition. The first two (4.2 and 4.3) must be justified through the construction of d.l.s.s representing their left-hand sides and right-hand sides. As an illustration, here is a prooof of 4.3 in the case where $\mathbf{x}$, $\mathbf{y}$ are non-parallel.

For arbitrary non-parallel vectors $\mathbf{x}$, $\mathbf{y}$, we construct (cf. figure V) a parallelogram $ABCD$ in which $\overrightarrow{AB}$ and $\overrightarrow{DC}$ represent $\mathbf{x}$ while $\overrightarrow{BC}$ and $\overrightarrow{AD}$ represent $\mathbf{y}$.

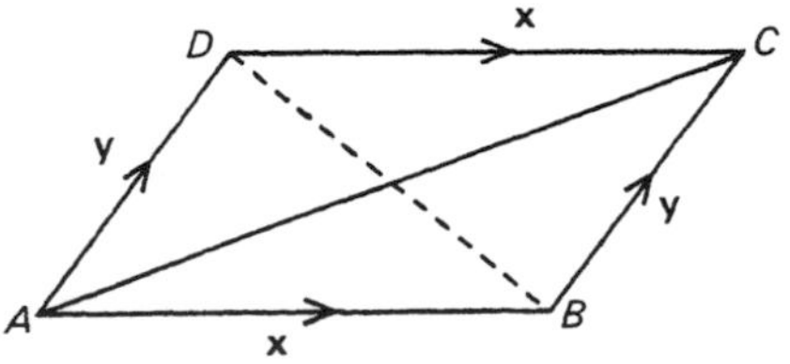

Figure V

Using 4.1 in the triangle ABC, we obtain

$$[\overrightarrow{AC}] = [\overrightarrow{AB}]+[\overrightarrow{BC}] = \mathbf{x}+\mathbf{y}.$$

But similarly, from the triangle ADC,

$$[\overrightarrow{AC}] = [\overrightarrow{AD}]+[\overrightarrow{DC}] = \mathbf{y}+\mathbf{x}.$$

Hence, as required, $\mathbf{x}+\mathbf{y} = \mathbf{y}+\mathbf{x}$ (both being the vector represented by $\overrightarrow{AC}$).

"The parallelogram law of vector addition" is the name given to the fact, brought to light in this discussion, that $\mathbf{x}+\mathbf{y}$ is represented by the diagonal $\overrightarrow{AC}$ of a parallelogram whose adjacent sides $\overrightarrow{AB}$ and $\overrightarrow{AD}$ represent $\mathbf{x}$ and $\mathbf{y}$.

While we are concerned with the parallelogram of figure V, it is instructive to express $[\overrightarrow{DB}]$ in terms of $\mathbf{x}$ and $\mathbf{y}$. We have:

$$\begin{aligned}[\overrightarrow{DB}] &= [\overrightarrow{DC}]+[\overrightarrow{CB}] && \text{(cf. 4.1)}\\ &= [\overrightarrow{DC}]-[\overrightarrow{BC}] && \text{(cf. 3.1(iii) and the definition of subtraction)}\\ &= \mathbf{x}-\mathbf{y}.\end{aligned}$$

There follow now some very general remarks, of significance here and in other places where addition, or some other similar operation, is defined on a given set of objects. The intention behind the generality of the remarks is to convey certain ideas once and for all, and thus to avoid the need for a separate cxplanation of these ideas on each of the several occasions when they are relevant.

(*a*) Let S be a set of objects any two of which may be added: i.e. $x+y$ has a meaning for all $x, y \in S$. We say that S is **closed** under addition if and only if $x+y \in S$ for every $x, y \in S$. So, for example, E_3 is closed under addition because in every case the result of adding two vectors in E_3 is also a vector in E_3. On the other hand, the set $\{1, 2, 3, 4, 5\}$ is *not* closed under addition since (for example) $3+4$ lies outside the set.

Leaving asidc the illustrations, one can give an exactly parallel explanation of the meaning of "closed" with addition replaced by multiplication or by any other such operation.

(*b*) Wherever in mathematics an associative law holds for addition (or multiplication or...), brackets may be omitted in writing down a long sum (or product or...): i.e. no ambiguity is introduced by omission of brackets. Further, whenever a commutative law holds for addition (or multiplication or...), one can change with impunity the order of the things being added (or multiplied or...). So, for example, if $\mathbf{x}$, $\mathbf{y}$, $\mathbf{z}$ denote vectors in E_3, "$\mathbf{x}+\mathbf{y}+\mathbf{z}$" is unambiguous and means the same as "$\mathbf{z}+\mathbf{x}+\mathbf{y}$".

(*c*) If objects in the set S can be added, if S contains a zero object, and if each object in S has a negative in S, then subtraction may be defined on S by

$$x-y = x+(-y).$$

We have already used this idea in the case $S = E_3$.

(*d*) When a hitherto unfamiliar operation (like addition or multiplication) is introduced, we shall describe it as a "no-catch operation" if all its properties (basic properties and their consequences) conform to patterns that are familiar from the elementary algebra of real numbers—i.e. if it can justly be said that "what looks true is true". Frequently much energy can be saved by giving students an assurance that a newly introduced operation is a no-catch operation.

This assurance can be given in the case of addition of vectors. A simple example of something that looks true and is true is the fact that if **a**, **b** are given vectors in E_3, the equation

$$\mathbf{a}+\mathbf{x} = \mathbf{b}$$

has precisely one solution, namely $\mathbf{x} = \mathbf{b}-\mathbf{a}$.

Returning from generalities to the system E_3 and its applications to geometry, we conclude this section with a simple but very important formula.

4.6 For any points A and B, $[\overrightarrow{AB}] = \mathbf{r}_B - \mathbf{r}_A$.

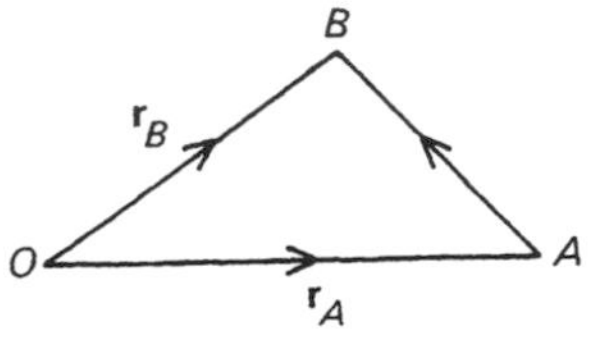

Figure VI

Proof. We refer to figure VI. From the triangle OAB (where O is the origin),

$$[\overrightarrow{OA}]+[\overrightarrow{AB}] = [\overrightarrow{OB}]; \qquad \text{i.e. } \mathbf{r}_A+[\overrightarrow{AB}] = \mathbf{r}_B.$$

Hence $[\overrightarrow{AB}] = \mathbf{r}_B-\mathbf{r}_A$ (cf. above remark about equations of the form $\mathbf{a}+\mathbf{x} = \mathbf{b}$).

5. Multiplication of a vector by a scalar

In this context "scalar" means "real number" (positive, negative, or zero). We use $\mathbb{R}$ to denote the set of all real numbers.

Let $\mathbf{x}$ be any vector in E_3 and λ any scalar. Then $\lambda\mathbf{x}$ is defined to be the vector with magnitude and direction as follows:

(i) $|\lambda\mathbf{x}| = |\lambda||\mathbf{x}|$; and

(ii) (if $|\lambda\mathbf{x}| \neq 0$) $\lambda\mathbf{x}$ has the same direction as $\mathbf{x}$ if $\lambda > 0$ but has direction opposite to that of $\mathbf{x}$ if $\lambda < 0$.

So, for example, if $\mathbf{x}$ has magnitude 2 and direction vertically upwards, then:

$5\mathbf{x}$ has magnitude 10 and direction vertically upwards;

$(-3)\mathbf{x}$ has magnitude 6 and direction vertically downwards;

$0\mathbf{x}$ is $\mathbf{0}$.

From the definition of "$\lambda\mathbf{x}$" it is clear that:

5.1 $\lambda\mathbf{x} = \mathbf{0} \Leftrightarrow \lambda = 0 \quad \text{or} \quad \mathbf{x} = \mathbf{0} \quad (\lambda \in \mathbb{R}, \mathbf{x} \in E_3)$.

The operation of multiplying a vector by a scalar is a "no-catch operation" (cf. remark (*d*) in §4). The absence of catches stems from the basic properties of the operation, which are as follows.

5.2 $\lambda(\mathbf{x}+\mathbf{y}) = \lambda\mathbf{x}+\lambda\mathbf{y} \qquad (\lambda \in \mathbb{R}; \mathbf{x}, \mathbf{y} \in E_3)$.

5.3 $(\lambda+\mu)\mathbf{x} = \lambda\mathbf{x}+\mu\mathbf{x} \qquad (\lambda, \mu \in \mathbb{R};\ \mathbf{x} \in E_3)$.

5.4 $\lambda(\mu\mathbf{x}) = (\lambda\mu)\mathbf{x} \qquad (\lambda, \mu \in \mathbb{R};\ \mathbf{x} \in E_3)$.

5.5 $1\mathbf{x} = \mathbf{x} \qquad (\mathbf{x} \in E_3)$.

The last of these properties, 5.5, is obvious from the definition of "$\lambda\mathbf{x}$". The results 5.3 and 5.4, about scalar multiples of a given vector **x**, are easily seen to be true in the case where λ and μ are both positive, though a complete proof involves the slightly tedious task of covering all other cases. And 5.2 depends (as will soon be appreciated by a student who attempts a proof) on the properties of similar triangles.

Let us turn our attention now to the topic of parallel vectors. It scarcely needs to be explained that two nonzero vectors **x**, **y** are parallel if and only if they have the same direction or opposite directions. More artificially, and purely because it turns out to be convenient, we regard **0** as parallel to every vector.

Let **x** be a given nonzero vector.

If α is a positive scalar, then the vector

$$\frac{\alpha}{|\mathbf{x}|}\mathbf{x}$$

has the same direction as **x** $\left(\dfrac{\alpha}{|\mathbf{x}|}\text{ being a positive scalar}\right)$ and has magnitude α (since $|\lambda\mathbf{x}| = \lambda|\mathbf{x}|$ if $\lambda > 0$).

This simple observation can be exploited to write down any specified vector parallel to **x** in the form of a scalar multiple of **x**. E.g. the vector with magnitude 5 and the same direction as **x** is $\dfrac{5}{|\mathbf{x}|}\mathbf{x}$, and hence the vector with magnitude 5 and direction opposite to that of **x** is $-\dfrac{5}{|\mathbf{x}|}\mathbf{x}$. More generally,

(i) if **y** has the same direction as **x**, then $\mathbf{y} = \dfrac{|\mathbf{y}|}{|\mathbf{x}|}\mathbf{x}$,

(ii) if **y** has direction opposite to that of **x**, then $\mathbf{y} = -\dfrac{|\mathbf{y}|}{|\mathbf{x}|}\mathbf{x}$, and

(iii) if $\mathbf{y} = \mathbf{0}$, then $\mathbf{y} = 0\mathbf{x}$.

Thus every vector parallel to **x** is expressible as a scalar multiple of **x**. The converse being obviously true, we conclude that:

5.6 If **x** is a nonzero vector, then a vector **y** is parallel to **x** if and only if **y** is expressible as a scalar multiple of **x**.

It is also clear from the above discussion that:

5.7 If $\mathbf{x}$ is a nonzero vector, then the unit vector with the same direction as $\mathbf{x}$ is $\dfrac{1}{|\mathbf{x}|}\mathbf{x}$.

The last proposition in this section is a fundamental preliminary to the use of vectorial methods in coordinate geometry.

5.8 Let O be the origin and let $\mathbf{i}$ be a unit vector whose direction is the positive direction along the x-axis. Then, for every point P on the x-axis, $[\overrightarrow{OP}] = x_P\mathbf{i}$, where, in the standard way, the coordinate x_P of P is the "signed distance" from O to P along the x-axis (positive when $\overrightarrow{OP}$ has the same direction as $\mathbf{i}$, negative when $\overrightarrow{OP}$ has the same direction as $-\mathbf{i}$).

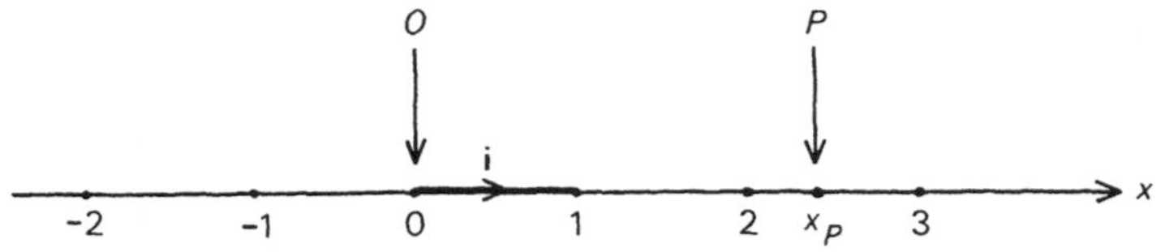

Figure VII

Proving the truth of 5.8 is simply a matter of checking that $[\overrightarrow{OP}]$ and $x_P\mathbf{i}$ are the same vector (a) when $x_P > 0$, (b) when $x_P < 0$, and (c) when $x_P = 0$.

While 5.8 has been worded so as to be about "the x-axis", there is, of course, an exactly corresponding story about any coordinate axis, whatever it is called.

To conclude this section, we introduce a very important item of vocabulary. Let $\mathbf{x}_1, \mathbf{x}_2, \ldots, \mathbf{x}_n$ denote vectors. Any vector expressible in the form

$$\lambda_1\mathbf{x}_1 + \lambda_2\mathbf{x}_2 + \ldots + \lambda_n\mathbf{x}_n \qquad \text{(with } \lambda_1, \lambda_2, \ldots, \lambda_n \text{ scalars)}$$

is said to be (expressible as) a **linear combination** of $\mathbf{x}_1, \mathbf{x}_2, \ldots, \mathbf{x}_n$. (In the trivial case $n = 1$, where we would be talking about "a linear combination of $\mathbf{x}_1$", that would simply mean a scalar multiple of $\mathbf{x}_1$.)

The phrase "linear combination" gets used in other and more general contexts, where its meaning is easily inferred from the usage just explained.

6. Section formula and collinear points

Consider first three distinct collinear points A, P, B.

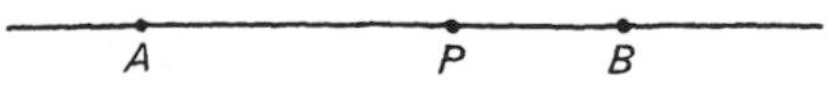

Figure VIII

The vectors $[\overrightarrow{AP}]$ and $[\overrightarrow{PB}]$ are parallel and so (cf. 5.6)

$$[\overrightarrow{AP}] = \lambda[\overrightarrow{PB}]$$

for some scalar λ, uniquely determined by A, P and B. This scalar λ is called the **position ratio** of P with respect to A and B, and it is usually denoted by AP/PB. From its definition, one soon sees that this position ratio is positive if and only if P lies between A and B.

The **section formula**, to which we now come, gives us the position vector of a point specified by its position ratio with respect to two fixed points.

6.1 Let A, B be given distinct points and let P be the point on the line AB such that $AP/PB = m/n$. Then

$$\mathbf{r}_P = \frac{m\mathbf{r}_B + n\mathbf{r}_A}{m+n} \qquad \left[\text{i.e. } \frac{1}{m+n}(m\mathbf{r}_B + n\mathbf{r}_A)\right].$$

Proof. In view of the meaning of "AP/PB", we have

$$[\overrightarrow{AP}] = \frac{m}{n}[\overrightarrow{PB}], \quad \text{and hence} \quad n[\overrightarrow{AP}] = m[\overrightarrow{PB}];$$

$$\text{i.e. (by 4.6)} \qquad n(\mathbf{r}_P - \mathbf{r}_A) = m(\mathbf{r}_B - \mathbf{r}_P).$$

This simplifies to $(m+n)\mathbf{r}_P = m\mathbf{r}_B + n\mathbf{r}_A$, from which the result follows.

The special case $m = n = 1$ gives us:

6.2 For any two distinct points A and B, the mid-point of AB has position vector $\frac{1}{2}(\mathbf{r}_A + \mathbf{r}_B)$.

So far in this section we have discussed situations where three points were known to be collinear. On the other hand, in many problems we can use position vectors to prove that three specified points—X, Y, Z, say—are collinear. This can be done by showing that $[\overrightarrow{XY}]$ is a scalar multiple of $[\overrightarrow{YZ}]$: it then follows that $[\overrightarrow{XY}]$ and $[\overrightarrow{YZ}]$ are parallel and hence that X, Y, Z are collinear.

Worked example. Let $OABC$ be a parallelogram; let P be the point on the diagonal AC such that $AP/PC = \frac{1}{4}$; and let Q be the point on the side AB such that $AQ/QB = \frac{1}{3}$. Prove that O, P, Q are collinear and determine the ratio OP/PQ.

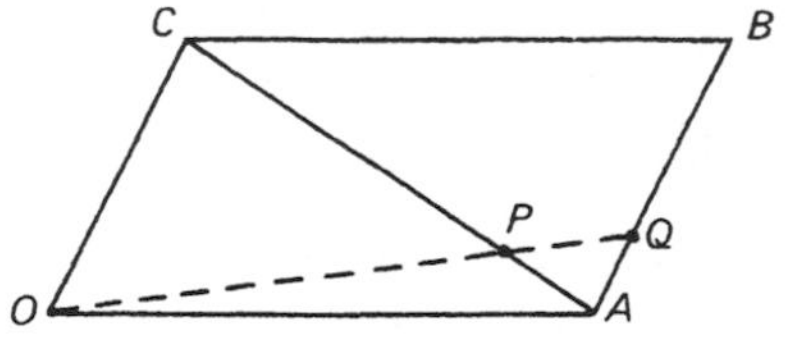

Figure IX

Solution. Take O as origin, and write $\mathbf{a}$ for $\mathbf{r}_A$, $\mathbf{b}$ for $\mathbf{r}_B$, etc. Since $OABC$ is a parallelogram, $\mathbf{b} = \mathbf{a}+\mathbf{c}$. (This may be obtained from 3.2 and 4.1; alternatively it comes straight from the "parallelogram law of vector addition" mentioned in §4.)

By the section formula, $\mathbf{p} = \frac{1}{5}(4\mathbf{a}+\mathbf{c})$,

while $\mathbf{q} = \frac{1}{4}(3\mathbf{a}+\mathbf{b})$
$= \frac{1}{4}(4\mathbf{a}+\mathbf{c})$ (since $\mathbf{b} = \mathbf{a}+\mathbf{c}$).

Hence (cf. 4.6)

$$[\overrightarrow{PQ}] = \mathbf{q}-\mathbf{p} = (\tfrac{1}{4}-\tfrac{1}{5})(4\mathbf{a}+\mathbf{c}) = \tfrac{1}{20}(4\mathbf{a}+\mathbf{c}) = \tfrac{1}{4}\mathbf{p} = \tfrac{1}{4}[\overrightarrow{OP}].$$

From this it follows that O, P, Q are collinear, with $OP/PQ = 4$ (since $[\overrightarrow{OP}] = 4[\overrightarrow{PQ}]$).

7. Centroids of a triangle and a tetrahedron

Consider first an arbitrary triangle ABC. Let D, E, F be the mid-points of the sides BC, CA, AB, respectively. The medians of the triangle are the lines AD, BE, CF. We shall now show, by vectorial methods, that these three lines are concurrent (cf. figure X(a)).

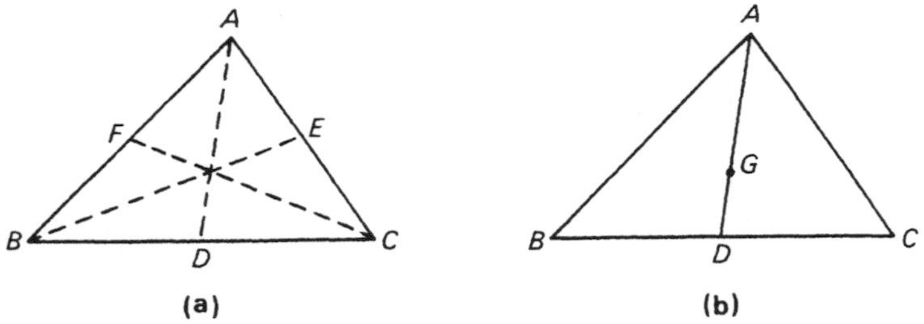

Figure X

Let G be defined as the point on the median AD such that $AG/GD = 2$ (cf. figure X(b)). Using position vectors relative to an arbitrary origin, we have:

$$\mathbf{r}_D = \tfrac{1}{2}(\mathbf{r}_B+\mathbf{r}_C);$$

and hence, by the section formula, $\mathbf{r}_G = \frac{1}{3}(2\mathbf{r}_D+\mathbf{r}_A) = \frac{1}{3}(\mathbf{r}_A+\mathbf{r}_B+\mathbf{r}_C)$.

Because this expression for $\mathbf{r}_G$ is completely symmetrical in A, B and C, we would obtain the same answer if we calculated the position vectors of the points on the medians BE and CF corresponding to G. Therefore, in fact, the point G lies on all three medians.

This establishes that the medians of the arbitrary triangle ABC are concurrent. The point G which the three medians pass through is called the **centroid** of the triangle ABC. For future reference, let us record the fact that:

7.1 The centroid of the triangle ABC has position vector $\frac{1}{3}(\mathbf{r}_A+\mathbf{r}_B+\mathbf{r}_C)$.

Moving from 2-dimensional to 3-dimensional geometry, we now undertake an analogous discussion of an arbitrary tetrahedron $ABCD$. Let G_1, G_2, G_3, G_4 be the centroids of the triangular faces BCD, ACD, ABD, ABC, respectively. We call the four lines AG_1, BG_2, CG_3, DG_4 (each joining a vertex to the centroid of the opposite face) the medians of the tetrahedron $ABCD$; and we shall prove that these four lines are concurrent.

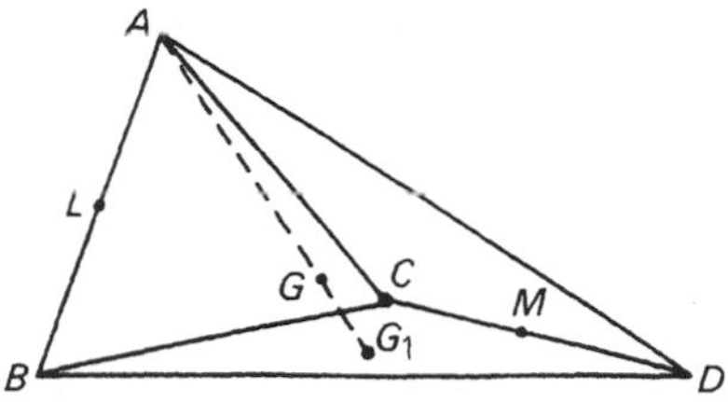

Figure XI

For this purpose, introduce the point G on AG_1 such that $AG/GG_1 = 3$. By 7.1, $\mathbf{r}_{G_1} = \frac{1}{3}(\mathbf{r}_B+\mathbf{r}_C+\mathbf{r}_D)$, and hence, by the section formula,

$$\mathbf{r}_G = \tfrac{1}{4}(3\mathbf{r}_{G_1}+\mathbf{r}_A) = \tfrac{1}{4}(\mathbf{r}_A+\mathbf{r}_B+\mathbf{r}_C+\mathbf{r}_D).$$

In view of the symmetry of this expression in A, B, C, D, it follows that the point G lies also on the other three medians of the tetrahedron $ABCD$. Thus the four medians are concurrent at G, which we call the centroid of the tetrahedron.

The 3-dimensional analogue of 7.1 is:

7.2 The centroid of the tetrahedron $ABCD$ has position vector $\frac{1}{4}(\mathbf{r}_A+\mathbf{r}_B+\mathbf{r}_C+\mathbf{r}_D)$.

Continuing the above discussion of the arbitrary tetrahedron $ABCD$, we can find other interesting lines that pass through the centroid G.

Let L and M be the respective mid-points of AB and CD (which form a pair of "opposite" edges of the tetrahedron). Then $\mathbf{r}_L = \frac{1}{2}(\mathbf{r}_A+\mathbf{r}_B)$ and $\mathbf{r}_M = \frac{1}{2}(\mathbf{r}_C+\mathbf{r}_D)$; and hence $\mathbf{r}_G = \frac{1}{2}(\mathbf{r}_L+\mathbf{r}_M)$—which shows that G is the mid-point of LM. In particular, the line LM passes through G. Similarly, the join of the mid-points of the opposite edges AC and BD passes through G; and the same goes for the join of the mid-points of the other pair of opposite edges, AD and BC.

We have now shown that seven interesting lines are concurrent at G—AG_1 and the other three medians, and LM and the two other similar lines. This is quite a formidable proposition in 3-dimensional geometry, and thus the simplicity of our proof of it is a striking demonstration of the power and elegance of vectorial methods of tackling geometrical problems. Further examples of this power and elegance may be seen later in the chapter and in some of the exercises which follow it.

8. Coordinates and components

The use of vectors is a helpful device in coordinate geometry. Conversely, as this section will show, the introduction of coordinates in 3-dimensional space opens up important new lines of thought in our consideration of the system E_3.

In a standard 2-dimensional coordinate system (such as is encountered at a fairly early stage in mathematical education) one starts with two mutually perpendicular coordinate axes (the x-axis and the y-axis), which are directed lines intersecting at the origin O. The coordinates (x_P, y_P) of a typical point P are described by dropping perpendiculars from P to the coordinate axes, so completing a rectangle $OLPM$ as in figure XII: x_P is the coordinate of L on the x-axis (i.e. the signed distance from O to L along the x-axis), and similarly y_P is the coordinate of M on the y-axis.

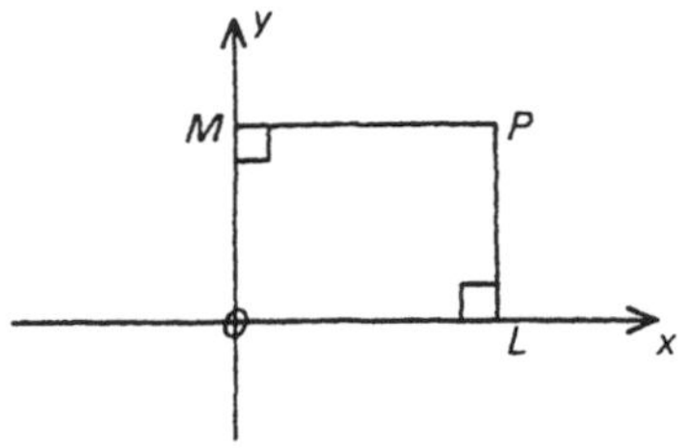

Figure XII

The corresponding story in three dimensions starts with three mutually perpendicular coordinate axes (the x-axis, the y-axis and the z-axis), which are directed lines concurrent at the origin O. To describe the coordinates of a typical point P, one introduces the feet L, M, N of the perpendiculars from P to the x-, y-, z-axes, respectively. The discussion is clarified by considering not just L, M and N, but the cuboid (see figure XIII) that has O, P, L, M, N as five of its vertices and all its edges parallel to the coordinate axes. This cuboid is the analogue of the rectangle $OLPM$ in the 2-dimensional discussion: its remaining vertices H, J, K are the feet of the perpendiculars from P to planes xOy, yOz, zOx, respectively. The coordinates of P are defined to be (x_P, y_P, z_P), where x_P is the coordinate of L on the x-axis, y_P is the coordinate of M on the y-axis, and z_P is the coordinate of N on the z-axis.

Newcomers to 3-dimensional coordinates may find it helpful to think of the plane xOy as horizontal and of the z-axis as pointing vertically upwards. Moreover, it is natural to think of the plane xOy (the x, y-plane, as it is often called) as the coordinatized 2-dimensional space considered earlier (as in figure XII), now embedded in a larger space. Against this background one can think of the coordinates of P as follows: P is vertically above or below H; x_P

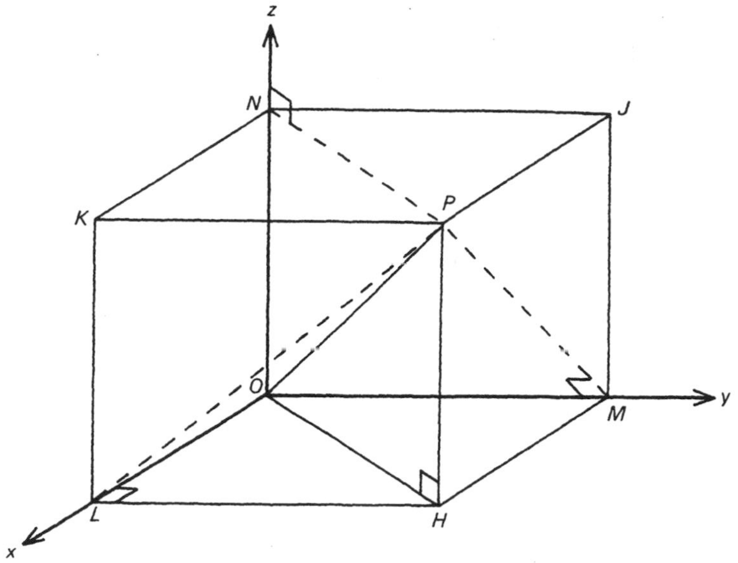

Figure XIII

and y_P are the coordinates of H in the x, y-plane; and z_P tells us how far P is above or below H.

Through thinking in this way, it is easy to convince oneself of the fact that each triple (α, β, γ) of real numbers is the coordinates of one and only one point in 3-dimensional space.

Returning to figure XIII and using the theorem of Pythagoras, we obtain

$$\begin{aligned} |OP|^2 &= |OH|^2 + |HP|^2 \qquad \text{(since } \angle OHP \text{ is a right angle)} \\ &= |OL|^2 + |LH|^2 + |HP|^2 \qquad \text{(since } \angle OLH \text{ is a right angle)} \\ &= |OL|^2 + |OM|^2 + |ON|^2 \qquad (OLHM \text{ and } OHPN \text{ being rectangles)} \\ &= x_P^2 + y_P^2 + z_P^2. \end{aligned}$$

Therefore:

8.1 For every point P, $|OP| = \sqrt{(x_P^2 + y_P^2 + z_P^2)}$.

We shall always use the symbols **i**, **j**, **k** for the unit vectors whose directions are the positive directions along the x-, y- and z-axes, respectively. An important fact is:

8.2 For every point P, $\mathbf{r}_P = x_P\mathbf{i} + y_P\mathbf{j} + z_P\mathbf{k}$.

Proof. We refer again to figure XIII, which we use as before to illustrate the coordinates of a typical point P.

By 5.8, since x_P is the coordinate of L on the x-axis, $[\overrightarrow{OL}] = x_P\mathbf{i}$. Similarly, $[\overrightarrow{OM}] = y_P\mathbf{j}$ and $[\overrightarrow{ON}] = z_P\mathbf{k}$. But

$$[\overrightarrow{OP}] = [\overrightarrow{OL}]+[\overrightarrow{LH}]+[\overrightarrow{HP}] \qquad \text{(cf. 4.1)}$$
$$= [\overrightarrow{OL}]+[\overrightarrow{OM}]+[\overrightarrow{ON}] \qquad \text{(cf. 3.2, } OLHM \text{ and } OHPN \text{ being rectangles).}$$

Therefore $\mathbf{r}_P = x_P\mathbf{i}+y_P\mathbf{j}+z_P\mathbf{k}$, as stated.

As well as revealing the connection between position vectors and coordinates, 8.2 leads via the next proposition (8.3) to some major developments in the study of E_3.

8.3 Let $\mathbf{x}$ be an arbitrary vector in E_3. Then $\mathbf{x}$ can be expressed in precisely one way as a linear combination of $\mathbf{i}$, $\mathbf{j}$, $\mathbf{k}$.

Proof. There are two things to prove: (1) that $\mathbf{x}$ can be expressed as a linear combination of $\mathbf{i}$, $\mathbf{j}$, $\mathbf{k}$; (2) that there is only one way in which this can be done. We take these two things one at a time.

(1) There is a point P whose position vector is $\mathbf{x}$, and, by 8.2, we have

$$\mathbf{x} = \mathbf{r}_P = x_P\mathbf{i}+y_P\mathbf{j}+z_P\mathbf{k}.$$

Thus $\mathbf{x}$ is expressible as a linear combination of $\mathbf{i}$, $\mathbf{j}$, $\mathbf{k}$.

(2) To show that there cannot be a second, different way of expressing $\mathbf{x}$ as a linear combination of $\mathbf{i}$, $\mathbf{j}$, $\mathbf{k}$, let us suppose that $\mathbf{x} = \alpha\mathbf{i}+\beta\mathbf{j}+\gamma\mathbf{k}$ and show that α, β, γ must coincide with x_P, y_P, z_P, respectively. For this purpose, introduce the point Q whose coordinates are (α, β, γ). Then we have

$$\mathbf{r}_Q = \alpha\mathbf{i}+\beta\mathbf{j}+\gamma\mathbf{k} \qquad \text{(by 8.2)}$$
$$= \mathbf{x} = \mathbf{r}_P.$$

Therefore Q coincides with P, and so $\alpha = x_P$, $\beta = y_P$, and $\gamma = z_P$.

It follows that there is only one way to express $\mathbf{x}$ as a linear combination of $\mathbf{i}$, $\mathbf{j}$, $\mathbf{k}$; and the proof of the stated result is complete.

To explore the implications of 8.3, consider an arbitrary vector $\mathbf{x}$ in E_3. We now know that there are scalars α, β, γ, uniquely determined by $\mathbf{x}$, such that

$$\mathbf{x} = \alpha\mathbf{i}+\beta\mathbf{j}+\gamma\mathbf{k}.$$

From now on we regard "$\alpha\mathbf{i}+\beta\mathbf{j}+\gamma\mathbf{k}$" as a standard way of expressing and denoting $\mathbf{x}$. Observe that it exhibits $\mathbf{x}$ split up as the sum of a vector parallel to the x-axis, a vector parallel to the y-axis, and a vector parallel to the z-axis. We call the uniquely determined scalars α, β, γ the **components** of $\mathbf{x}$ (with respect to $\mathbf{i}$, $\mathbf{j}$, $\mathbf{k}$). Moreover, we use the triple (α, β, γ) as an alternative notation for $\mathbf{x}$, so that (α, β, γ), if it stands for a vector, will mean $\alpha\mathbf{i}+\beta\mathbf{j}+\gamma\mathbf{k}$.

Notice that the triple (α, β, γ) may now be used to denote the coordinates of a certain point (P, say) *and* to denote a certain vector $\mathbf{x}$ (the vector $\alpha\mathbf{i}+\beta\mathbf{j}+\gamma\mathbf{k}$).

The point P and the vector $\mathbf{x}$ are closely related, since (cf. 8.2) $\mathbf{x}$ is the position vector of P. No confusion should arise from having these two possible meanings for the triple (α, β, γ); and in fact it will be found helpful in practice to have the same notation for the two associated things that specify a point P—the coordinates of P and the position vector of P.

A vector $\mathbf{x}$ represented as a triple (α, β, γ) is said to be expressed in **component form**. The next results give a formula for the magnitude of a vector expressed in component form and show how very easy it is to add vectors, or to subtract them or multiply them by scalars, when they are in component form.

8.4 The vector (α, β, γ) has magnitude $\sqrt{(\alpha^2+\beta^2+\gamma^2)}$;

$$\text{i.e.}\quad |\alpha\mathbf{i}+\beta\mathbf{j}+\gamma\mathbf{k}| = \sqrt{(\alpha^2+\beta^2+\gamma^2)}.$$

Proof. Let $\mathbf{x}$ be the vector (α, β, γ) and P the point with coordinates (α, β, γ). Then $\mathbf{x} = \mathbf{r}_P = [\overrightarrow{OP}]$. Hence

$$|\mathbf{x}| = |OP| = \sqrt{(\alpha^2+\beta^2+\gamma^2)} \qquad \text{(by 8.1)},$$

as claimed.

8.5 Let $(\alpha_1, \beta_1, \gamma_1)$, $(\alpha_2, \beta_2, \gamma_2)$ and (α, β, γ) denote vectors and let λ be a scalar. Then

(i) $(\alpha_1, \beta_1, \gamma_1)+(\alpha_2, \beta_2, \gamma_2) = (\alpha_1+\alpha_2, \beta_1+\beta_2, \gamma_1+\gamma_2)$,
(ii) $(\alpha_1, \beta_1, \gamma_1)-(\alpha_2, \beta_2, \gamma_2) = (\alpha_1-\alpha_2, \beta_1-\beta_2, \gamma_1-\gamma_2)$,
(iii) $\lambda(\alpha, \beta, \gamma) = (\lambda\alpha, \lambda\beta, \lambda\gamma)$.

One could sum these three facts up by saying that the operations in question may be carried out 'componentwise'. To prove the assertions is extremely easy: e.g., in the case of (i),

$$\begin{aligned}\text{left-hand side} &= \alpha_1\mathbf{i}+\beta_1\mathbf{j}+\gamma_1\mathbf{k}+\alpha_2\mathbf{i}+\beta_2\mathbf{j}+\gamma_2\mathbf{k}\\ &= (\alpha_1+\alpha_2)\mathbf{i}+(\beta_1+\beta_2)\mathbf{j}+(\gamma_1+\gamma_2)\mathbf{k} = \text{right-hand side.}\end{aligned}$$

Worked example. Let A, B, C be the points with coordinates $(3, -2, 4)$, $(4, 0, 2)$, and $(7, 6, -4)$, respectively. (*a*) Prove that A, B, C are collinear, and find the value of AB/BC. (*b*) Obtain the unit vector whose direction is that of the d.l.s. $\overrightarrow{AB}$.

Solution. [The solution to part (*a*) illustrates, amongst other things, an earlier remark about how helpful it is to be able to use the same notation for the coordinates of a point and for its position vector.]

(*a*) $[\overrightarrow{AB}] = \mathbf{r}_B-\mathbf{r}_A = (4,0,2)-(3,-2,4) = (1,2,-2)$ (cf. 8.5(ii));
and $[\overrightarrow{BC}] = \mathbf{r}_C-\mathbf{r}_B = (7,6,-4)-(4,0,2) = (3,6,-6)$
$= 3(1,2,-2)$ (cf. 8.5(iii)).

Hence $[\overrightarrow{AB}] = \frac{1}{3}[\overrightarrow{BC}]$, and from this it follows that A, B, C are collinear, with $AB/BC = \frac{1}{3}$.

(*b*) By 8.4, the vector $[\overrightarrow{AB}]$ has magnitude $\sqrt{(1^2+2^2+(-2)^2)} = 3$. Hence (cf. 5.7) the required unit vector is

$$\tfrac{1}{3}[\overrightarrow{AB}] = \tfrac{1}{3}(1,2,-2) = (\tfrac{1}{3}, \tfrac{2}{3}, -\tfrac{2}{3}) \qquad [\text{i.e. } \tfrac{1}{3}\mathbf{i}+\tfrac{2}{3}\mathbf{j}-\tfrac{2}{3}\mathbf{k}].$$

The finale in this section is the general 3-dimensional distance formula.

8.6 For any points A and B,

$$|AB| = \sqrt{[(x_B-x_A)^2+(y_B-y_A)^2+(z_B-z_A)^2]}.$$

Proof. $[\overrightarrow{AB}] = \mathbf{r}_B-\mathbf{r}_A = (x_B, y_B, z_B)-(x_A, y_A, z_A)$
$= (x_B-x_A, y_B-y_A, z_B-z_A).$

The distance $|AB|$ is the magnitude of this vector, and hence the stated result follows by 8.4.

9. Scalar products

As a preliminary, let us make clear what is meant by the *angle between* two nonzero vectors $\mathbf{x}$ and $\mathbf{y}$. For this purpose, introduce (cf. figure XIV) the points P and Q with position vectors $\mathbf{x}$ and $\mathbf{y}$. Then, quite simply, the angle between $\mathbf{x}$ and $\mathbf{y}$ is the measure in the range 0 to π inclusive (i.e. 0° to 180° inclusive) of $\angle POQ$. No questions of sign convention arise; moreover it is clear that the choice of origin used in the definition is immaterial.

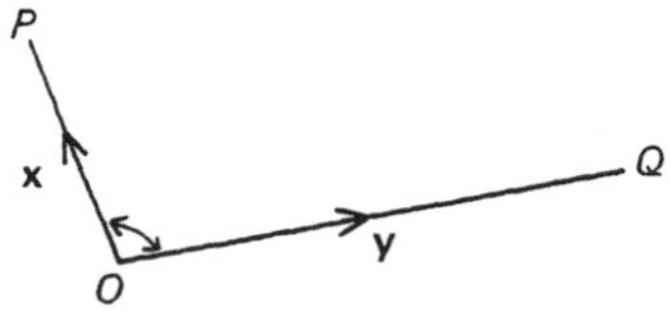

Figure XIV

The extreme possible values, 0 and π, arise when $\mathbf{x}$ and $\mathbf{y}$ are parallel—0 when they have the same direction and π when they have opposite directions.

We shall now define (for arbitrary $\mathbf{x}, \mathbf{y} \in E_3$) something called the **scalar product** of $\mathbf{x}$ and $\mathbf{y}$. As the name suggests, this is a scalar (depending on $\mathbf{x}$ and $\mathbf{y}$). It is denoted by $\mathbf{x}\,.\,\mathbf{y}$ (the dot being mandatory), and it is defined as follows:

(1) If $\mathbf{x}$ and $\mathbf{y}$ are nonzero, then $\mathbf{x}\,.\,\mathbf{y} = |\mathbf{x}||\mathbf{y}| \cos\theta$, where θ is the angle between $\mathbf{x}$ and $\mathbf{y}$;

(2) If $\mathbf{x}$ or $\mathbf{y}$ is $\mathbf{0}$ (or if both are $\mathbf{0}$), then $\mathbf{x}\,.\,\mathbf{y} = 0$. (Case (2) must be given separately because the angle between $\mathbf{x}$ and $\mathbf{y}$ does not exist if either vector is zero.)

There are several things in physics that provide very strong motivation for the definition of the scalar product of two vectors. Soon (see 9.3 below) we shall obtain a simple formula for $\mathbf{x} \cdot \mathbf{y}$ in terms of the components of $\mathbf{x}$ and $\mathbf{y}$; and the simplicity of this formula confirms that the scalar product of two vectors is a natural thing for mathematicians to consider.

From the definition it is clear that:

9.1 In all cases, $\mathbf{x} \cdot \mathbf{y} = \mathbf{y} \cdot \mathbf{x}$ $(\mathbf{x}, \mathbf{y} \in E_3)$.

Next we observe that in case (1) of the definition (where $\mathbf{x}, \mathbf{y}$ are nonzero and $\mathbf{x} \cdot \mathbf{y}$ is given by $|\mathbf{x}||\mathbf{y}| \cos \theta$) the factors $|\mathbf{x}|$ and $|\mathbf{y}|$ are certainly nonzero, while the $\cos \theta$ factor is zero if and only if θ is a right angle. Therefore:

9.2 If $\mathbf{x}$ and $\mathbf{y}$ are nonzero vectors, then $\mathbf{x} \cdot \mathbf{y} = 0$ if and only if $\mathbf{x}$ and $\mathbf{y}$ are perpendicular to each other.

We shall now derive the simple expression for $\mathbf{x} \cdot \mathbf{y}$ in terms of the components of $\mathbf{x}$ and $\mathbf{y}$.

9.3 Let $\mathbf{x} = (x_1, x_2, x_3)$ and $\mathbf{y} = (y_1, y_2, y_3)$ be arbitrary vectors in E_3. Then

$$\mathbf{x} \cdot \mathbf{y} = x_1 y_1 + x_2 y_2 + x_3 y_3.$$

Proof. We shall deal with the case where both $\mathbf{x}$ and $\mathbf{y}$ are nonzero, the other cases being trivial.

Let P and Q be the points with position vectors $\mathbf{x}$ and $\mathbf{y}$, so that P has coordinates (x_1, x_2, x_3) and Q has coordinates (y_1, y_2, y_3).

Figure XV

We apply the cosine formula to the triangle OPQ, O being the origin (cf. figure XV). The formula tells us that

$$\begin{aligned} |PQ|^2 &= |OP|^2 + |OQ|^2 - 2|OP||OQ| \cos(\angle POQ) \\ &= |\mathbf{x}|^2 + |\mathbf{y}|^2 - 2(\mathbf{x} \cdot \mathbf{y}). \end{aligned}$$

Hence, by 8.4 and 8.6,

$$(y_1 - x_1)^2 + (y_2 - x_2)^2 + (y_3 - x_3)^2 = (x_1^2 + x_2^2 + x_3^2) + (y_1^2 + y_2^2 + y_3^2) - 2(\mathbf{x} \cdot \mathbf{y}).$$

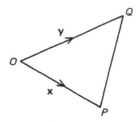

On expanding the left-hand side and then cancelling the $x_1^2, \ldots, y_3^2$ terms from both sides, we are left with

$$-2x_1y_1 - 2x_2y_2 - 2x_3y_3 = -2(\mathbf{x}\,.\,\mathbf{y}),$$

and from this the stated result at once follows.

Worked example (illustrating how 9.3 makes it easy to find the angle between two vectors given in component form). Find the angle between the vectors $(1, 0, -1)$ and $(7, -1, -2)$ (i.e. the vectors $\mathbf{i}-\mathbf{k}$ and $7\mathbf{i}-\mathbf{j}-2\mathbf{k}$).

Solution. Call the given vectors $\mathbf{x}$ and $\mathbf{y}$, and let θ be the angle between them. By the definition of $\mathbf{x}\,.\,\mathbf{y}$,

$$\mathbf{x}\,.\,\mathbf{y} = |\mathbf{x}||\mathbf{y}| \cos \theta.$$

But also, by 9.3, $\mathbf{x}\,.\,\mathbf{y} = 1 \times 7 + 0 \times (-1) + (-1) \times (-2) = 9$, while (by 8.4) $|\mathbf{x}| = \sqrt{2}$ and $|\mathbf{y}| = \sqrt{54} = 3\sqrt{6}$. Hence

$$9 = \sqrt{2} \times 3\sqrt{6} \times \cos \theta = 3\sqrt{12} \cos \theta = 6\sqrt{3} \cos \theta,$$

and so $\cos \theta = 9/6\sqrt{3} = \sqrt{3}/2$.

It follows that $\theta = \pi/6$ (or 30°).

The formula 9.3 enables us to prove mechanically the following "distributive laws".

9.4 For all $\mathbf{x}, \mathbf{y}, \mathbf{z} \in E_3$,
(i) $\mathbf{x}\,.\,(\mathbf{y}+\mathbf{z}) = \mathbf{x}\,.\,\mathbf{y}+\mathbf{x}\,.\,\mathbf{z}$ and (ii) $(\mathbf{y}+\mathbf{z})\,.\,\mathbf{x} = \mathbf{y}\,.\,\mathbf{x}+\mathbf{z}\,.\,\mathbf{x}$.

9.5 For all $\mathbf{x}, \mathbf{y} \in E_3$ and $\lambda \in \mathbb{R}$, $\mathbf{x}\,.\,(\lambda\mathbf{y}) = (\lambda\mathbf{x})\,.\,\mathbf{y} = \lambda(\mathbf{x}\,.\,\mathbf{y})$.

It is, of course, true that each half of 9.4 follows immediately from the other half because of the commutative property 9.1. As a sample of the ease with which these propositions can be proved, here is a proof of 9.4(i).

Let $\mathbf{x} = (x_1, x_2, x_3)$, $\mathbf{y} = (y_1, y_2, y_3)$, $\mathbf{z} = (z_1, z_2, z_3)$. Then

$$\begin{aligned} \text{left-hand side} &= (x_1, x_2, x_3)\,.\,(y_1+z_1, y_2+z_2, y_3+z_3) \\ &= x_1(y_1+z_1)+x_2(y_2+z_2)+x_3(y_3+z_3) \qquad \text{(by 9.3)} \\ &= (x_1y_1+x_2y_2+x_3y_3)+(x_1z_1+x_2z_2+x_3z_3) \\ &= \mathbf{x}\,.\,\mathbf{y}+\mathbf{x}\,.\,\mathbf{z} \qquad \text{(by 9.3).} \end{aligned}$$

It is fair to say that the scalar product operation is another example of a "no-catch operation", provided one bears in mind the basic fact that "$\mathbf{x}\,.\,\mathbf{y}$" always denotes a scalar. The student should realize, in particular, that the propositions 9.4 and 9.5 can be used (along with 9.1) to prove the correctness of natural-looking expansions of relatively complicated scalar products—e.g.

$$(\mathbf{a}+\mathbf{b})\,.\,(\mathbf{c}-\mathbf{d}) = \mathbf{a}\,.\,\mathbf{c}+\mathbf{b}\,.\,\mathbf{c}-\mathbf{a}\,.\,\mathbf{d}-\mathbf{b}\,.\,\mathbf{d},$$
$$\text{and} \quad (2\mathbf{x}-3\mathbf{y})\,.\,(\mathbf{x}+4\mathbf{y}) = 2(\mathbf{x}\,.\,\mathbf{x})+5(\mathbf{x}\,.\,\mathbf{y})-12(\mathbf{y}\,.\,\mathbf{y}).$$

Obviously the angle between any nonzero vector $\mathbf{x}$ and itself is 0, and so (for $\mathbf{x} \neq \mathbf{0}$) $\mathbf{x}.\mathbf{x} = |\mathbf{x}||\mathbf{x}| \cos 0 = |\mathbf{x}|^2$. When $\mathbf{x} = \mathbf{0}$, it is still true that $\mathbf{x}.\mathbf{x} = |\mathbf{x}|^2$. Thus:

9.6 For every vector $\mathbf{x}$, $\mathbf{x}.\mathbf{x} = |\mathbf{x}|^2$.

Worked example. (This involves expansion of a complicated scalar product and the use of 9.6 to find a length.) In a tetrahedron $OABC$, $|OA| = 2$, $|OB| = 4$, $|OC| = 3$; $\angle COA$ is a right angle, while both $\angle BOC$ and $\angle AOB$ are 60°; and G is the centroid of the tetrahedron. Find $|OG|$.

Solution. We use position vectors relative to O as origin, and we write $\mathbf{a}$ for $\mathbf{r}_A$, $\mathbf{b}$ for $\mathbf{r}_B$, etc. By 7.2, the position vector of G is $\mathbf{g} = \frac{1}{4}(\mathbf{a}+\mathbf{b}+\mathbf{c})$. The required length $|OG|$ is $|\mathbf{g}|$, and so, using 9.6, we have

$$|OG|^2 = \mathbf{g}.\mathbf{g} = \tfrac{1}{16}(\mathbf{a}+\mathbf{b}+\mathbf{c}).(\mathbf{a}+\mathbf{b}+\mathbf{c})$$
$$= \tfrac{1}{16}[(\mathbf{a}.\mathbf{a})+(\mathbf{b}.\mathbf{b})+(\mathbf{c}.\mathbf{c})+2(\mathbf{a}.\mathbf{b})+2(\mathbf{b}.\mathbf{c})+2(\mathbf{a}.\mathbf{c})].$$

Now $\mathbf{a}.\mathbf{a} = |\mathbf{a}|^2 = |OA|^2 = 4$; and similarly $\mathbf{b}.\mathbf{b} = 16$, $\mathbf{c}.\mathbf{c} = 9$. Further, $\mathbf{a}.\mathbf{b} = |OA||OB| \cos(\angle AOB) = 2 \times 4 \times \frac{1}{2} = 4$. Similarly $\mathbf{b}.\mathbf{c}$ works out to be 6, while $\mathbf{a}.\mathbf{c}$ is 0 since $\angle AOC$ is a right angle (cf. 9.2). Hence

$$|OG|^2 = \tfrac{1}{16}(4+16+9+2 \times 4+2 \times 6+2 \times 0) = \tfrac{49}{16}.$$

Therefore $|OG| = \frac{7}{4}$.

Several geometrical applications of scalar products hinge on 9.2, which enables the perpendicularity of two lines to be translated into an algebraic statement (that a certain scalar product is zero). A good illustration is the very simple proof which we now give of the concurrence of the altitudes of a triangle.

Since it is obvious that the altitudes of a right-angled triangle are concurrent, we consider a non-right-angled triangle ABC. Introduce the altitudes through A and B, and let them intersect at H (cf. figure XVI). Take H as origin, and let $\mathbf{a}$, $\mathbf{b}$, $\mathbf{c}$ be the position vectors of A, B, C, respectively (relative to H).

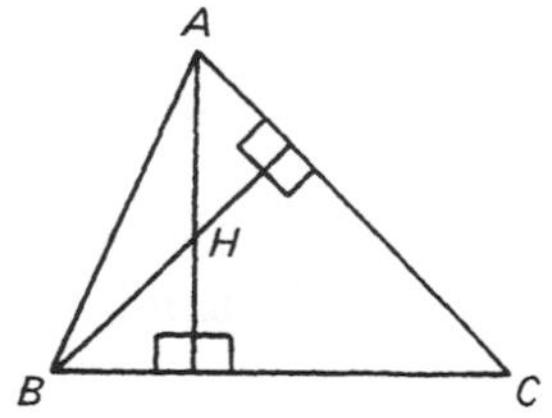

Figure XVI

The vectors $[\overrightarrow{HA}]$ and $[\overrightarrow{BC}]$ are perpendicular to each other. Therefore $[\overrightarrow{HA}].[\overrightarrow{BC}] = 0$, i.e. $\mathbf{a}.(\mathbf{c}-\mathbf{b}) = 0$, i.e. $\mathbf{a}.\mathbf{c}-\mathbf{a}.\mathbf{b} = 0$,

$$\text{i.e.} \quad \mathbf{a}.\mathbf{c} = \mathbf{a}.\mathbf{b}.$$

Similarly, $[\overrightarrow{HB}].[\overrightarrow{AC}] = 0$, i.e. $\mathbf{b}.(\mathbf{c}-\mathbf{a}) = 0$; and hence

$$\mathbf{b}.\mathbf{c} = \mathbf{a}.\mathbf{b}.$$

Now $[\overrightarrow{HC}].[\overrightarrow{AB}] = \mathbf{c}.(\mathbf{b}-\mathbf{a}) = \mathbf{b}.\mathbf{c}-\mathbf{a}.\mathbf{c}$; and this is zero since, as seen above, both $\mathbf{b}.\mathbf{c}$ and $\mathbf{a}.\mathbf{c}$ are equal to $\mathbf{a}.\mathbf{b}$. Therefore $[\overrightarrow{HC}]$ is perpendicular to $[\overrightarrow{AB}]$, and hence it is apparent that CH (produced) is the third altitude of the triangle.

This shows that the three altitudes of the triangle ABC are concurrent (at H).

10. Postscript

Although the discussion of E_3 and its applications to geometry can be taken much further, enough has already been said about this subject-matter for this chapter to serve its purpose in a book on linear algebra—to give the reader some familiarity with the system E_3 and to prepare the way for generalizations which are important in linear algebra. Before we move on to a different topic, it remains only to make some miscellaneous tidy-up remarks.

(*a*) Right from our first introduction of vectors in this chapter we have worked in three dimensions and have discussed the system E_3 containing vectors with all possible directions in 3-dimensional space. Instead, though, one can decide at the outset to stick to two dimensions as in elementary plane geometry. In that case, one restricts attention to a plane Ω and considers vectors with all the directions that exist in Ω. The system of vectors so conceived is useful for purely 2-dimensional geometrical purposes; and algebraically it is a perfectly satisfactory self-contained system, to which, naturally, we give the name E_2.

In due course in a discussion of E_2 one would introduce x- and y-axes in the plane Ω, and thus it is natural to identify the "2-dimensional universe" Ω with the x, y-plane of coordinatized 3-dimensional space. Accordingly E_2 is identified with a specific subset of E_3, namely the set of vectors in E_3 that arise as position vectors of points in the x, y-plane—i.e. the set of vectors that are expressible as linear combinations of $\mathbf{i}$ and $\mathbf{j}$.

(*b*) In E_3, one describes vectors $\mathbf{x}_1, \mathbf{x}_2, \mathbf{x}_3, \ldots$ as **coplanar** if and only if the representations of these vectors starting at the origin are coplanar. (This is another place where one's choice of origin is immaterial.) For example, the vectors $\mathbf{i}+\mathbf{j}, \mathbf{i}+2\mathbf{j}, \mathbf{i}+3\mathbf{j}$ are coplanar since the representations of these vectors starting at the origin all lie in the x, y-plane; on the other hand, the vectors $\mathbf{i}, \mathbf{j}, \mathbf{k}$ are clearly non-coplanar.

Suppose that $\mathbf{u}$, $\mathbf{v}$, $\mathbf{w}$ are three vectors which are not coplanar. It can be shown that an arbitrary vector $\mathbf{x}$ is expressible in precisely one way as a linear combination $\alpha\mathbf{u}+\beta\mathbf{v}+\gamma\mathbf{w}$ of $\mathbf{u}$, $\mathbf{v}$ and $\mathbf{w}$. So the idea of components can be developed with respect to any triad of non-coplanar vectors (not just with respect to the special triad $\mathbf{i}, \mathbf{j}, \mathbf{k}$). It must be noted, however, that nice formulae like 8.4 and 9.3 do not apply to components with respect to an arbitrary non-coplanar triad: they rest on the special properties of $\mathbf{i}, \mathbf{j}, \mathbf{k}$.

(*c*) The final remark of § 2 referred to a one-to-one correspondence between the vectors in E_3 and the points in 3-dimensional space. Ideas encountered since then bring this same correspondence to light in the form

$$\text{vector } \mathbf{x} \leftrightarrow \text{point with position vector } \mathbf{x},$$

or (to express exactly the same thing in another way)

$$\text{vector } (\alpha, \beta, \gamma) \leftrightarrow \text{point with coordinates } (\alpha, \beta, \gamma).$$

Sometimes it is helpful to blur or forget the distinction between a vector and the corresponding point, and accordingly to use $\mathbf{x}$ as a symbol for "the point with position vector $\mathbf{x}$". This is especially so in the discussion of geometrical transformations of 2- or 3-dimensional space. For, if we identify each point with its position vector, such a geometrical transformation can be represented as an algebraic transformation of E_2 or E_3. A straightforward example should make the general idea clear. Consider the transformation of 3-dimensional space in which every point is moved out radially from the origin O to twice its original distance from O: this transformation can be described algebraically by the very simple formula

$$\mathbf{x} \mapsto 2\mathbf{x} \qquad (\mathbf{x} \in E_3),$$

meaning that, for each $\mathbf{x} \in E_3$, the point with position vector $\mathbf{x}$ goes to the point with position vector $2\mathbf{x}$ under the transformation.

EXERCISES ON CHAPTER ONE

1. If the vector **x** has magnitude 4, what are (*a*) the vector with magnitude 8 and the same direction as **x**, (*b*) the vector with magnitude 2 and direction opposite to that of **x**, (*c*) the unit vectors parallel to **x**?

2. Let ABC be a triangle. Let D be the point on BC such that $BD/DC = \frac{1}{2}$ and E the point on AC such that $AE/EC = \frac{1}{3}$; and let F be the mid-point of AD. Using position vectors, show that B, F, E are collinear, and find the value of BF/FE.

3. Let ABC be a triangle, and let D, E, F be points on the sides BC, CA, AB, respectively, such that $BD/DC = CE/EA = AF/FB$. Prove that the centroids of the triangles ABC and DEF coincide.

4. Let $ABCD$ be a quadrilateral, and let M_1, M_2, M_3, M_4 be the mid-points of the sides AB, BC, CD, DA, respectively. By using position vectors relative to an arbitrary origin, show that $M_1M_2M_3M_4$ is a parallelogram.

Let O be a point not in the plane of $ABCD$, and let G_1, G_2, G_3, G_4 be the centroids of the triangles OCD, ODA, OAB, OBC, respectively. Taking O as origin, prove that the lines $M_1G_1, M_2G_2, M_3G_3, M_4G_4$ are concurrent. (Hint: consider the point P on M_1G_1 such that $M_1P/PG_1 = \frac{3}{2}$.)

5. Let **x** and **y** be non-parallel vectors. Prove that the vectors $\mathbf{x}+\mathbf{y}$ and $\mathbf{x}-\mathbf{y}$ are perpendicular to each other if and only if $|\mathbf{x}| = |\mathbf{y}|$. Do this (*a*) by considering a parallelogram whose sides represent **x** and **y**, (*b*) by using scalar products.

6. Let ABC be a non-right-angled triangle with centroid G. Let **a**, **b**, **c** be the position vectors of the vertices relative to the circumcentre O of the triangle. (This is the point where the perpendicular bisectors of the sides meet, i.e. the point equidistant from the three vertices.) By using the result in question 5, show that the point with position vector $\mathbf{a}+\mathbf{b}+\mathbf{c}$ coincides with the orthocentre H of the triangle. Deduce that O, G, H are collinear and that, if these three points do not coincide, $OG/GH = \frac{1}{2}$. (The result remains true, and is easily proved, in the case of a right-angled triangle. The line OGH is called the *Euler line* of triangle ABC.)

7.* Let OAB be a triangle. Take O as origin, and let A, B have position vectors **a**, **b**, respectively. Let P be the point on OA such that $OP/PA = 2$; let Q be the point on the line AB such that $AQ/QB = -\frac{3}{4}$; and let X be the point where the lines PQ and OB intersect. Find the position vector of X and the value of PX/XQ.

8. Let A, B, C be the points with coordinates $(-3, 4, -6)$, $(0, 1, 3)$, $(1, 0, 6)$, respectively. Prove that A, B, C are collinear, and find the ratio AB/BC. Find also the coordinates of the point D on the line ABC such that $AD/DC = \frac{1}{3}$.

9. Show that the points with coordinates $(1, 2, 0)$, $(4, -1, 2)$ and $(-2, 2, 4)$ are the vertices of a triangle (i.e. are not collinear), and find the coordinates of the centroid of this triangle.

10. What is the vector of magnitude 14 with the same direction as $2\mathbf{i}-3\mathbf{j}+6\mathbf{k}$?

11. Find the angle between the vectors $(7, -4, 4)$ and $(1, -1, 4)$.

12. Let **x** be an arbitrary vector. Show that the components of **x** with respect to **i**, **j**, **k** are $\mathbf{x}.\mathbf{i}$, $\mathbf{x}.\mathbf{j}$, $\mathbf{x}.\mathbf{k}$. Deduce that

$$(\mathbf{x}.\mathbf{i})^2+(\mathbf{x}.\mathbf{j})^2+(\mathbf{x}.\mathbf{k})^2 = |\mathbf{x}|^2.$$

Now suppose that **x** is nonzero and that α, β, γ are the angles between **x** and the vectors **i**, **j**, **k**, respectively. Show that

$$\cos^2\alpha+\cos^2\beta+\cos^2\gamma = 1.$$

13. Let $\mathbf{u}$ and $\mathbf{v}$ denote unit vectors the angle between which is $\pi/3$, and let $\mathbf{w} = 3\mathbf{u} + 5\mathbf{v}$. Find $|\mathbf{w}|$ by considering $\mathbf{w}.\mathbf{w}$.

14.* Let $OABC$ be a regular tetrahedron whose edges are taken to have unit length. Let G be the centroid of the tetrahedron, and let H be the point on the line OG such that G is the mid-point of OH. Using position vectors relative to O, prove that the three lines HA, HB, HC are mutually perpendicular.

15. Let OAB be a triangle. Using position vectors relative to O, prove that the point C on AB such that $AC/CB = |OA|/|OB|$ lies on the bisector of $\angle AOB$.

16.* (i) Let $\mathbf{x}, \mathbf{y}, \mathbf{z}$ be vectors such that $\mathbf{x}, \mathbf{y}$ are non-parallel while $\mathbf{x}, \mathbf{y}, \mathbf{z}$ are coplanar. By a suitable geometrical construction, show that $\mathbf{z}$ is expressible as a linear combination of $\mathbf{x}$ and $\mathbf{y}$.

(ii) Suppose that the origin O lies in the plane of triangle ABC, but not on any of its sides. Let $\mathbf{a}, \mathbf{b}, \mathbf{c}$ be the position vectors of A, B, C. By applying part (i) with $\mathbf{x} = [\overrightarrow{AC}]$, $\mathbf{y} = [\overrightarrow{BC}]$ and $\mathbf{z} = [\overrightarrow{OC}]$, prove that there are scalars α, β, γ such that

$$\alpha\mathbf{a}+\beta\mathbf{b}+\gamma\mathbf{c} = \mathbf{0} \quad \text{and} \quad \alpha+\beta+\gamma = 1.$$

Let the lines OA, OB, OC cut the lines BC, CA, AB at D, E, F, respectively. Show that the point on BC with position ratio γ/β with respect to B and C coincides with D. Deduce that

$$(BD/DC) \times (CE/EA) \times (AF/FB) = 1. \qquad \text{(This is } \textit{Ceva's theorem.})$$

Prove also that

$$(OA/AD)+(OB/BE)+(OC/CF) = -2.$$

CHAPTER TWO

MATRICES

11. Introduction

The purpose of this chapter is to give an account of the elementary properties of matrices—the rectangular arrays of numbers (or other coefficients) that arise from systems of linear equations and in other contexts. In particular, the chapter will introduce operations such as addition and multiplication of matrices and will consider the properties of these operations.

When we come to discuss matrices and related ideas in linear algebra, we always have in mind a system of *scalars*—numbers or number-like objects which may arise as coefficients in equations or in other ways. In elementary work the system of scalars is likely to be $\mathbb{R}$ (the set of real numbers) or perhaps $\mathbb{C}$ (the set of complex numbers), but, from a more advanced point of view, these are just two among many possibilities. Naturally, one wants to construct a body of theory that takes account of these many possibilities and is applicable to the broadest possible variety of sensible systems of scalars. This generality can, without complicating matters, be built into our study of linear algebra right from the start—simply by declaring that we are using as set of available scalars a system F (which could be $\mathbb{R}$ or $\mathbb{C}$ or some other system). The symbol F will be reserved for this purpose throughout the rest of the book; and, whenever the symbol F appears, the reader will understand, without any further explanation being necessary, that F stands for the (general) system of scalars in use.

As suggested above, we want a body of theory that applies to any *sensible* system of scalars. So it is not our intention that F may stand for any algebraic system whatever. Indeed all the generality that we want can be achieved while stipulating that the system F must have all the basic nice properties of $\mathbb{R}$ and $\mathbb{C}$. To be precise, F must be what mathematicians call a *field*. This is not the place to explain in full detail what this technical term "field" means. Suffice it to say that, in demanding that F be a field, we are demanding in particular that: (i) it is possible to add, subtract and multiply the "scalars" in F; (ii) these operations obey the same basic "laws of algebra" as real numbers (e.g. it must be the case that $\alpha(\beta+\gamma) = \alpha\beta+\alpha\gamma$ for all $\alpha, \beta, \gamma \in F$); and (iii) each nonzero scalar in F has a reciprocal ($1/\alpha$ or α^{-1}) in F.

Let no one be put off by this slightly technical digression: the truth is that one will lose nothing if one simply regards F as standing for "$\mathbb{R}$ or $\mathbb{C}$ or some other similar system".

Students who are familiar with the notion of "a field" should consider one further point. In certain fields it is the case that $1+1 = 0$. Such fields are said to have characteristic two. In a few places in our work in linear algebra it is necessary to exclude such fields, i.e. to stipulate that our field of scalars F does not have characteristic two. This we indicate by the abbreviated statement "char $F \neq 2$".

12. Basic nomenclature for matrices

A matrix is a rectangular array of scalars : e.g.

$$\begin{bmatrix} 3 & 1 & -1 \\ -1 & 0 & 4 \end{bmatrix}.$$

The scalars appearing in the matrix are called its **entries**. Each horizontal line of entries is called a **row** of the matrix; each vertical line of entries is called a **column** of the matrix. In any matrix we number the rows from top to bottom and the columns from left to right: so the 1st row means the top row, and the 1st column means the column at the extreme left.

A matrix with m rows and n columns is called an $m \times n$ matrix or a matrix of **type** $m \times n$. So the form of an $m \times n$ matrix is

$$\begin{array}{r} \text{1st row} \rightarrow \\ \\ \\ i\text{th row} \rightarrow \\ \\ m\text{th row} \rightarrow \end{array} \begin{bmatrix} \alpha_{11} & \alpha_{12} & \cdots & \alpha_{1k} & \cdots & \alpha_{1n} \\ \alpha_{21} & \alpha_{22} & \cdots & \alpha_{2k} & \cdots & \alpha_{2n} \\ & & & \vdots & & \\ \alpha_{i1} & \alpha_{i2} & \cdots & \alpha_{ik} & \cdots & \alpha_{in} \\ & & & \vdots & & \\ \alpha_{m1} & \alpha_{m2} & \cdots & \alpha_{mk} & \cdots & \alpha_{mn} \end{bmatrix}.$$

$$\begin{array}{ccc} \uparrow & \uparrow & \uparrow \\ \text{1st} & k\text{th} & n\text{th} \\ \text{column} & \text{column} & \text{column} \end{array}$$

Notice the use of double subscripts to indicate the positions of the entries. The first subscript tells which row the entry is in, while the second subscript tells which column the entry is in. Thus, in the above matrix, α_{ik} is the entry in the ith row and the kth column: we call this the (i, k)th entry of the matrix.

Remarks. (*a*) We shall normally denote matrices by capital letters (A, B, C, etc.).

(*b*) A *real matrix* is a matrix whose entries are all real numbers. A *complex matrix* is one whose entries are all complex numbers. More generally, a matrix whose entries are scalars from the system F is often described as a matrix *over* F. We shall denote by $F_{m \times n}$ the set of all $m \times n$ matrices over F.

(*c*) Let us consider matrices with only one row or only one column.

A $1 \times n$ matrix is of the form

$$[\alpha_1 \quad \alpha_2 \quad \dots \quad \alpha_n].$$

Such a matrix is called a row matrix or a row vector or, quite simply, a row.

An $m \times 1$ matrix is of the form

$$\begin{bmatrix} \alpha_1 \\ \alpha_2 \\ \vdots \\ \alpha_m \end{bmatrix}.$$

Such a matrix is called a column matrix or a column vector or, quite simply, a column.

For obvious reasons of economy of space, we bring in the alternative notation $\operatorname{col}(\alpha_1, \alpha_2, \dots, \alpha_m)$ for the column matrix displayed above. Correspondingly, $\operatorname{row}(\alpha_1, \alpha_2, \dots, \alpha_n)$ is an alternative notation for the row matrix displayed above.

(*d*) A **square matrix** is a matrix in which the number of rows is the same as the number of columns—i.e. a matrix of type $n \times n$ for some positive integer n. In an $n \times n$ matrix the **main diagonal** means the diagonal line of entries consisting of the $(1, 1)$th, $(2, 2)$th, $\dots$, (n, n)th entries (cf. top left to bottom right).

(*e*) The **zero matrix** in $F_{m \times n}$ is the $m \times n$ matrix whose every entry is the zero scalar. The zero matrix is denoted by $O_{m,n}$, or simply by O if the type is obvious from the context.

(*f*) We use the notation

$$[\alpha_{ik}]_{m \times n}$$

as an abbreviation for the $m \times n$ matrix whose (i, k)th entry is α_{ik}.

(*g*) It is important to establish a clear meaning for the statement "$A = B$" that two matrices A and B are equal. This will mean that (i) A and B are of the same type ($m \times n$, say) and (ii) the (i, k)th entry of A equals the (i, k)th entry of B for all relevant i, k. ("All relevant i, k" is a convenient shorthand for "all i in the range $1, 2, \dots, m$ and all k in the range $1, 2, \dots, n$".)

Observe that if $A = [\alpha_{ik}]_{m \times n}$ and $B = [\beta_{ik}]_{m \times n}$ are $m \times n$ matrices, then the matrix equation "$A = B$" is equivalent to mn scalar equations, namely $\alpha_{11} = \beta_{11}, \alpha_{12} = \beta_{12}, \dots, \alpha_{mn} = \beta_{mn}$.

13. Addition and subtraction of matrices

It should be noted at the outset that a meaning is given to the sum $A+B$ of two matrices A and B only when A and B are of the same type. The same goes for $A-B$.

The definition of matrix addition is as follows. Let $A = [\alpha_{ik}]_{m\times n}$ and $B = [\beta_{ik}]_{m\times n}$ be matrices in $F_{m\times n}$. Then $A+B$ is the $m \times n$ matrix whose (i, k)th entry is in every case $\alpha_{ik}+\beta_{ik}$.

So in practice each entry of $A+B$ is obtained by adding the corresponding entries of A and B—a fact that may be summarized by saying that matrix addition is performed "entrywise". For example

$$\begin{bmatrix} 3 & 1 & -1 \\ -1 & 0 & 4 \end{bmatrix} + \begin{bmatrix} 4 & 3 & 1 \\ 0 & -5 & 2 \end{bmatrix} = \begin{bmatrix} 7 & 4 & 0 \\ -1 & -5 & 6 \end{bmatrix}.$$

The negative, $-A$, of a matrix $A = [\alpha_{ik}]_{m\times n}$ is defined to be the matrix of type $m \times n$ whose (i, k)th entry is in every case $-\alpha_{ik}$; i.e. $-A$ is the matrix obtained from A by replacing each entry by its negative.

In the light of remark (*c*) in §4, it is now scarcely necessary to say that subtraction of matrices in $F_{m\times n}$ is defined by the equation

$$A-B = A+(-B) \qquad (A, B \in F_{m\times n}).$$

As one quickly sees, this definition means that in practice the difference $A-B$ of two given matrices A, B can be obtained by "entrywise subtraction". For example

$$\begin{bmatrix} 3 & 1 & -1 \\ -1 & 0 & 4 \end{bmatrix} - \begin{bmatrix} 4 & 3 & 1 \\ 0 & -5 & 2 \end{bmatrix} = \begin{bmatrix} -1 & -2 & -2 \\ -1 & 5 & 2 \end{bmatrix}.$$

The operations of addition and subtraction of matrices in $F_{m\times n}$ are "no-catch" operations (cf. remark (*d*) in §4). The underlying reason for this pleasant state of affairs is the truth of the following four basic properties of matrix addition.

13.1 (The associative property)

$$(A+B)+C = A+(B+C) \qquad (\text{for all } A, B, C \in F_{m\times n}).$$

13.2 (The commutative property)

$$A+B = B+A \qquad (\text{for all } A, B \in F_{m\times n}).$$

13.3 $A+O_{m,n} = A$ (for all $A \in F_{m\times n}$).

13.4 $A+(-A) = O_{m,n}$ (for all $A \in F_{m\times n}$).

These are almost obvious. To provide a formal proof of any of them, one must fall back on the meaning of a statement that two matrices are equal (cf. remark (*g*) in §12). As an illustration, here is a proof of 13.2.

Let $A = [\alpha_{ik}]_{m\times n}$, $B = [\beta_{ik}]_{m\times n}$ be arbitrary matrices in $F_{m\times n}$. Then both $A+B$ and $B+A$ are $m \times n$ matrices. Further, for all relevant i, k,

$$(i, k)\text{th entry of } A+B = \alpha_{ik}+\beta_{ik} = (i, k)\text{th entry of } B+A.$$

Hence $A+B = B+A$; and this proves the result.

Several elementary propositions mentioned in the next few pages can be proved by the same approach—i.e. by showing that the two matrices alleged to be equal are of the same type and have equal (i, k)th entries for all relevant i, k. As in the case of 13.2, such proofs are often very straightforward, and so most of them will be omitted without comment.

In more advanced work the consideration of (i, k)th entries is potentially very messy and complicated. Therefore, once the basic facts of the algebra of matrices have been established, the recommended tactics for dealing with further problems are to use the established basic facts and to avoid, whenever possible, having to work with (i, k)th entries.

14. Multiplication of a matrix by a scalar

Let A be a matrix in $F_{m\times n}$—say $A = [\alpha_{ik}]_{m\times n}$; and let $\lambda \in F$. Then λA is defined to be the $m \times n$ matrix whose (i, k)th entry is $\lambda\alpha_{ik}$.

So, for example, if $A = \begin{bmatrix} 3 & 1 & -1 \\ -1 & 0 & 4 \end{bmatrix}$, then $4A = \begin{bmatrix} 12 & 4 & -4 \\ -4 & 0 & 16 \end{bmatrix}$.

The fundamental properties of the operation of multiplying a matrix by a scalar are as follows.

14.1 $\lambda(A+B) = \lambda A+\lambda B \qquad (A, B\in F_{m\times n};\ \lambda\in F).$

14.2 $(\lambda+\mu)A = \lambda A+\mu A \qquad (A\in F_{m\times n};\ \lambda, \mu\in F).$

14.3 $\lambda(\mu A) = (\lambda\mu)A \qquad (A\in F_{m\times n};\ \lambda, \mu\in F).$

14.4 $1A = A \qquad (A\in F_{m\times n}).$

These fundamental properties and their consequences enable the operation to be designated as yet another "no-catch" operation.

At this point the student should be able to discern that the systems $F_{m\times n}$ and E_3 have something significant in common. One might express it by saying that they are two examples of the same mathematical phenomenon—the phenomenon of a set of objects that can be added together and multiplied by scalars, these operations obeying all natural-looking laws. This is an important observation. It ties in with the hint dropped in the second paragraph of §1, about generalization of the idea of a system of vectors; and, like that paragraph, it points forward to chapter 5, where we take the far-reaching step of studying in general the "mathematical phenomenon" of which both $F_{m\times n}$ and E_3 are examples.

15. Multiplication of matrices

From this section onwards, frequent use will be made of the $\sum$-notation exemplified by $\sum_{j=1}^{n} \alpha_j$, which is a short-hand for the sum $\alpha_1 + \alpha_2 + \ldots + \alpha_n$.

Our concern in this section is with defining, in certain circumstances, the product AB of matrices A and B. We define this product only when

$$\text{number of columns of } A = \text{number of rows of } B,$$

i.e. only when each row of A has the same number of entries as each column of B. When this condition holds, the types of A and B must be $l \times m$ and $m \times n$ (for some positive integers l, m, n); and then, as part of the ensuing formal definition will state, AB will be a matrix of type $l \times n$. So the pattern formed by the types of A, B and AB (in cases where AB is defined) is as follows.

$$\begin{array}{ccccc} A & & B & \qquad & AB \\ l \times m & \frown & m \times n & & l \times n \\ \uparrow & & \quad\uparrow & & \uparrow \quad \uparrow \end{array}$$

As a step towards the main definition, we first give a meaning to the product XY where X is a $1 \times m$ row matrix and Y is an $m \times 1$ column matrix (both with entries in F): if $X = \text{row}(x_1, x_2, \ldots, x_m)$ and $Y = \text{col}(y_1, y_2, \ldots, y_m)$, we define XY to be the scalar

$$x_1 y_1 + x_2 y_2 + \ldots + x_m y_m, \qquad \text{i.e. } \sum_{j=1}^{m} x_j y_j.$$

This done, we can give the main definition—that of AB, where $A \in F_{l \times m}$ and $B \in F_{m \times n}$. Suppose that $A = [\alpha_{ik}]_{l \times m}$ and $B = [\beta_{ik}]_{m \times n}$. Then AB is defined to be the $l \times n$ matrix whose (i, k)th entry is

$$(i\text{th row of } A) \times (k\text{th column of } B) \quad [\text{in the sense of the previous paragraph}],$$

$$\text{i.e. } \text{row}(\alpha_{i1}, \alpha_{i2}, \ldots, \alpha_{im}) \times \text{col}(\beta_{1k}, \beta_{2k}, \ldots, \beta_{mk}),$$

$$\text{i.e. } \sum_{j=1}^{m} \alpha_{ij} \beta_{jk}.$$

As an illustration, consider

$$A = \begin{bmatrix} 3 & 1 & -1 \\ -1 & 0 & 4 \end{bmatrix} \quad \text{and} \quad B = \begin{bmatrix} 2 & 1 & 3 \\ -1 & 3 & 4 \\ 0 & 2 & 5 \end{bmatrix}.$$

The student should test his grasp of the definition by verifying that

$$AB = \begin{bmatrix} 5 & 4 & 8 \\ -2 & 7 & 17 \end{bmatrix}.$$

(E.g. the (1, 2)th entry is obtained as

$$\begin{aligned}(\text{1st row of } A) \times (\text{2nd column of } B) &= \text{row}(3,1,-1) \times \text{col}(1,3,2)\\ &= 3 \times 1 + 1 \times 3 + (-1) \times 2 = 4.)\end{aligned}$$

In this particular instance, the product BA does not exist, since the number of columns of B does not equal the number of rows of A.

Remarks. (*a*) Consider a typical system of simultaneous linear equations

$$\left\{\begin{matrix} \alpha_{11}x_1 + \alpha_{12}x_2 + \dots + \alpha_{1n}x_n = \kappa_1 \\ \dots \\ \alpha_{m1}x_1 + \alpha_{m2}x_2 + \dots + \alpha_{mn}x_n = \kappa_m \end{matrix}\right\}$$

(a system of m equations in the n "unknowns" $x_1, x_2, \dots, x_n$). Let $A = [\alpha_{ik}]_{m \times n}$, which, naturally, we call the **coefficient matrix** of the system; and let $X = \text{col}(x_1, x_2, \dots, x_n)$ and $K = \text{col}(\kappa_1, \dots, \kappa_m)$.

The system is exactly equivalent to the single matrix equation

$$AX = K.$$

To understand this claim, simply observe that AX is an $m \times 1$ column whose entries are precisely the left-hand sides of the m given equations. Since systems of linear equations represent the primary motivation for considering matrices, it is a cause for satisfaction that our definition of matrix multiplication enables a system of equations to be expressed in such a concise and (as we shall see) useful form.

(*b*) Further light is shed on the definition of matrix multiplication by supposing that we have two systems of linear equations

$$x_i = \alpha_{i1}y_1 + \alpha_{i2}y_2 + \dots + \alpha_{im}y_m \qquad (1 \leqslant i \leqslant l) \qquad (1)$$

$$\text{and} \quad y_j = \beta_{j1}z_1 + \beta_{j2}z_2 + \dots + \beta_{jn}z_n \qquad (1 \leqslant j \leqslant m), \qquad (2)$$

system (1) giving $x_1, x_2, \dots, x_l$ in terms of $y_1, y_2, \dots, y_m$, and system (2) giving $y_1, y_2, \dots, y_m$ in terms of $z_1, z_2, \dots, z_n$. Let X, Y, Z denote the columns $\text{col}(x_1, \dots, x_l)$, $\text{col}(y_1, \dots, y_m)$, $\text{col}(z_1, \dots, z_n)$, respectively; and let $A = [\alpha_{ik}]_{l \times m}$ and $B = [\beta_{ik}]_{m \times n}$, which are the coefficient matrices of the two systems.

In the right-hand side of each equation in the system (1) we may substitute for each y_j the expression for y_j given by (2); and hence we can obtain a system of equations expressing $x_1, \dots, x_l$ in terms of $z_1, \dots, z_n$. The details may look forbiddingly complicated, but it is not difficult to discover that the coefficient matrix of the resulting system is the product AB. It is helpful to summarize the situation by the diagram

$$\begin{bmatrix} z_1 \\ \vdots \\ z_n \end{bmatrix} \xrightarrow{B} \begin{bmatrix} y_1 \\ \vdots \\ y_m \end{bmatrix} \xrightarrow{A} \begin{bmatrix} x_1 \\ \vdots \\ x_l \end{bmatrix}, \qquad \begin{bmatrix} z_1 \\ \vdots \\ z_n \end{bmatrix} \xrightarrow{AB} \begin{bmatrix} x_1 \\ \vdots \\ x_l \end{bmatrix}$$

where, by each labelled arrow, we indicate the coefficient matrix of the system of equations that express one set of "variables" in terms of another. There is more than a suggestion of a connection with the idea of the composition of two mappings; and one possible view of the definition of matrix multiplication is that it is designed to cause that connection to exist between matrix multiplication and mapping composition.

(*c*) Alert readers may complain of an inconsistency in the meaning given to XY, where X is a $1 \times m$ row matrix and Y is an $m \times 1$ column matrix. On the one hand, we stated (as the first step towards the definition of matrix multiplication) that XY is a certain scalar. On the other hand, the later general definition of matrix multiplication implies that XY is a 1×1 matrix whose sole entry is the aforementioned scalar. However, because 1×1 matrices behave algebraically like the scalars that are their sole entries

$$(\text{e.g. } [\alpha]+[\beta] = [\alpha+\beta] \quad \text{and} \quad [\alpha][\beta] = [\alpha\beta]),$$

it is customary to identify the 1×1 matrix $[\alpha]$ and the scalar α, i.e. to ignore the difference between them. This convention removes the difficulty outlined above.

16. Properties and non-properties of matrix multiplication

It would be entirely wrong to imagine that matrix multiplication is a "no-catch operation". Accordingly, it is important to know which properties matrix multiplication possesses and which it does not possess; and every handler of matrices must learn to avoid certain pitfalls.

Before we come to details about these matters, we shall first establish an important general fact which we need for the proof of a result in this section and also in later chapters.

Consider an arbitrary matrix $A = [\alpha_{ik}]_{m \times n}$, and think about forming the sum T of all mn entries of A. The many ways of doing this include:

(*a*) forming the sum of the entries in each row of A and then adding together all the row totals so obtained;

(*b*) forming the sum of all the entries in each column of A and then adding together all the column totals so obtained.

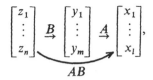

Using method (*a*), we have

$$\text{sum of entries of } s\text{th row} = \sum_{t=1}^{n} \alpha_{st},$$

and hence

$$T = \sum_{s=1}^{m} \left(\sum_{t=1}^{n} \alpha_{st} \right).$$

On the other hand, method (*b*) gives

$$\text{sum of entries of } t\text{th column} = \sum_{s=1}^{m} \alpha_{st},$$

and hence

$$T = \sum_{t=1}^{n} \left(\sum_{s=1}^{m} \alpha_{st} \right).$$

Comparing the two versions, we see that:

16.1 $$\sum_{s=1}^{m} \left(\sum_{t=1}^{n} \alpha_{st} \right) = \sum_{t=1}^{n} \left(\sum_{s=1}^{m} \alpha_{st} \right).$$

This is an important general fact. It means in practice that two summation signs of the form $\sum_{s=1}^{m}$ and $\sum_{t=1}^{n}$, where m, n are (finite) constants, can be validly interchanged; and of course this applies irrespective of what symbols are used for the "dummy variables" (s and t in the above discussion). Application of this principle is called *interchange of the order of summation.* We use it in the proof of the next proposition, which tells us that matrix multiplication is associative.

16.2 Let $A \in F_{l \times m}$, $B \in F_{m \times n}$, $C \in F_{n \times p}$. Then $(AB)C = A(BC)$.

Proof. Let $A = [\alpha_{ik}]_{l \times m}$, $B = [\beta_{ik}]_{m \times n}$, $C = [\gamma_{ik}]_{n \times p}$.
Both $(AB)C$ and $A(BC)$ are $l \times p$ matrices. Further, for all relevant i, k,

$$(i, k)\text{th entry of } (AB)C = \sum_{t=1}^{n} ((i, t)\text{th entry) of } AB) \times \gamma_{tk}$$

$$= \sum_{t=1}^{n} \left(\sum_{s=1}^{m} \alpha_{is}\beta_{st} \right) \gamma_{tk}$$

$$= \sum_{t=1}^{n} \left(\sum_{s=1}^{m} \alpha_{is}\beta_{st}\gamma_{tk} \right) \qquad (\gamma_{tk} \text{ being independent of } s)$$

$$= \sum_{s=1}^{m} \left(\sum_{t=1}^{n} \alpha_{is}\beta_{st}\gamma_{tk} \right) \qquad (\text{on interchanging the order of summation})$$

$$= \sum_{s=1}^{m} \alpha_{is}\left(\sum_{t=1}^{n} \beta_{st}\gamma_{tk}\right) \qquad (\alpha_{is} \text{ being independent of } t)$$

$$= \sum_{s=1}^{m} \alpha_{is} \times ((s,k)\text{th entry of } BC)$$

$$= (i,k)\text{th entry of } A(BC).$$

Therefore $(AB)C = A(BC)$.

The general comment on associative laws made in remark (*b*) of §4 applies to matrix multiplication in particular. Moreover, positive integral powers of a square matrix A can be satisfactorily defined by:

$$A^r = \underbrace{A \times A \times A \times \ldots \times A}_{r \text{ factors}} \qquad (r \text{ a positive integer}).$$

And fairly obviously the following index laws hold:

16.3 For $A \in F_{n \times n}$ and r, s positive integers,
(i) $A^r A^s = A^{r+s}$, (ii) $(A^r)^s = A^{rs}$.

The next two propositions give further important positive properties of matrix multiplication. They may be proved by the standard basic method of comparing types and (i, k)th entries of matrices alleged to be equal.

16.4 (i) $A(B+C) = AB+AC \qquad (A \in F_{l \times m};\ B, C \in F_{m \times n})$.
(ii) $(A+B)C = AC+BC \qquad (A, B \in F_{l \times m};\ C \in F_{m \times n})$.

16.5 $A(\lambda B) = (\lambda A)B = \lambda(AB) \qquad (A \in F_{l \times m}, B \in F_{m \times n}, \lambda \in F)$.

It should be appreciated that 16.4 and 16.5 guarantee the truth of more elaborate "distributive" properties, such as

$$A(B+C+D) = AB+AC+AD, \qquad A(2B-3C) = 2AB-3AC, \text{ etc.}$$

Having seen that matrix multiplication does have some nice major properties, we turn attention to the operation's more tricky aspects. Foremost among these is the fact that:

16.6 Matrix multiplication is not commutative: i.e. the matrix products AB and BA may be different.

There is, it should be realized, more than one sense in which the matrix products AB and BA may be "different". (i) As in an illustration in §15 (where A was of type 2×3 and B of type 3×3), it is possible for one of the products AB, BA to exist while the other does not. (ii) It is possible for AB, BA both to exist but to be of different types—e.g. if A and B are of types 2×3 and 3×2, respectively. (iii) Even if AB and BA both exist and are of the same type (which happens when both A and B are of type $n \times n$ for some n), AB and BA may be different. An illustration of this last possibility is obtained by taking $A = \begin{bmatrix} 1 & -1 \\ 1 & -1 \end{bmatrix}$ and $B = \begin{bmatrix} 1 & 1 \\ 1 & 1 \end{bmatrix}$. In this case, $AB = \begin{bmatrix} 0 & 0 \\ 0 & 0 \end{bmatrix} = O$, while $BA = \begin{bmatrix} 2 & -2 \\ 2 & -2 \end{bmatrix}$.

This illustration brings to light a second tricky point:

16.7 The matrix product AB may be zero when neither A nor B is a zero matrix.

Remarks. (*a*) It is of course true that if one of the factors in a matrix product AB is a zero matrix, then AB will be zero. But 16.7 points out that the converse of this is not true.

(*b*) Because of 16.6 the instruction "multiply A by B" is ambiguous when A and B are matrices. (Is one to form the product AB or the product BA?) This difficulty is overcome by introducing the words **premultiply** and **postmultiply**: to premultiply A by B is to form the product BA, and to postmultiply A by B is to form the product AB.

(*c*) In some cases the matrix products AB and BA are equal. When they are equal, we say that the matrices A and B *commute* (with each other). For example, any two powers of a square matrix C commute with each other since (cf. 16.3) C^rC^s and C^sC^r are both equal to C^{r+s}.

(*d*) The opening paragraph of this section referred to pitfalls to be avoided. The "pitfalls" are the ramifications of 16.6 and 16.7, and there follows a brief indication of some of them.

(i) The index law $(AB)^r = A^rB^r$ does not hold for square matrices. In particular, $(AB)^2$, which means $ABAB$, need not equal A^2B^2, which means $AABB$.

(ii) In expanding products such as $(A+B)(C+D)$, care must be taken to preserve the distinction between the matrix products XY and YX. E.g. one might, without thinking, rewrite $(A+B)^2$ as $A^2+2AB+B^2$, but in fact this would be wrong. Correct would be

$$(A+B)^2 = (A+B)(A+B) = A(A+B)+B(A+B) = A^2+AB+BA+B^2,$$

which is different from $A^2+2AB+B^2$ unless A and B commute. Likewise, $(A-B)(A+B)$ is in general different from A^2-B^2.

(iii) From $A^2 = O$ (or $A^3 = O$, etc.) it does not follow that $A = O$.

(iv) Given that $AB = AC$ and that A is a nonzero matrix, one cannot simply cancel and assert that $B = C$. It is true that $A(B-C) = O$, but (cf. 16.7) from this and the fact that $A \neq O$ it does not follow that $B-C = O$.

17. Some special matrices and types of matrices

Throughout the section we shall be concerned with matrices over the arbitrary field F.

(1) A very useful notation is the **Kronecker delta** symbol δ_{ik}, which is defined to mean 1 if $i = k$ and to mean 0 if $i \neq k$. So, for example δ_{33} is 1 and δ_{52} is 0. Observe that:

17.1 $\sum_{t=1}^{n} \delta_{st}\alpha_t = \alpha_s \qquad (1 \leqslant s \leqslant n)$

(there being only one term in the sum that can possibly be nonzero).

We now define the **identity** $n \times n$ **matrix** to be the $n \times n$ matrix whose (i, k)th entry is δ_{ik}. This matrix is denoted by I_n (or simply by I if the intended size is clear from the context). So, for example,

$$I_2 = \begin{bmatrix} 1 & 0 \\ 0 & 1 \end{bmatrix} \quad \text{and} \quad I_3 = \begin{bmatrix} 1 & 0 & 0 \\ 0 & 1 & 0 \\ 0 & 0 & 1 \end{bmatrix}.$$

The matrix I_n is special because it behaves like a "one" in multiplication and is the only $n \times n$ matrix that always does so. More precisely:

17.2 (i) $AI_n = A$ for every matrix A with n columns.

(ii) $I_n B = B$ for every matrix B with n rows.

(iii) If U is an $n \times n$ matrix such that $AU = A$ for every matrix A with n columns, then $U = I_n$.

(iv) If V is an $n \times n$ matrix such that $VB = B$ for every matrix B with n rows, then $V = I_n$.

Proof. Let $A = [\alpha_{ik}]_{m \times n}$. Then both AI_n and A are $m \times n$ matrices. Further, for all relevant i, k,

$$\begin{aligned} (i,k)\text{th entry of } AI_n &= \sum_{j=1}^{n} \alpha_{ij}\delta_{jk} = \alpha_{ik} \qquad \text{(cf. 17.1)} \\ &= (i,k)\text{th entry of } A. \end{aligned}$$

Hence $AI_n = A$.

Part (i) follows; and part (ii) is proved similarly.

For part (iii), simply note that if U satisfies the condition stated, then

$$\begin{aligned} I_n &= I_n U \qquad \text{(by the stated condition)} \\ &= U \qquad \text{(by part (ii))}. \end{aligned}$$

And part (iv) is similarly proved.

(2) A *diagonal matrix* is a square matrix with the property that all the entries not on the main diagonal are zero. Thus an $n \times n$ diagonal matrix is of the form

$$\begin{bmatrix} \alpha_1 & 0 & 0 & \dots & 0 \\ 0 & \alpha_2 & 0 & \dots & 0 \\ 0 & 0 & \alpha_3 & \dots & 0 \\ & \dots & & \ddots & \\ 0 & 0 & 0 & \dots & \alpha_n \end{bmatrix},$$

for which we use the abbreviation $\operatorname{diag}(\alpha_1, \alpha_2, \dots, \alpha_n)$. (Here each of $\alpha_1, \dots, \alpha_n$ may be zero or nonzero.)

The effect of pre- or postmultiplying by a diagonal matrix is easily discovered:

17.3 Let D be the diagonal matrix $\operatorname{diag}(\alpha_1, \alpha_2, \dots, \alpha_n)$, let B be any matrix with n rows, and let C be any matrix with n columns. Then (i) DB is the matrix obtained from B by multiplying the first row by α_1, the second row by α_2, etc.; and (ii) CD is the matrix obtained from C by multiplying the first column by α_1, the second column by α_2, etc.

(Here the phrase "multiplying by α_i" is used in the sense of §14: e.g. multiplying the first row by α_1 means multiplying all of its entries by α_1.)

The simple rule for multiplying two $n \times n$ diagonal matrices may be regarded as a particular case of 17.3. The rule is:

17.4 $\operatorname{diag}(\alpha_1, \alpha_2, \dots, \alpha_n) \times \operatorname{diag}(\beta_1, \beta_2, \dots, \beta_n) = \operatorname{diag}(\alpha_1\beta_1, \alpha_2\beta_2, \dots, \alpha_n\beta_n)$.

It follows in particular from 17.4 that:

17.5 Any two diagonal matrices commute with each other.

(3) In discussing matrices in $F_{m \times n}$, it is sometimes helpful to consider, for each relevant i, k, the matrix which has

$$\left\{\begin{array}{l} (i,k)\text{th entry equal to } 1 \\ \text{all other entries equal to } 0 \end{array}\right\}.$$

We denote the matrix just described by E_{ik}. Thus, for example, in $F_{2 \times 3}$, E_{21} means

$$\begin{bmatrix} 0 & 0 & 0 \\ 1 & 0 & 0 \end{bmatrix}.$$

In $F_{m \times n}$ the mn matrices E_{ik} $(1 \leqslant i \leqslant m,\ 1 \leqslant k \leqslant n)$ are termed *matrix units*. Notice that an arbitrary matrix $A = [\alpha_{ik}]_{m \times n}$ in $F_{m \times n}$ can be expressed as a linear combination of the mn matrix units: for

$$A = \alpha_{11}E_{11} + \alpha_{12}E_{12} + \alpha_{13}E_{13} + \dots + \alpha_{mn}E_{mn}.$$

And it is fairly obvious that this last equation gives the only way in which the arbitrary A can be expressed as a linear combination of the mn matrix units.

In the case of square $n \times n$ matrices the product of any two matrix units can be formed, and the answer is given in all cases by the following simple formula, whose proof is left as an exercise.

17.6 In $F_{n \times n}$, $E_{st}E_{uv} = \delta_{tu}E_{sv}$.

18. Transpose of a matrix

Let $A = [\alpha_{ik}]_{m \times n}$ be an arbitrary matrix in $F_{m \times n}$. The **transpose** of A, denoted by A^T, is defined to be the $n \times m$ matrix whose (i, k)th entry is α_{ki}.

E.g. if $A = \begin{bmatrix} 1 & 2 & 3 \\ 4 & 5 & 6 \end{bmatrix}$, then $A^T = \begin{bmatrix} 1 & 4 \\ 2 & 5 \\ 3 & 6 \end{bmatrix}$.

It is sometimes said that the effect of transposing (i.e. of forming A^T from A) is to turn rows into columns and vice versa. Imprecise though this remark is, it does bring out one important feature of the relationship between A and A^T.

The following four propositions give the basic properties of transposes.

18.1 $(A^T)^T = A \qquad (A \in F_{m \times n})$.

18.2 $(\lambda A)^T = \lambda A^T \qquad (\lambda \in F,\ A \in F_{m \times n})$.

18.3 $(A+B)^T = A^T + B^T \qquad (A, B \in F_{m \times n})$.

18.4 $(AB)^T = B^T A^T \qquad (A \in F_{l \times m},\ B \in F_{m \times n})$.

Of these, the first three are very easy to prove. Here is a proof of the much less trivial 18.4, which, for obvious reasons, is known as the *reversal rule* for transposes.

Let $A = [\alpha_{ik}]_{l \times m}$, $B = [\beta_{ik}]_{m \times n}$.

Both $(AB)^T$ and $B^T A^T$ are matrices of type $n \times l$. Further, for all relevant i, k,

$$\begin{aligned} (i,k)\text{th entry of } B^T A^T &= (i\text{th row of } B^T) \times (k\text{th column of } A^T) \\ &= \text{row}(\beta_{1i}, \beta_{2i}, \dots, \beta_{mi}) \times \text{col}(\alpha_{k1}, \alpha_{k2}, \dots, \alpha_{km}) \\ &= \sum_{j=1}^{m} \beta_{ji}\alpha_{kj} = \sum_{j=1}^{m} \alpha_{kj}\beta_{ji} \\ &= (k, i)\text{th entry of } AB \\ &= (i, k)\text{th entry of } (AB)^T. \end{aligned}$$

This proves the result.

It should be appreciated that, by suitably using 18.2 and 18.3, one can obtain the corresponding fact about the transpose of any linear combination of matrices: e.g. $(2A-B+3C)^T = 2A^T - B^T + 3C^T$. Moreover, through repeated application, the reversal rule 18.4 extends to longer products: e.g. $(ABC)^T = C^T B^T A^T$.

Two special kinds of matrices can be described by means of transposes. A matrix A is said to be

(1) **symmetric** if and only if $A^T = A$,

(2) **skew-symmetric** if and only if $A^T = -A$.

So, for example the 3×3 matrices

$$\begin{bmatrix} 0 & 1 & 2 \\ 1 & 4 & 3 \\ 2 & 3 & 5 \end{bmatrix} \text{ and } \begin{bmatrix} 0 & 1 & 2 \\ -1 & 0 & 3 \\ -2 & -3 & 0 \end{bmatrix}$$

are symmetric and skew-symmetric, respectively.

Remarks. (*a*) It is clear from the definition that every symmetric or skew-symmetric matrix must be square.

(*b*) It should be noted that (rather obviously):

18.5 Every diagonal matrix (I_n, in particular) is symmetric.

(*c*) Consider an arbitrary real skew-symmetric matrix $A = [\alpha_{ik}]_{n \times n}$. Since $A^T = -A$, we have, for each relevant i,

$$(i, i)\text{th entry of } A^T = (i, i)\text{th entry of } (-A),$$
$$\text{i.e. } \alpha_{ii} = -\alpha_{ii}, \qquad \text{i.e. } \alpha_{ii} = 0.$$

Thus every entry on the main diagonal of a real skew-symmetric matrix is zero. (*Technical note*: this extends from matrices over $\mathbb{R}$ to matrices over any field F such that char $F \neq 2$.)

Worked example. Let A be an arbitrary $m \times n$ matrix. Show that $I_n + A^T A$ is symmetric.

The student should pause to check that the product $A^T A$ exists and is an $n \times n$ matrix; $I_n + A^T A$ also exists, therefore. This having been noted, the solution proceeds as follows.

$$\begin{aligned} (I_n + A^T A)^T &= I_n^T + (A^T A)^T && \text{(cf. 18.3)} \\ &= I_n + A^T (A^T)^T && \text{(cf. 18.5 for the first term and 18.4 for the second)} \\ &= I_n + A^T A && \text{(cf. 18.1).} \end{aligned}$$

Thus $I_n + A^T A$ is symmetric.

19. First considerations of matrix inverses

For reasons that can be seen in perspective later (e.g. once chapter 6 has been read) the discussion of matrix inverses is restricted to *square* matrices. This section contains just the tentative beginnings of that discussion. The scene is set by the following definitions.

Let A denote a matrix in $F_{n \times n}$.

(1) A matrix P in $F_{n \times n}$ is said to be a **left inverse** of A if and only if $PA = I_n$.

(2) A matrix R in $F_{n \times n}$ is said to be a **right inverse** of A if and only if $AR = I_n$.

(3) A matrix Q in $F_{n \times n}$ is said to be an **inverse** of A if and only if it is both a left inverse and a right inverse of A, i.e. if and only if $QA = AQ = I_n$.

(4) The matrix A is described as **nonsingular** if an inverse of A exists, and **singular** if no inverse of A exists.

Remarks. (*a*) Definition (4) relates to an arbitrary *square* matrix A. Accordingly, whenever a matrix is described as nonsingular, it is implied that the matrix is square.

(*b*) In chapter 3 it will be proved that the three notions "left inverse", "right inverse" and "inverse", as defined above, coincide: for, in fact, any left inverse of the square matrix A must also be a right inverse, and vice versa. However, that is far from evident at the present stage and must not yet be assumed.

(*c*) We can focus more clearly on nonsingular matrices and their inverses once we realize that:

19.1 A nonsingular matrix has just one inverse.

Proof. Let A be an arbitrary nonsingular $n \times n$ matrix. Suppose that Q and S both denote inverses of A. We shall prove that Q and S must coincide.

We have:
$$\begin{aligned} Q = QI_n &= QAS && \text{(since } S \text{ is an inverse of } A) \\ &= I_nS && \text{(since } Q \text{ is an inverse of } A) \\ &= S. \end{aligned}$$

The truth of 19.1 follows.

Because of 19.1 we may legitimately speak of *the* inverse of a nonsingular matrix A; and we denote that one and only inverse by A^{-1}. (The notations "$1/A$" and "I/A" should not be used. Worse still would be the use of "B/A" for the product of the matrix B and the inverse of A: one could not tell whether $A^{-1}B$ or BA^{-1} was intended.)

(*d*) At this point we can define another special type of square matrix. A square matrix A is described as **orthogonal** if and only if $A^TA = AA^T = I$. Another way of putting this is that an orthogonal matrix is a nonsingular matrix whose inverse is equal to its transpose. It is easily verified that all 2×2 matrices of the form $\begin{bmatrix} \cos\theta & -\sin\theta \\ \sin\theta & \cos\theta \end{bmatrix}$ are orthogonal. Later (especially in chapter 9) it will be seen that real orthogonal matrices come to the fore through certain very natural geometrical considerations.

(*e*) Many nonzero square matrices are singular. This is apparent from the following useful little result.

19.2 If the square matrix A has a zero row, then A has no right inverse (and hence A is singular). (By a "zero row" we mean a row every entry of which is zero.)

Proof. If $A(\in F_{n\times n})$ has zero ith row, then (by the definition of matrix multiplication) the ith row is zero in AR for every $R \in F_{n\times n}$; and therefore there is no matrix $R \in F_{n\times n}$ such that $AR = I$. This proves the stated result.

(*f*) Two important major questions in linear algebra now confront us: for a given square matrix A, (1) how can one tell whether A is nonsingular, and (2) if A is nonsingular, how can its inverse be found? These are not easy questions, but in chapters 3 and 4 we shall obtain answers to them through two different approaches. Meanwhile we can dispose of certain special cases from first principles: for if, given a square matrix A, we can produce (by any means) a matrix Q such that $AQ = QA = I$, then of course we can conclude that A is nonsingular and that $A^{-1} = Q$. There follows an account of two special cases—diagonal matrices and 2×2 matrices—and then a worked example showing a technique applicable when a certain kind of information is available about the matrix in question.

19.3 The diagonal matrix $D = \text{diag}(\alpha_1, \alpha_2, \ldots, \alpha_n)$ is nonsingular if and only if all of $\alpha_1, \alpha_2, \ldots, \alpha_n$ are nonzero; and when all of $\alpha_1, \alpha_2, \ldots, \alpha_n$ are nonzero, $D^{-1} = \text{diag}(1/\alpha_1, 1/\alpha_2, \ldots, 1/\alpha_n)$.

Proof. First suppose that all of $\alpha_1, \ldots, \alpha_n$ are nonzero. Then we can introduce the matrix $E = \text{diag}(1/\alpha_1, 1/\alpha_2, \ldots, 1/\alpha_n)$. By 17.4,

$$DE = ED = \text{diag}(1, 1, \ldots, 1) = I.$$

Hence in this case D is nonsingular and $D^{-1} = E$.

On the other hand, if any α_i is zero, then D has a zero row and so is singular (by 19.2).

The whole proposition is now proved.

19.4 Let $A = \begin{bmatrix} a & b \\ c & d \end{bmatrix}$ be an arbitrary matrix in $F_{2\times 2}$. Then A is nonsingular if and only if $ad - bc \neq 0$; and if $ad - bc \neq 0$,

$$A^{-1} = \frac{1}{ad-bc}\begin{bmatrix} d & -b \\ -c & a \end{bmatrix}.$$

Proof. Let $k = ad-bc$, and let $B = \begin{bmatrix} d & -b \\ -c & a \end{bmatrix}$. By direct calculation, we find that $AB = BA = kI$.

In the case $k \neq 0$, it follows that $AC = CA = I$, where $C = (1/k)B$, and hence that A is nonsingular with inverse $(1/k)B$.

Consider the remaining case where $k = 0$ and, therefore, $AB = O$. Suppose, with a view to obtaining a contradiction, that A^{-1} exists. Then we have

$$O = A^{-1}O = A^{-1}AB = IB = B,$$

and hence all of a, b, c, d are zero: i.e. $A = O$. This (cf. 19.2) contradicts the supposition that A^{-1} exists. Therefore, in the case $k = 0$, A must be singular.

The result stated is now fully proved.

In the notation of 19.4 we call the scalar $ad-bc$ the **determinant** of A. Chapter 4 will outline a generalization: the determinant of any larger square matrix can be defined as a certain scalar calculable from the entries of the matrix, and in every case the determinant is zero if and only if the matrix is singular.

Worked example. The square matrix A is such that

$$A^2+2A-2I = O.$$

Prove that A is nonsingular, and find A^{-1} in terms of A.

Solution. The given equation can be re-written as $A^2+2A = 2I$. Hence $A(A+2I) = (A+2I)A = 2I$; and so $AQ = QA = I$, where $Q = \frac{1}{2}(A+2I) = I+\frac{1}{2}A$. It follows that A is nonsingular and that $A^{-1} = I+\frac{1}{2}A$.

Further to this example, it should be explained that, when A is a matrix in $F_{n\times n}$, any matrix expressible in the form

$$\alpha_r A^r + \alpha_{r-1}A^{r-1} + \ldots + \alpha_2 A^2 + \alpha_1 A + \alpha_0 I_n$$

(where each $\alpha_i \in F$) is described as a *polynomial* in A. Because any two powers of A commute, it can be seen that any two polynomials in A commute. The generalization of the above worked example is that if O can be expressed as a polynomial in A with the coefficient of I nonzero, then A is nonsingular and A^{-1} is expressible as a polynomial in A.

20. Properties of nonsingular matrices

First there are some straightforward results telling us that matrices constructed in certain obvious ways from given nonsingular matrices are also nonsingular and giving information about the inverses of these further nonsingular matrices.

20.1 If A is a nonsingular matrix, then A^{-1} is also nonsingular and $(A^{-1})^{-1} = A$.

20.2 If A is a nonsingular matrix and λ is a nonzero scalar, then λA is also nonsingular and $(\lambda A)^{-1} = (1/\lambda)A^{-1}$.

20.3 If A is a nonsingular matrix, then A^T is also nonsingular and $(A^T)^{-1} = (A^{-1})^T$.

20.4 If A and B are nonsingular matrices in $F_{n \times n}$, then AB is also nonsingular and $(AB)^{-1} = B^{-1}A^{-1}$.

Each of these results is easily proved just by verifying that the alleged inverse fits the bill. For example, here is a proof of 20.4.

Suppose that A and B are nonsingular matrices in $F_{n \times n}$. Let $Q = B^{-1}A^{-1}$. Then

$$(AB)Q = ABB^{-1}A^{-1} = AIA^{-1} = AA^{-1} = I$$

and similarly $Q(AB) = I$. It follows that AB is nonsingular with inverse $Q(=B^{-1}A^{-1})$; and this establishes the stated result.

The fact that $(AB)^{-1} = B^{-1}A^{-1}$ when A and B are nonsingular matrices in $F_{n \times n}$ is called the **reversal rule** for matrix inverses. Either by adapting the above argument or by repeated application of 20.4, we can obtain a generalization of 20.4 to products of arbitrary length, namely:

20.5 If $A_1, A_2, \ldots, A_{k-1}, A_k$ are all nonsingular matrices in $F_{n \times n}$, then $A_1 A_2 \ldots A_{k-1} A_k$ is also nonsingular, and

$$(A_1 A_2 \ldots A_{k-1} A_k)^{-1} = A_k^{-1} A_{k-1}^{-1} \ldots A_2^{-1} A_1^{-1}.$$

It remains to point out that certain simplifications and lines of argument are open to us when the nonsingularity of a matrix is given. For example, suppose we know that

$$AB = AC, \qquad (*)$$

A, B, C being matrices. In §16 we remarked that, in general, one may not jump to the conclusion that $B = C$. However, if A is given to be nonsingular, we can premultiply both sides of (*) by A^{-1} to obtain $A^{-1}AB = A^{-1}AC$, and hence in this special case the conclusion $B = C$ legitimately emerges.

Parallel remarks could be made about the equation $BA = CA$. Notice, though, that if we are given that $AB = CA$ (A, B, $C \in F_{n \times n}$) and that A is nonsingular, then we can deduce that $B = A^{-1}CA$; but, because matrix multiplication is non-commutative, there is in general no reason to expect $A^{-1}CA$ to be equal to C.

A useful general observation is:

20.6 (i) If $A \in F_{m \times m}$, B and $C \in F_{m \times n}$, and A is nonsingular, then

$$AB = C \text{ is equivalent to } B = A^{-1}C.$$

(ii) If $A \in F_{n \times n}$, B and $C \in F_{m \times n}$, and A is nonsingular, then

$$BA = C \text{ is equivalent to } B = CA^{-1}.$$

Proof of (i). Suppose that $A \in F_{m \times m}$, $B \in F_{m \times n}$, $C \in F_{m \times n}$ and that A is nonsingular.

If $AB = C$, then, on premultiplying by A^{-1}, we obtain

$$A^{-1}AB = A^{-1}C, \quad \text{i.e. } IB = A^{-1}C, \quad \text{i.e. } B = A^{-1}C;$$

and conversely if $B = A^{-1}C$, then $AB = AA^{-1}C = IC = C$.

This proves (i); and (ii) may be proved similarly.

The following worked example concerns an **idempotent** matrix. A square matrix A is described as idempotent if and only if $A^2 = A$. When this is the case, every positive power of A coincides with A: e.g. $A^3 = A^2A = AA = A^2 = A$, and hence $A^4 = A^3A = AA = A^2 = A$; etc. A simple nontrivial example of an idempotent matrix is $\begin{bmatrix} \frac{1}{2} & \frac{1}{2} \\ \frac{1}{2} & \frac{1}{2} \end{bmatrix}$.

Worked example. Suppose that the square matrix A is idempotent and that $A \neq I$. Prove that A is singular.

Solution. Suppose that A is nonsingular. Then (since $A^2 = A$) we have $A^{-1}A^2 = A^{-1}A = I$. But $A^{-1}A^2 = A^{-1}AA = IA = A$. Therefore $A = I$, which contradicts what is given.

It follows that A must be singular.

21. Partitioned matrices

It is often helpful to think of a given matrix as "partitioned" into a number of "submatrices" by horizontal and/or vertical dividing lines. For example, a 4×6 matrix

$$\left[\begin{array}{cc|ccc|c} \alpha_{11} & \alpha_{12} & \alpha_{13} & \alpha_{14} & \alpha_{15} & \alpha_{16} \\ \alpha_{21} & \alpha_{22} & \alpha_{23} & \alpha_{24} & \alpha_{25} & \alpha_{26} \\ \hline \alpha_{31} & \alpha_{32} & \alpha_{33} & \alpha_{34} & \alpha_{35} & \alpha_{36} \\ \alpha_{41} & \alpha_{42} & \alpha_{43} & \alpha_{44} & \alpha_{45} & \alpha_{46} \end{array}\right]$$

may be partitioned into 6 submatrices as indicated by the lines drawn in above, and accordingly it may be represented as

$$\begin{bmatrix} X_1 & X_2 & X_3 \\ X_4 & X_5 & X_6 \end{bmatrix},$$

where $X_1 = \begin{bmatrix} \alpha_{11} & \alpha_{12} \\ \alpha_{21} & \alpha_{22} \end{bmatrix}$, $X_2 = \begin{bmatrix} \alpha_{13} & \alpha_{14} & \alpha_{15} \\ \alpha_{23} & \alpha_{24} & \alpha_{25} \end{bmatrix}$, etc.

A "partitioned matrix" simply means a matrix thought of or represented in this way. In any partitioned matrix, the submatrices such as $X_1, X_2, X_3, X_4, X_5, X_6$ in our illustration are termed **blocks**.

Consider two matrices A, B of the same type that are identically partitioned (i.e. partitioned by dividing lines occurring in the same places in both matrices), so that corresponding blocks in the two matrices are of the same type. E.g. we might have

$$A = \begin{bmatrix} X_1 & X_2 & X_3 \\ X_4 & X_5 & X_6 \end{bmatrix}, \qquad B = \begin{bmatrix} Y_1 & Y_2 & Y_3 \\ Y_4 & Y_5 & Y_6 \end{bmatrix},$$

where, for each j, X_j is of the same type as Y_j. One readily sees in any such case the sum $A+B$ is obtainable by adding "blockwise". E.g. in the above specific case

$$A+B = \begin{bmatrix} X_1+Y_1 & X_2+Y_2 & X_3+Y_3 \\ X_4+Y_4 & X_5+Y_5 & X_6+Y_6 \end{bmatrix}.$$

It is equally obvious that one can work blockwise in multiplying a partitioned matrix by a scalar or in subtracting two identically partitioned matrices of the same type.

A rather less obvious fact is that, provided certain details are in order, two partitioned matrices can be correctly multiplied by treating the blocks as though they were entries. Before this process of "block multiplication" can be used to obtain a matrix product AB, it is necessary that the partitioning of the columns of A correspond exactly to the partitioning of the rows of B: i.e. if there are vertical partitions in A after the ath, bth, cth,... columns, then in B there must be horizontal partitions after the ath, bth, cth,... rows. When this is the case, we say that A, B are **conformably partitioned** for the product AB.

A simple example should serve to make the idea clear. Consider

$$A = \left[\begin{array}{c|c} X_1 & X_2 \\ \hline X_3 & X_4 \end{array}\right] \begin{array}{l} \updownarrow r \\ \updownarrow l-r, \end{array} \qquad B = \left[\begin{array}{c|c} Y_1 & Y_2 \\ \hline Y_3 & Y_4 \end{array}\right] \begin{array}{l} \updownarrow s \\ \updownarrow m-s. \end{array}$$
$$\qquad \overleftrightarrow{s} \quad \overleftrightarrow{m-s} \qquad\qquad\qquad \overleftrightarrow{t} \quad \overleftrightarrow{n-t}$$

These matrices A, B are of types $l \times m$, $m \times n$, respectively; the dimensions of the blocks are indicated; and A, B are conformably partitioned for the product

AB because the partitioning after the sth column in A corresponds exactly with the partitioning after the sth row in B. Treating blocks as entries (while (cf. 16.6) avoiding confusion between the matrix products XY and YX), we obtain

$$AB = \begin{bmatrix} X_1Y_1 + X_2Y_3 & X_1Y_2 + X_2Y_4 \\ X_3Y_1 + X_4Y_3 & X_3Y_2 + X_4Y_4 \end{bmatrix},$$

and this is in fact correct.

To begin to understand why it is correct, the student is recommended first to check its correctness at the (1, 1)th entry—by writing down the (1, 1)th entry of $X_1Y_1 + X_2Y_3$ in terms of the entries of A and B and verifying that it is the same as the (1, 1)th entry of AB. This piece of work should give the student insight into the block multiplication process—enough to perceive why the above example is correct at all entries (not just the (1, 1)th) and to see that the process will work in other cases (where the pattern of the partitioning may be different). Because of the horrendous notational complications that have to be negotiated, it is inadvisable to try to write out a formal general proof of the correctness of the block multiplication procedure.

Turning to a specially easy case, consider $A \in F_{l\times m}$ and $B \in F_{m\times n}$, where A is not subdivided but B is partitioned into its individual columns $C_1, C_2, \ldots, C_n$, so that $B = [C_1\ C_2 \ldots C_n]$. Then

$$\begin{aligned} AB &= A[C_1 \quad C_2 \quad \ldots \quad C_n] \\ &= [AC_1 \quad AC_2 \quad \ldots \quad AC_n] \quad \text{(by block multiplication).} \end{aligned}$$

Thus:

21.1 The kth column of the matrix product AB is

$$A \times (k\text{th column of } B).$$

(This, it should be said, can be satisfactorily verified without reference to the generalities of block multiplication.)

The next two results are offshoots of 21.1.

21.2 The (i, k)th entry of a matrix product ABC is

$$(i\text{th row of } A) \times B \times (k\text{th column of } C).$$

Proof. (i, k)th entry of ABC [i.e. of $A(BC)$]

$$\begin{aligned} &= (i\text{th row of } A) \times (k\text{th column of } BC) \\ &= (i\text{th row of} A) \times B \times (k\text{th column of } C) \qquad \text{(by 21.1).} \end{aligned}$$

21.3 Suppose that A, B are matrices in $F_{m\times n}$ with the property that $AX = BX$ for every $X \in F_{n\times 1}$. Then $A = B$.

Proof. Let $E_1, E_2, \ldots, E_n$ denote the columns of I_n. Then

$$\begin{aligned} A = AI_n &= A \times [E_1 \quad E_2 \ldots E_n] \\ &= [AE_1 \quad AE_2 \ldots AE_n] \qquad \text{(cf. 21.1)} \\ &= [BE_1 \quad BE_2 \ldots BE_n] \qquad \text{(by the given property of } A, B) \\ &= B \times [E_1 \quad E_2 \ldots E_n] = BI_n \\ &= B, \end{aligned}$$

and the result is proved.

Finally, another easy case of block multiplication gives us:

21.4 Let $A \in F_{m \times n}$ have columns $C_1, C_2, \ldots, C_n$, and let $X = \text{col}(x_1, x_2, \ldots, x_n)$, a column in $F_{n \times 1}$. Then

$$AX = x_1C_1 + x_2C_2 + \ldots + x_nC_n.$$

This is soon verified once X is dressed up as a proper partitioned matrix by writing it in the apparently pedantic form

$$\begin{bmatrix} x_1I_1 \\ x_2I_1 \\ \vdots \\ x_nI_1 \end{bmatrix}.$$

EXERCISES ON CHAPTER TWO

1. If $A = \begin{bmatrix} 1 & \alpha \\ 2 & \beta \end{bmatrix}$, $B = \begin{bmatrix} \alpha & \beta \\ -1 & 1 \end{bmatrix}$ and $C = \begin{bmatrix} 5 & \delta \\ 0 & 1 \end{bmatrix}$, and if $A+\gamma B = C$, find the values of the scalars α, β, γ, δ.

2. Locate and correct the two errors in the right-hand side of the equation

$$\begin{bmatrix} 4 & 0 & 2 \\ 1 & -1 & -2 \end{bmatrix} \begin{bmatrix} 3 & 1 & 0 \\ -1 & 1 & 3 \\ 0 & 5 & 1 \end{bmatrix} = \begin{bmatrix} 12 & 14 & 2 \\ 2 & -10 & -7 \end{bmatrix}.$$

3. The matrices A and B in $F_{n\times n}$ are such that A, B, and $A+B$ are all idempotent (i.e. satisfy $X^2 = X$). Prove that $AB = -BA$, and deduce, by considering ABA, that $AB = O$.

4. The matrices A, B in $F_{n\times n}$ are such that both $A+B$ and $A-B$ are idempotent. Prove that $A = A^2+B^2$, and deduce that B^2 commutes with A. By considering $A = \frac{1}{4}\begin{bmatrix} x & 1 \\ 1 & 1 \end{bmatrix}$ and $B = \frac{1}{4}\begin{bmatrix} y & -1 \\ -1 & -1 \end{bmatrix}$, where x and y are numbers to be chosen, show that A^2 need not commute with B.

5. The *trace* of a square matrix means the sum of the entries on its main diagonal. Prove that, for all $A, B \in F_{n\times n}$, AB and BA have equal traces. Deduce that if $A, B \in F_{n\times n}$ are such that $AB-BA$ is a scalar multiple of I_n, then $AB-BA$ must be O.

6. In $\mathbb{R}_{n\times n}$, C and D are diagonal matrices, the entries on whose main diagonals are all positive; and A is such that $C^2A = AD^2$. Prove that $CA = AD$.

7.* Investigate the circumstances under which the matrix $A = [\alpha_{ik}]_{n\times n}$ in $F_{n\times n}$ commutes with the $n \times n$ matrix unit E_{ik}. Hence or otherwise prove that the only matrices in $F_{n\times n}$ that commute with all matrices in $F_{n\times n}$ are the scalar multiples of I_n.

8. A square matrix is described as *lower triangular* if all its entries above the main diagonal are zero; and if, in addition, all the entries on the main diagonal are zero, the matrix is described as *strictly lower triangular*. Show that, in $F_{3\times 3}$,

(i) the set of lower triangular matrices is closed under multiplication,

(ii) the cube of every strictly lower triangular matrix is O. (These results generalize to $F_{n\times n}$, with "cube" replaced by "nth power" in (ii).)

9. Let A, B be symmetric matrices in $F_{n\times n}$. Show that:

(i) AB is symmetric if and only if A and B commute;

(ii) $AB-BA$ is skew-symmetric.

10. Let X be a real $n \times 1$ column and S a real $n \times n$ skew-symmetric matrix. Show that the 1×1 matrix X^TSX is skew-symmetric, and deduce that $X^TSX = 0$.

11. In $\mathbb{R}_{n\times n}$, show that the set of orthogonal matrices is (i) closed under multiplication and (ii) closed under the taking of inverses [meaning that if A is in the set, so also is A^{-1}].

12. In $\mathbb{R}_{n\times n}$, A is an orthogonal matrix and the matrix B is such that AB is skew-symmetric. Show that (i) BA is also skew-symmetric; (ii) if B is orthogonal, then $(AB)^2 = (BA)^2 = -I$.

13. (i) Show that if A, B, P are matrices in $F_{n\times n}$ such that $AP = PB$ and P is nonsingular, then, for every positive integer m, $A^m = PB^mP^{-1}$.

(ii) Let $A = \begin{bmatrix} 4 & -2 \\ 3 & -1 \end{bmatrix}$ and $B = \text{diag}(2, 1)$. Show that a real 2×2 matrix P satisfies the equation $AP = PB$ if and only if P is of the form $\begin{bmatrix} a & 2b \\ a & 3b \end{bmatrix}$. By considering the case $a = b = 1$ and by using part (i), find a general formula for A^m (m a positive integer).

14. The square matrix A is such that $(A+2I)^2 = O$. Prove that A is nonsingular and express A^{-1} in terms of A. Show also that $A+I$ is nonsingular.

15. The square matrix A is such that $I+A$ is nonsingular. Show that $(I+A)^{-1}$ commutes with A.

16. Let J be the $n \times n$ real matrix with every entry equal to 1, so that $J^2 = nJ$; and let α, β be real numbers such that $\alpha \neq 0$ and $\alpha \neq -n\beta$. Prove that $\alpha I_n + \beta J$ is nonsingular by finding an inverse for it of the form $(1/\alpha)(I_n + \gamma J)$. Deduce the inverse of the matrix

$$\begin{bmatrix} 3 & 2 & 2 & 2 \\ 2 & 3 & 2 & 2 \\ 2 & 2 & 3 & 2 \\ 2 & 2 & 2 & 3 \end{bmatrix}.$$

17. The matrices A and B in $F_{n\times n}$ are such that $AB = A$. Show that:
(i) if $B \neq I$, then A is singular;
(ii) if $B^2 = O$, then $A = O$;
(iii) if $B^3 = O$ (but B^2 is not necessarily O), then $A = O$;
(iv)* if $BA = B$, then A and B are idempotent and $(A-B)^2 = O$.

18. In $\mathbb{R}_{n\times n}$, P is a nonsingular matrix and the matrix A is such that P^TAP is symmetric. Prove that A is symmetric.

19. A square matrix A is described as *nilpotent* if some positive power of A is zero. (E.g. all strictly lower triangular matrices: see question 8.) Prove that every nilpotent matrix is singular.

20. In $F_{n\times n}$, the nonsingular matrices A and B are such that $A+B$ is also nonsingular and $(A+B)^{-1} = A^{-1}+B^{-1}$. Let C stand for AB^{-1}. Show that $C+C^{-1} = -I$, and deduce that $C^3 = I$.

21. In the following statements, A and B stand for $n \times n$ matrices over F. Classify each statement as true or false, and justify each answer by giving a proof or a counterexample, as appropriate.
(i) If $AB^2 = A^2B$ and A and B are nonsingular, then $A = B$.
(ii) If A and B are nonsingular, then $A+B$ is nonsingular.
(iii) If $AB = O$ but neither A nor B is O, then A and B are both singular.
(iv) If there exists a nonzero column X such that $AX = O$, then A is singular.
(v)* If $ABA = O$ and B is nonsingular, then $A^2 = O$.

22. Let $A = \begin{bmatrix} a & b \\ c & d \end{bmatrix}$ be an arbitrary 2×2 matrix over F; let $k = ad-bc$ (the determinant of A); and let $t = a+d$. Prove that

$$A^2 - tA + kI = O.$$

* Deduce that if $A^3 = O$ (so that $k = 0$—cf. question 19 and 19.4), then $A^2 = O$.

23.* In $F_{(2n)\times(2n)}$, M is the nonsingular matrix $\begin{bmatrix} I_n & A \\ B & C \end{bmatrix}$, partitioned after its nth row and nth column; and M^{-1}, similarly partitioned, is $\begin{bmatrix} W & X \\ Y & Z \end{bmatrix}$. Prove that $C - BA$ is nonsingular and that its inverse is Z, and express W, X, Y in terms of A, B, C.

24. Let A and B be the matrices $\begin{bmatrix} 1 & 1 \\ 0 & 1 \end{bmatrix}$ and $\begin{bmatrix} 1 & 1 & 0 \\ 0 & 1 & 0 \\ 0 & 0 & 3 \end{bmatrix}$, respectively. Find general formulae for A^n and B^n.

CHAPTER THREE

ELEMENTARY ROW OPERATIONS

22. Introduction

The title of the chapter refers to operations of three standard types which, for various constructive purposes, we may carry out on the rows of a matrix. The operations in question are:

(I) Interchange of two rows;

(II) Multiplication of a row by a nonzero scalar;

(III) Addition to one row of a scalar multiple of another row.

These operations are called **elementary row operations** (e.r.o.s for short)—of types (I), (II), (III), respectively.

There will be a standing notational convention henceforth that R_i will denote the ith row of the matrix being operated on. Accordingly, we use shorthands for e.r.o.s as follows:

(i) $R_i \leftrightarrow R_j$ denotes the interchange of rows i and j $(i \neq j)$;

(ii) $R_i \rightarrow \lambda R_i$ denotes the multiplication of the ith row by the scalar λ $(\lambda \neq 0)$;

(iii) $R_i \rightarrow R_i + \lambda R_j$ denotes the addition to the ith row of λ times the jth row $(i \neq j)$.

Let us illustrate by performing certain e.r.o.s on the matrix

$$\begin{bmatrix} 1 & 2 & 3 & 4 \\ 5 & 6 & 7 & 8 \\ 9 & 10 & 11 & 12 \end{bmatrix}.$$

If we start on each occasion with this matrix and perform (separately) the e.r.o.s $R_1 \leftrightarrow R_2$, $R_2 \rightarrow 4R_2$, $R_3 \rightarrow R_3 + 6R_1$, the results are, respectively,

$$\begin{bmatrix} 5 & 6 & 7 & 8 \\ 1 & 2 & 3 & 4 \\ 9 & 10 & 11 & 12 \end{bmatrix}, \begin{bmatrix} 1 & 2 & 3 & 4 \\ 20 & 24 & 28 & 32 \\ 9 & 10 & 11 & 12 \end{bmatrix}, \begin{bmatrix} 1 & 2 & 3 & 4 \\ 5 & 6 & 7 & 8 \\ 15 & 22 & 29 & 36 \end{bmatrix}.$$

There is a discernible resemblance between e.r.o.s on matrices and the operations one carries out in handling a system of linear equations. For example, a type (III) e.r.o. resembles the adding of some multiple of one equation to another in the system. This resemblance is no accident, and indeed it is a major reason for taking an interest in e.r.o.s. Not surprisingly, therefore, our discussion of e.r.o.s will enable us to make major advances in our study of systems of linear equations, and we shall focus our attention on that topic later in the chapter (sections 27 and 28). Prior to that (§24) come explanations of how to use e.r.o.s to transform any given matrix into a matrix of standard simple form. And this and other technicalities about e.r.o.s lead (as shown in §26) to the solving of important problems identified in chapter 2: e.g. we shall develop a systematic method for finding the inverse of a nonsingular matrix and prove that a left inverse (of a square matrix) must also be a right inverse (and vice versa).

Throughout the chapter the field of scalars in the general discussion will be the arbitrary F.

23. Some generalities concerning elementary row operations

We often use a symbol such as θ to denote an e.r.o.; and then $\theta(A)$ means the matrix that results from performing the e.r.o. θ on the matrix A.

When we speak of "performing the sequence of e.r.o.s $\theta_1, \theta_2, \ldots, \theta_s$ on A", we mean performing θ_1 on A, then performing θ_2 on the result of that first stage, then performing θ_3 on the result of that second stage, and so on until finally θ_s is performed. The matrix finally produced by this multi-stage process is denoted by

$$\theta_s\theta_{s-1}\ldots\theta_2\theta_1(A).$$

Here we may think of $\theta_s\theta_{s-1}\ldots\theta_2\theta_1$ as a composite operation, whose performance consists of the s steps described above (θ_1 being performed first, and θ_s last).

It should be noted that the order of the e.r.o.s in a sequence matters. For example, the composite operation $\theta_2\theta_1$ (meaning θ_1 followed by θ_2) may well have a different effect from the composite operation $\theta_1\theta_2$ (meaning θ_2 followed by θ_1). A specific illustration of this is obtained by taking θ_1 to be $R_1 \leftrightarrow R_2$ and θ_2 to be $R_1 \to 2R_1$: in this case $\theta_2\theta_1(I_2)$ is $\begin{bmatrix} 0 & 2 \\ 1 & 0 \end{bmatrix}$, but $\theta_1\theta_2(I_2)$ is $\begin{bmatrix} 0 & 1 \\ 2 & 0 \end{bmatrix}$.

For each e.r.o. θ we specify, as detailed in the following table, an e.r.o. called the *inverse* of θ and denoted by θ^{-1}.

e.r.o. θ	its inverse θ^{-1}
$R_i \leftrightarrow R_j \ (i \neq j)$	$R_i \leftrightarrow R_j$
$R_i \rightarrow \lambda R_i \ (\lambda \neq 0)$	$R_i \rightarrow (1/\lambda) R_i$
$R_i \rightarrow R_i + \lambda R_j \ (i \neq j)$	$R_i \rightarrow R_i - \lambda R_j$

It is easy to see that:

23.1 In every case the composite operations $\theta^{-1}\theta$ and $\theta\theta^{-1}$ leave any matrix unaltered: i.e. for any matrix A, $\theta^{-1}\theta(A)$ and $\theta\theta^{-1}(A)$ are just A.

Another way of expressing 23.1 would be to say that, in every case, the e.r.o.s θ and θ^{-1} undo (or reverse) each other's effects. Observe that, more generally:

23.2 For any sequence $\theta_1, \theta_2, \ldots, \theta_s$ of e.r.o.s, the composite operation $\theta_s\theta_{s-1}\ldots\theta_2\theta_1$ is reversed by the composite operation $\theta_1^{-1}\theta_2^{-1}\ldots\theta_{s-1}^{-1}\theta_s^{-1}$: i.e. after performing the sequence of e.r.o.s $\theta_1, \theta_2, \ldots, \theta_{s-1}, \theta_s$, one can get back to where one started by then performing the sequence $\theta_s^{-1}, \theta_{s-1}^{-1}, \ldots, \theta_2^{-1}, \theta_1^{-1}$.

At this point we can introduce the useful notion of **row equivalence**. Let A, B denote matrices. We say that A is row equivalent to B (written $A \sim B$) if and only if there is a sequence of e.r.o.s $\theta_1, \theta_2, \ldots, \theta_s$ that transforms A into B (i.e. is such that $B = \theta_s\ldots\theta_2\theta_1(A)$).

Suppose it is the case that $A \sim B$. Then there is some sequence of e.r.o.s $\theta_1, \theta_2, \ldots, \theta_s$ that transforms A into B; and (cf. 23.2) the sequence of e.r.o.s $\theta_s^{-1}, \ldots, \theta_2^{-1}, \theta_1^{-1}$ transforms B into A, so that it is also true that $B \sim A$.

From this discussion it is seen that row equivalence is a *symmetric* relation: i.e. if $A \sim B$ then also $B \sim A$ (and vice versa, as is seen by interchanging the roles of A and B), so that, in fact, the statements "$A \sim B$" and "$B \sim A$" are synonymous.

It is also true that row equivalence is what is called a *transitive* relation: i.e. if $A \sim B$ and $B \sim C$, then $A \sim C$. This is immediately apparent from the definition of row equivalence. Because of this property of row equivalence, we can present a sequence of the form

$$A \sim B_1 \sim B_2 \sim B_3 \sim \ldots \sim B_n \sim C,$$

meaning that A is row equivalent to B_1, B_1 to B_2, B_2 to $B_3, \ldots,$ and B_n to C, and indicating that consequently (by virtue of the transitive property) A is row equivalent to C.

(*Technical note*: Readers who know what is meant by "an equivalence relation" will recognize that row equivalence is such a relation. To justify this assertion, it remains to prove that row equivalence possesses the reflexive property—i.e. that every matrix is row equivalent to itself. This is clearly true because the arbitrary matrix A is transformed into itself by the trivial e.r.o. $R_1 \rightarrow 1R_1$.)

Our final observation in this section is that:

23.3 If A and B are matrices with the same number of rows and θ is an e.r.o., then $\theta([A \quad B]) = [\theta(A) \quad \theta(B)]$.

This means that when we have a matrix partitioned by a vertical dividing line, the performing of a given e.r.o. on both of the submatrices is identical in effect to the performing of that same e.r.o. on the entire matrix. One soon satisfies oneself that this is true for each of the three types of e.r.o.

24. Echelon matrices and reduced echelon matrices

Some important terminology will now be introduced.

(1) In any nonzero row, the **leading entry** is the first nonzero entry (i.e. the left-most nonzero entry). For example, in the row

$$[0 \quad 0 \quad 4 \quad 0 \quad 6]$$

the 4 is the leading entry.

(2) An **echelon matrix** means a matrix with the following two properties:

(i) the zero rows, if any, occur at the bottom;

(ii) if $R_1, R_2, \dots, R_l$ are the nonzero rows (in order from the top), then the leading entry of R_{i+1} occurs strictly farther right than the leading entry of R_i ($i = 1, 2, 3, \dots, l-1$).

Examples of echelon matrices include

$$\begin{bmatrix} 1 & 2 & 3 & 4 & 5 \\ 0 & 6 & 7 & 8 & 9 \\ 0 & 0 & 0 & 10 & 11 \\ 0 & 0 & 0 & 0 & 12 \end{bmatrix} \quad \text{and} \quad \begin{bmatrix} 0 & 4 & -1 & 0 & 5 \\ 0 & 0 & 2 & 1 & 0 \\ 0 & 0 & 0 & 0 & 8 \\ 0 & 0 & 0 & 0 & 0 \\ 0 & 0 & 0 & 0 & 0 \end{bmatrix}.$$

The zigzag lines drawn through these matrices are there purely to help the student see clearly the pattern of leading entries found farther and farther to the right as one moves down each matrix.

A zero matrix is deemed to be an echelon matrix. It certainly satisfies condition (i), and condition (ii) is, as it were, true by default for a zero matrix. Similarly any 1-rowed matrix is an echelon matrix.

(3) A **reduced echelon matrix** is an echelon matrix satisfying the following two further conditions:

(*a*) in each nonzero row the leading entry is 1;

(*b*) in each column that contains the leading entry of a row, all other entries are zero.

An example of a reduced echelon matrix is

$$\begin{bmatrix} \textcircled{1} & 0 & 2 & 0 & -4 \\ 0 & \textcircled{1} & 7 & 0 & 0 \\ 0 & 0 & 0 & \textcircled{1} & 3 \\ 0 & 0 & 0 & 0 & 0 \end{bmatrix}.$$

Once again a zigzag has been drawn to make it clear that the matrix is an echelon matrix. In addition the leading 1s of the nonzero rows have been circled, to highlight them and to make it easy to check that condition (*b*) of the definition is satisfied.

A zero matrix is deemed to be a reduced echelon matrix. (The conditions (*a*) and (*b*) may be held to be true by default for any zero matrix.)

The rest of this section is devoted to showing how any given matrix can (by a straightforward and systematic use of e.r.o.s) be transformed into an echelon matrix or (by a second process which is a modification of the first) be transformed into a reduced echelon matrix. Before details are revealed, it might be suggested that the first of these processes is redundant. As every reduced echelon matrix is in particular an echelon matrix, the second process fulfils the stated purpose of the first (the production of an echelon matrix by e.r.o.s). So, one might ask, why not just forget the first process and concentrate on the second process only? The answer is that the second process entails more work than the first and that there are many occasions when we merely want to produce an echelon matrix (not necessarily a reduced echelon matrix) and when, therefore, the extra work entailed in the second process would be a waste of effort. So it is helpful to have, and be separately aware of, the first process, which will produce an echelon matrix efficiently, without any superfluous work.

The uses of these processes are many. Applications of the second process to the theory of matrix inverses and of systems of linear equations will be seen later in this chapter.

In the general discussion A will denote an arbitrary given matrix which is to be transformed by e.r.o.s to an echelon matrix or (when we come to the second process) to a reduced echelon matrix.

There is one simple idea that is used repeatedly, and the following fragment of an example will make this idea clear. Consider

$$\begin{bmatrix} * & * & 4 & * & * & * & * \\ 0 & 0 & 2 & * & * & * & * \\ * & * & 3 & * & * & * & * \\ * & * & -1 & * & * & * & * \\ * & * & 6 & * & * & * & * \end{bmatrix},$$

where (and this is standard practice) the stars denote entries which could be anything and whose precise values do not concern us. We focus attention on the 2 which is the leading entry of the second row, and we shall consider the use of this entry as what is termed a **pivot**. This means that while the entry itself is unaltered, it is used, in type (III) e.r.o.s, to change some or all of the entries in its column to zeros. In the current illustration, the e.r.o.s.

$$R_1 \to R_1 - 2R_2, \quad R_3 \to R_3 - \tfrac{3}{2}R_2, \quad R_4 \to R_4 + \tfrac{1}{2}R_2, \quad R_5 \to R_5 - 3R_2$$

do this job, i.e. change to zeros the other entries in the column containing the entry chosen as pivot.

There is nothing special about this specific illustration. Indeed any nonzero entry in any position in any matrix could be used as a pivot to create (through the use of type (III) e.r.o.s) zeros in the column containing that entry. The advantage of using a leading entry as a pivot (as in the illustration) is that the e.r.o.s performed do not change any entries to the left of the pivot.

In using a pivot, we need not change all other entries in its column to zero. In the first of the two processes to be described, each pivot will be used to change to zeros the entries that lie *below* the pivot.

We are now ready to describe that first process—the transformation of the arbitrary matrix A to an echelon matrix by means of e.r.o.s. We may leave aside all cases in which A itself is an echelon matrix (including the case $A = O$ and the case where A has just one row): for in all such cases the trivial e.r.o. $R_1 \to 1R_1$ transforms A into an echelon matrix (by leaving it alone).

There may be some zero columns at the left of A. If so, we can ignore these columns since all entries in them will remain zero no matter what e.r.o.s we perform. Therefore we focus attention on the leftmost column of A that contains a nonzero entry. If the top entry of this column is nonzero, that is fine: otherwise we perform a type (I) e.r.o. to bring a nonzero entry to the top of the column. So we are now looking at a matrix of the form

$$\left[\begin{array}{ccc|c|cccc}
0 & \cdots & 0 & \alpha_1 & * & * & \cdots & * \\ \hline
0 & \cdots & 0 & * & & & & \\
0 & \cdots & 0 & * & & & & \\
 & \cdots & & \vdots & & \ast & & \\
0 & \cdots & 0 & * & & & &
\end{array}\right], \quad \text{where } \alpha_1 \neq 0.$$

By using the α_1 as a pivot, we change to zero all entries below it, so producing a matrix of the form

$$\left[\begin{array}{ccc|c|cccc}
0 & \cdots & 0 & \alpha_1 & * & * & \cdots & * \\ \hline
0 & \cdots & 0 & 0 & & & & \\
0 & \cdots & 0 & 0 & & A_1 & & \\
 & \cdots & & \vdots & & & & \\
0 & \cdots & 0 & 0 & & & &
\end{array}\right].$$

If the submatrix A_1 is an echelon matrix, then the entire matrix is now in echelon form. Otherwise, we leave row 1 alone now and apply the procedure of the previous paragraph (our basic unit of procedure, as we shall call it) to the submatrix consisting of rows 2, 3, 4, ... In effect this means applying the procedure to A_1, since none of the e.r.o.s performed will affect the zero submatrix lying to the left of A_1. The result of this second stage is a matrix of the form

$$\left[\begin{array}{ccc|c|ccc|c|ccc} 0 & \dots & 0 & \alpha_1 & * & \dots & * & * & * & \dots & * \\ \hline 0 & \dots & 0 & 0 & 0 & \dots & 0 & \alpha_2 & * & \dots & * \\ \hline 0 & \dots & 0 & 0 & 0 & \dots & 0 & 0 & & & \\ & \dots & & \vdots & & \dots & & \vdots & & A_2 & \\ 0 & \dots & 0 & 0 & 0 & \dots & 0 & 0 & & & \end{array}\right].$$

Already the beginnings of echelon form are apparent. If A_2 is an echelon matrix, then echelon form has been reached. Otherwise, we leave rows 1 and 2 alone and apply our basic unit of procedure to the submatrix consisting of rows 3, 4, ... (i.e., in effect, to A_2). We continue in this way, repeating the basic unit of procedure as often as necessary, applying it each time to a matrix with one row fewer than the time before. The whole process terminates in the production of an echelon matrix when the residual bottom right-hand block (like A_1, A_2) is an echelon matrix. In the most protracted cases this happens only when that bottom right-hand block is reduced to just one row, but often it happens sooner.

The specific numerical example which follows should convince the student that (despite the length of the above general description) the process is a straightforward one. It is the sort of mechanical process where a moderate amount of practice leads to great fluency.

For demonstration example, let us begin with the matrix

$$A = \begin{bmatrix} 0 & 2 & 4 & -1 & -8 \\ 1 & 1 & 1 & 3 & 5 \\ 1 & 0 & -1 & 6 & 6 \\ 1 & 2 & 3 & 5 & -2 \end{bmatrix}.$$

The working goes as follows. E.r.o.s that have been performed are indicated at the side, along with explanatory comments.

$$A \sim \begin{bmatrix} 1 & 1 & 1 & 3 & 5 \\ 0 & 2 & 4 & -1 & -8 \\ 1 & 0 & -1 & 6 & 6 \\ 1 & 2 & 3 & 5 & -2 \end{bmatrix}$$

($R_1 \leftrightarrow R_2$) (to bring the intended pivot to the top)

$$\sim \begin{bmatrix} 1 & 1 & 1 & 3 & 5 \\ 0 & 2 & 4 & -1 & -8 \\ 0 & -1 & -2 & 3 & 1 \\ 0 & 1 & 2 & 2 & -7 \end{bmatrix}$$

($R_3 \rightarrow R_3 - R_1$, then $R_4 \rightarrow R_4 - R_1$) (using the (1, 1)th entry as a pivot to produce zeros below it)

$$\sim \begin{bmatrix} 1 & 1 & 1 & 3 & 5 \\ 0 & 1 & 2 & 2 & -7 \\ 0 & -1 & -2 & 3 & 1 \\ 0 & 2 & 4 & -1 & -8 \end{bmatrix}$$

($R_2 \leftrightarrow R_4$) (not strictly necessary, but it is arithmetically easier to have 1 (rather than 2) as pivot)

$$\sim \begin{bmatrix} 1 & 1 & 1 & 3 & 5 \\ 0 & 1 & 2 & 2 & -7 \\ 0 & 0 & 0 & 5 & -6 \\ 0 & 0 & 0 & -5 & 6 \end{bmatrix}$$

($R_3 \rightarrow R_3 + R_2$, then $R_4 \rightarrow R_4 - 2R_2$) (using the (2, 2)th entry as a pivot to produce zeros below it)

$$\sim \begin{bmatrix} 1 & 1 & 1 & 3 & 5 \\ 0 & 1 & 2 & 2 & -7 \\ 0 & 0 & 0 & 5 & -6 \\ 0 & 0 & 0 & 0 & 0 \end{bmatrix}$$

($R_4 \rightarrow R_4 + R_3$) (using the (3, 4)th entry as a pivot to make the entry below it zero).

And this is an echelon matrix.

Remarks. (*a*) The process just described and illustrated is called **echelon reduction** (of the given matrix A). Because this process may be used to transform any matrix whatever by a sequence of e.r.o.s into an echelon matrix, it follows that:

24.1 Every matrix is row equivalent to an echelon matrix.

(*b*) There are, in general, many ways of carrying out the echelon reduction of a given matrix; and the echelon matrix obtained at the end of the process is not uniquely determined by the matrix one starts with. However, as knowledge to be gained in chapter 5 will reveal, all echelon matrices row equivalent to a given matrix A have one thing in common: they all have the same number of nonzero rows. It will be seen in later chapters that this number is an important property of the matrix A.

(*c*) In general when a pivot is used, we perform a batch of e.r.o.s, one for each entry that is being changed to zero. The proper understanding of any such "batch of e.r.o.s" (cf. our interest in *sequences* of e.r.o.s) is that the e.r.o.s in the batch are performed consecutively (first one, then the second, then the third,...) and not simultaneously. It is good advice to beware of trying to do too many e.r.o.s in any one stage of the working, and there is a pitfall to be avoided: e.g. if one decides to carry out the programme $R_j \to \ldots$, then $R_i \to R_i + \lambda R_j$, it must be realized that the second of these e.r.o.s involves not the original R_j, but the "new" R_j produced by the first of the e.r.o.s.

As earlier remarks stated, there is a second process, a modification of the echelon reduction process, which will produce a *reduced* echelon matrix from an arbitrary given matrix A. The two necessary changes in the echelon reduction process are as follows.

(1) Every entry that will be a leading entry in the echelon matrix finally obtained (and thus every entry used as a pivot in the echelon reduction) must be made into a 1. There may be several satisfactory ways of achieving this, but certainly it can always be done by a type (II) e.r.o. on the row in which the entry occurs. (If this entry is originally α, multiply the row by $1/\alpha$ to change the entry to 1.)

(2) Whenever a pivot is used, *all* other entries in its column should be changed to zero.

As we work through the modified process, then, we create leading 1s and use them as pivots to eliminate all nonzero entries above and below them. As we proceed, the leading 1s that occupy our attention occur farther and farther right, and thus the reduced echelon pattern is spread across the matrix from left to right until the whole matrix becomes a reduced echelon matrix. It should be noted that at each stage our use of a leading 1 as pivot does not spoil the reduced echelon pattern that has already been built up to the left: for, as pointed out earlier, there is never any change in entries to the left of a pivot when that pivot is a leading entry.

All this will no doubt be more clearly understood in the light of a specific numerical example. Let us again take the matrix

$$A = \begin{bmatrix} 0 & 2 & 4 & -1 & -8 \\ 1 & 1 & 1 & 3 & 5 \\ 1 & 0 & -1 & 6 & 6 \\ 1 & 2 & 3 & 5 & -2 \end{bmatrix},$$

on which we have already carried out an echelon reduction. This time we carry out the modified process to transform A, by a sequence of e.r.o.s, to a reduced echelon matrix:

$$A \sim \begin{bmatrix} 1 & 1 & 1 & 3 & 5 \\ 0 & 2 & 4 & -1 & -8 \\ 1 & 0 & -1 & 6 & 6 \\ 1 & 2 & 3 & 5 & -2 \end{bmatrix} \qquad (R_1 \leftrightarrow R_2)$$

$$\sim\begin{bmatrix}1 & 1 & 1 & 3 & 5\\ 0 & 2 & 4 & -1 & -8\\ 0 & -1 & -2 & 3 & 1\\ 0 & 1 & 2 & 2 & -7\end{bmatrix}$$

($R_3 \rightarrow R_3 - R_1$, then $R_4 \rightarrow R_4 - R_1$) (so far there is no change as compared with the earlier echelon reduction)

$$\sim\begin{bmatrix}1 & 1 & 1 & 3 & 5\\ 0 & 1 & 2 & 2 & -7\\ 0 & -1 & -2 & 3 & 1\\ 0 & 2 & 4 & -1 & -8\end{bmatrix}$$

($R_2 \leftrightarrow R_4$) (this time it is essential to make the intended pivot a 1; $R_2 \rightarrow \frac{1}{2}R_2$ would be an alternative strategy)

$$\sim\begin{bmatrix}1 & 0 & -1 & 1 & 12\\ 0 & 1 & 2 & 2 & -7\\ 0 & 0 & 0 & 5 & -6\\ 0 & 0 & 0 & -5 & 6\end{bmatrix}$$

($R_1 \rightarrow R_1 - R_2$, then $R_3 \rightarrow R_3 + R_2$, then $R_4 \rightarrow R_4 - 2R_2$) (this time we clear above the pivot, as well as below it)

$$\sim\begin{bmatrix}1 & 0 & -1 & 1 & 12\\ 0 & 1 & 2 & 2 & -7\\ 0 & 0 & 0 & 1 & -1.2\\ 0 & 0 & 0 & -5 & 6\end{bmatrix}$$

($R_3 \rightarrow \frac{1}{5}R_3$) (making the next pivot a 1)

$$\sim\begin{bmatrix}1 & 0 & -1 & 0 & 13.2\\ 0 & 1 & 2 & 0 & -4.6\\ 0 & 0 & 0 & 1 & -1.2\\ 0 & 0 & 0 & 0 & 0\end{bmatrix}$$

($R_1 \rightarrow R_1 - R_3$, then $R_2 \rightarrow R_2 - 2R_3$, then $R_4 \rightarrow R_4 + 5R_3$) (again clearing above and below a pivot);

and this is a reduced echelon matrix.

Once satisfied that by this second process any matrix whatever can be transformed to a reduced echelon matrix, we can record the conclusion that:

24.2 Every matrix is row equivalent to a reduced echelon matrix.

It can be shown (though no attempt is made to prove it here) that there is only one reduced echelon matrix row equivalent to any given matrix A.

25. Elementary matrices

Throughout this section, the general discussion will deal with matrices with n rows.

For any e.r.o. θ, the **elementary matrix** corresponding to θ is the matrix $\theta(I_n)$ which results from performing θ on I_n.

For example, in the case $n = 3$, the elementary matrices corresponding to the e.r.o.s $R_1 \leftrightarrow R_2$, $R_2 \to 4R_2$, $R_3 \to R_3 + 6R_1$ are, respectively,

$$\begin{bmatrix} 0 & 1 & 0 \\ 1 & 0 & 0 \\ 0 & 0 & 1 \end{bmatrix}, \quad \begin{bmatrix} 1 & 0 & 0 \\ 0 & 4 & 0 \\ 0 & 0 & 1 \end{bmatrix}, \quad \begin{bmatrix} 1 & 0 & 0 \\ 0 & 1 & 0 \\ 6 & 0 & 1 \end{bmatrix}.$$

The following proposition is a key fact about elementary matrices.

25.1 Let θ be any e.r.o. (applicable to n-rowed matrices) and let P be the $(n \times n)$ elementary matrix corresponding to θ. Then, for every n-rowed matrix A, $\theta(A) = PA$. That is, θ has the same effect as premultiplication by P.

To prove 25.1, it is necessary to consider separately the three types of e.r.o. We shall look in detail at the case of a type (III) e.r.o.

Let θ be the arbitrary type (III) e.r.o. $R_i \to R_i + \lambda R_j$ (where $i \neq j$), and let P be the corresponding elementary matrix. Then $P = \theta(I_n) = I_n + \lambda E_{ij}$ (where E_{ij} is an $n \times n$ matrix unit, as defined in §17(3)).

Let A, with rows $R_1, R_2, \ldots, R_n$, be an arbitrary n-rowed matrix. Then

$$PA = (I_n + \lambda E_{ij})A = A + \lambda E_{ij}A$$

$$= A + \lambda E_{ij} \begin{bmatrix} R_1 \\ R_2 \\ \vdots \\ R_n \end{bmatrix} = \begin{bmatrix} R_1 \\ \vdots \\ R_{i-1} \\ R_i \\ R_{i+1} \\ \vdots \\ R_n \end{bmatrix} + \lambda \begin{bmatrix} O \\ \vdots \\ O \\ R_j \\ O \\ \vdots \\ O \end{bmatrix} \leftarrow \begin{pmatrix} i\text{th} \\ \text{row} \end{pmatrix} = \begin{bmatrix} R_1 \\ \vdots \\ R_{i-1} \\ R_i + \lambda R_j \\ R_{i+1} \\ \vdots \\ R_n \end{bmatrix};$$

and thus $PA = \theta(A)$.

This proves 25.1 for the case of a type (III) e.r.o. By the same approach of partitioning the arbitrary matrix A into its individual rows, one can soon show that 25.1 is true also for a type (I) e.r.o. There remains the case of a type (II) e.r.o., which is relatively trivial: for in this case the elementary matrix P is a diagonal matrix, and the result follows immediately from 17.3(i).

The rest of this section is devoted to consequences of 25.1. Among these, 25.3 and 25.4 will play a crucial role in our discussion of matrix inverses in §26.

25.2 Every elementary matrix is nonsingular.

Proof. Let P be an arbitrary $(n \times n)$ elementary matrix, corresponding to the e.r.o. θ.

Let Q be the elementary matrix corresponding to θ^{-1}. By 25.1, $Q(PI_n)$ is the matrix obtained by performing θ on I_n and then performing θ^{-1} on the result. So (cf. 23.1) $Q(PI_n) = I_n$, i.e. $QP = I_n$. But similarly

$$PQ = P(QI_n) = \theta\theta^{-1}(I_n) = I_n \qquad \text{(again by 25.1 and 23.1).}$$

Hence P is nonsingular (its inverse being Q).

The stated result follows.

While 25.1 was about a single e.r.o., the next result tells us about any sequence of e.r.o.s.

25.3 Let $\theta_1, \theta_2, \ldots, \theta_s$ be a sequence of e.r.o.s applicable to n-rowed matrices. Then there is a nonsingular $n \times n$ matrix Q such that, for every n-rowed matrix A,

$$\theta_s\theta_{s-1}\ldots\theta_2\theta_1(A) = QA.$$

That is, the performing of the sequence of e.r.o.s $\theta_1, \theta_2, \ldots, \theta_s$ has the same effect as premultiplication by a certain nonsingular matrix Q.

Proof. Let A be an arbitrary n-rowed matrix, and let $P_1, P_2, \ldots, P_s$ be the elementary matrices corresponding to $\theta_1, \theta_2, \ldots, \theta_s$, respectively. By 25.1, we successively deduce:

$$\theta_1(A) = P_1A; \qquad \theta_2\theta_1(A) = P_2P_1A; \qquad \theta_3\theta_2\theta_1(A) = P_3P_2P_1A;$$

and eventually

$$\theta_s\theta_{s-1}\ldots\theta_2\theta_1(A) = P_sP_{s-1}\ldots P_2P_1A = QA,$$

where $Q = P_sP_{s-1}\ldots P_2P_1$, which is nonsingular in view of 25.2 and 20.5. This establishes the result.

It follows at once from 25.3 that:

25.4 If the matrix A is row equivalent to the matrix B, then $B = QA$ for some nonsingular matrix Q.

26. Major new insights on matrix inverses

We begin this section by noting that:

26.1 If E is an $n \times n$ reduced echelon matrix with no zero row, then $E = I_n$.

Proof. Let E be an $n \times n$ reduced echelon matrix with no zero row. Then E contains n leading 1s (one in every row). Because these leading 1s are found further and further right as one moves down the matrix, it is apparent (there being n leading 1s and E having type $n \times n$) that the leading 1s occupy the positions on the main diagonal of E. Therefore (cf. condition (*b*) in the definition of "reduced echelon matrix") $E = I_n$.

It will be recalled (cf. 24.2) that any matrix can be transformed to a reduced echelon matrix by a sequence of e.r.o.s. The next result shows that, in the case of a square matrix, a simple feature of the reduced echelon matrix obtained by this process tells us whether or not the matrix is singular.

26.2 Let A be an arbitrary $n \times n$ matrix, and let it be transformed (by a sequence of e.r.o.s) to a reduced echelon matrix E (so that $A \sim E$).

(i) If E contains a zero row, then A has no right inverse and hence A is singular.

(ii) If E contains no zero row (so that $E = I_n$, by 26.1), then A is nonsingular.

Proof. Row equivalence being a symmetric relation (as explained in §23), we have $E \sim A$; and hence, by 25.4, $A = QE$ for some nonsingular matrix Q. Hence also $E = Q^{-1}A$ (cf. 20.6).

(i) Suppose that E contains a zero row. Then if A had a right inverse P, we would have

$$\begin{aligned} I &= Q^{-1}Q \qquad (Q \text{ being nonsingular}) \\ &= Q^{-1}APQ \qquad (AP = I,\ P \text{ being a right inverse of } A) \\ &= EPQ \qquad (\text{since } E = Q^{-1}A), \end{aligned}$$

showing that PQ is a right inverse of E—a contradiction by 19.2. Therefore in this case A has no right inverse and thus is singular.

(ii) If E contains no zero row (so that $E = I$), then we have $A = QE = QI = Q$; and so, since Q is nonsingular, A is nonsingular.

Both parts of the result are now proved.

An easily obtained corollary is:

26.3 If the $n \times n$ matrix A has a right inverse (in particular, if A is nonsingular), then $A \sim I_n$.

Proof. Suppose that A is an $n \times n$ matrix with a right inverse. By 24.2, there is a reduced echelon matrix E such that $A \sim E$. Since A has a right inverse, it follows from 26.2(i) that E does not contain a zero row. Hence, by 26.1, $E = I_n$. So $A \sim I_n$. This establishes the corollary.

As will now be revealed, a method of finding the inverse of a nonsingular matrix lies within our grasp.

26.4 Let A be an arbitrary nonsingular matrix. Let $\theta_1, \theta_2, \ldots, \theta_s$ be any sequence of e.r.o.s that transforms A into I. (*Note*: there is at least one such sequence, by 26.3). Then the same sequence of e.r.o.s transforms I into A^{-1}.

Proof. By 25.3, the performing of the sequence of e.r.o.s has the same effect as premultiplication by a certain nonsingular matrix Q. Since the sequence transforms A into I, $QA = I$. Hence, since A is nonsingular, $Q = IA^{-1} = A^{-1}$. Further, the sequence must transform I into the matrix

$$QI = Q = A^{-1},$$

as claimed.

We can now describe and understand a simple procedure which will determine whether a given $n \times n$ matrix A is nonsingular and will produce A^{-1} if A is nonsingular. We perform a sequence of e.r.o.s on the $n \times (2n)$ matrix $[A \quad I_n]$: this means (cf. 23.3) that we simultaneously perform the same sequence of e.r.o.s on A and on I_n. We choose the e.r.o.s so that, in the left-hand half, A is transformed into a reduced echelon matrix E. If E contains a zero row, then the conclusion to be drawn is that A is singular (cf. 26.2(i)). Consider the remaining case, where the reduced echelon matrix produced in the left-hand half is I_n and the nonsingularity of A is thus disclosed (cf. 26.2(ii)). In this case our sequence of e.r.o.s has transformed A into I_n: so (in view of 26.4) it will have transformed the right-hand half of $[A \quad I_n]$ into A^{-1}!

This, therefore, is a procedure which will produce A^{-1} whenever that inverse exists. It may be helpful to summarize the pattern of the procedure when A is nonsingular by the diagram

$$[A \quad I] \xrightarrow{\text{e.r.o.s}} [I \quad A^{-1}].$$

Worked example. Consider the matrix $A = \begin{bmatrix} 3 & 4 & 5 \\ 1 & -1 & 2 \\ 2 & 1 & 3 \end{bmatrix}$.

$$[A \quad I] = \left[\begin{array}{ccc|ccc} 3 & 4 & 5 & 1 & 0 & 0 \\ 1 & -1 & 2 & 0 & 1 & 0 \\ 2 & 1 & 3 & 0 & 0 & 1 \end{array}\right]$$

$$\sim \left[\begin{array}{ccc|ccc} 1 & -1 & 2 & 0 & 1 & 0 \\ 3 & 4 & 5 & 1 & 0 & 0 \\ 2 & 1 & 3 & 0 & 0 & 1 \end{array}\right] \quad (R_1 \leftrightarrow R_2)$$

$$\sim \left[\begin{array}{ccc|ccc} 1 & -1 & 2 & 0 & 1 & 0 \\ 0 & 7 & -1 & 1 & -3 & 0 \\ 0 & 3 & -1 & 0 & -2 & 1 \end{array}\right] \quad \begin{array}{l}(R_2 \rightarrow R_2 - 3R_1, \text{ then} \\ R_3 \rightarrow R_3 - 2R_1)\end{array}$$

$$\sim \left[\begin{array}{ccc|ccc} 1 & -1 & 2 & 0 & 1 & 0 \\ 0 & 1 & 1 & 1 & 1 & -2 \\ 0 & 3 & -1 & 0 & -2 & 1 \end{array}\right]$$ ($R_2 \to R_2 - 2R_3$) (a subtle way to create a 1 as the intended pivot—a way that delays the appearance of fractions in the working)

$$\sim \left[\begin{array}{ccc|ccc} 1 & 0 & 3 & 1 & 2 & -2 \\ 0 & 1 & 1 & 1 & 1 & -2 \\ 0 & 0 & -4 & -3 & -5 & 7 \end{array}\right]$$ ($R_1 \to R_1 + R_2$, then $R_3 \to R_3 - 3R_2$)

$$\sim \left[\begin{array}{ccc|ccc} 1 & 0 & 3 & 1 & 2 & -2 \\ 0 & 1 & 1 & 1 & 1 & -2 \\ 0 & 0 & 1 & \frac{3}{4} & \frac{5}{4} & -\frac{7}{4} \end{array}\right]$$ ($R_3 \to -\frac{1}{4}R_3$)

$$\sim \left[\begin{array}{ccc|ccc} 1 & 0 & 0 & -\frac{5}{4} & -\frac{7}{4} & \frac{13}{4} \\ 0 & 1 & 0 & \frac{1}{4} & -\frac{1}{4} & -\frac{1}{4} \\ 0 & 0 & 1 & \frac{3}{4} & \frac{5}{4} & -\frac{7}{4} \end{array}\right]$$ ($R_1 \to R_1 - 3R_3$, then $R_2 \to R_2 - R_3$)

Our success in transforming the left-hand half into I shows that the given matrix A is nonsingular. And A^{-1} is the final right-hand half; i.e.

$$A^{-1} = \frac{1}{4}\begin{bmatrix} -5 & -7 & 13 \\ 1 & -1 & -1 \\ 3 & 5 & -7 \end{bmatrix}.$$

Having developed one very satisfactory method for testing whether a given square matrix is nonsingular and for finding its inverse when it is nonsingular, let us turn our attention to another problem outstanding from chapter 2—the relationship between left and right inverses of a square matrix. This matter can now be swiftly cleared up, by means of results already proved in this section.

26.5 Let A be a square matrix.

(i) Any right inverse of A is also a left inverse of A.

(ii) Any left inverse of A is also a right inverse of A.

Proof. (i) Suppose that R is a right inverse of A. Because this right inverse exists, $A \sim I$ (by 26.3), and hence (by 25.4) there is a nonsingular matrix Q such that $QA = I$. This matrix Q is a left inverse of A, and

$$\begin{aligned} R &= QAR && \text{(since } QA = I\text{)} \\ &= Q && \text{(since } R \text{ is a right inverse of } A\text{, i.e. } AR = I\text{).} \end{aligned}$$

Therefore R is also a left inverse of A, and this proves part (i).

(ii) Suppose that P is a left inverse of A, so that $PA = I$. Then A is a right inverse of P, and so, by part (i), A is also a left inverse of P. So $AP = I$, which shows that P is a right inverse of A. Part (ii) is thus established.

It follows that:

26.6 If A is an $n \times n$ matrix and an $n \times n$ matrix Q is known to be either a left inverse or a right inverse of A, then A is nonsingular and $Q = A^{-1}$.

(One needs 19.1 to justify the last clause.)

These results are pleasing because they mean that we may simply forget all distinctions between the terms "left inverse", "right inverse" and "inverse" in square matrix theory. In particular, if $A, B \in F_{n \times n}$, the equations $AB = I$ and $BA = I$ imply each other. So, for example, if one can show that a square matrix A satisfies $A^T A = I$, that (because it implies that also $AA^T = I$) suffices to prove that A is orthogonal. Moreover, 26.6 enables us to prove that:

26.7 If $A, B \in F_{n \times n}$ and at least one of A, B is singular, then AB is also singular.

Proof. For $A, B \in F_{n \times n}$,

$$AB \text{ is nonsingular (with inverse } P \text{, say)} \Rightarrow \left\{\begin{matrix} A(BP) = (AB)P = I \quad \text{and} \\ (PA)B = P(AB) = I \end{matrix}\right\}$$

$\Rightarrow A$ and B are both nonsingular (with inverses BP, PA, respectively);

and from this the stated result clearly follows.

There is one last useful fact we can glean from the ideas of this section.

26.8 Every nonsingular matrix can be expressed as a product of elementary matrices.

Proof. Let A be an arbitrary nonsingular matrix. In view of 26.3 and the symmetric property of row equivalence, I is row equivalent to A, and thus there is a sequence of e.r.o.s $\theta_1, \theta_2, \ldots, \theta_s$ that transforms I into A. Let $P_1, P_2, \ldots, P_s$ be the elementary matrices corresponding to $\theta_1, \theta_2, \ldots, \theta_s$, respectively. Then, by 25.1 (cf. proof of 25.3)

$$\begin{aligned} \theta_s\theta_{s-1}\ldots\theta_2\theta_1(I) &= P_sP_{s-1}\ldots P_2P_1I, \\ \text{i.e.}\quad A &= P_sP_{s-1}\ldots P_2P_1. \end{aligned}$$

This equation displays the arbitrary nonsingular A expressed as a product of elementary matrices, and so the result is proved.

27. Generalities about systems of linear equations

It is helpful to establish a standard notation which we can use whenever we wish to discuss an arbitrary system of linear equations. For this purpose we shall return to the notation used in the brief allusion to systems of linear equations in §15 (remark (*a*)): that is, the arbitrary system of equations will be taken to be

$$\left\{\begin{matrix} \alpha_{11}x_1+\alpha_{12}x_2+\ldots+\alpha_{1n}x_n=\kappa_1 \\ \ldots \\ \alpha_{m1}x_1+\alpha_{m2}x_2+\ldots+\alpha_{mn}x_n=\kappa_m \end{matrix}\right\},$$

a system of m equations in n "unknowns" $x_1, x_2, \ldots, x_n$; the coefficient matrix $[\alpha_{ik}]_{m\times n}$ will be denoted by A; the column of unknowns, $\text{col}(x_1, x_2, \ldots, x_n)$, will be denoted by X; and the column of right-hand sides, $\text{col}(\kappa_1, \kappa_2, \ldots, \kappa_m)$, will be denoted by K; accordingly, it will be possible to write the system concisely as $AX = K$.

In future, then, whenever we wish to initiate a discussion of an arbitrary system of linear equations, we shall say, "Consider the arbitrary system of equations $AX = K$, where the notational details are as usual." That will mean that *all* the notational details of the system are to be precisely as above (including, for example, m for the number of equations, n for the number of unknowns, α_{ik} for the (i, k)th entry of the coefficient matrix).

The first matter to clarify in discussing the arbitrary system $AX = K$ (the notational details being as usual) is the variety of ways in which the word "solution" gets used with reference to the system. When we give specific values of the unknowns (e.g. $x_1 = 2, x_2 = 5, \ldots$) that satisfy all the equations in the system, we are giving what is called a **particular solution**. The implication of the word "particular" here is that this solution may be just one among many and that there may be other specific values of the unknowns that satisfy the equations. More interesting than a particular solution is the **general solution** of the system: by this we mean a sequence of statements of the form

$$x_1 = \ldots, x_2 = \ldots \quad , \ldots \quad , x_n = \ldots$$

precisely equivalent to the given system of equations, so that these statements indicate all possibilities for values of $x_1, x_2, \ldots, x_n$ satisfying the equations.

Of course, specifying $x_1, x_2, \ldots, x_n$ is the same as specifying the column X; and specifying all possibilities for $x_1, \ldots, x_n$ is the same as specifying all possibilities for the column X. So both a particular solution of the system and the general solution can be presented in the "matrix form"

$$X = \ldots.$$

A further natural thing to consider is the set

$$\{X \in F_{n\times 1} : AX = K\},$$

i.e. the set of all columns that satisfy $AX = K$. This is a subset of $F_{n\times 1}$, and we shall call it the **solution set** of the system.

Let us illustrate all this with a trivial example. Take the "system" consisting of the single equation

$$x_1 + x_2 = 4$$

in 2 unknowns. It is easy to write down several different particular solutions (e.g. $x_1 = 1, x_2 = 3$; or $x_1 = 6, x_2 = -2$). The general solution may be given as

$$x_1 = 4-\alpha,\ x_2 = \alpha \qquad (\alpha \in F).$$

Here α is what we call a **parameter**; and the claim being made is that the totality of particular solutions of the "system" is precisely the variety of things one can obtain from "$x_1 = 4-\alpha$, $x_2 = \alpha$" by giving α a specific value in F. (That claim is clearly true.) The matrix form of the general solution is

$$X = \begin{bmatrix} 4-\alpha \\ \alpha \end{bmatrix} \qquad (\alpha \in F);$$

and thus the solution set of the "system" is the set of all columns of the form $\begin{bmatrix} 4-\alpha \\ \alpha \end{bmatrix}$ with $\alpha \in F$.

In other examples, there may be two or more independent parameters. A trivial illustration comes from the system comprising the one equation

$$x_1 + x_2 + x_3 = 4$$

in 3 unknowns. Here the general solution may be given by

$$x_1 = 4-\alpha-\beta,\ x_2 = \beta,\ x_3 = \alpha \qquad (\alpha, \beta \in F).$$

The meaning of this statement is most concisely explained by saying that it defines the solution set as the set of all columns of the form $\begin{bmatrix} 4-\alpha-\beta \\ \alpha \\ \beta \end{bmatrix}$ with $\alpha, \beta \in F$.

Another possibility to be noted at this stage is that a system of equations may have no solutions at all, i.e. that its solution set may be $\emptyset$, the empty set. This is clearly the case for the system

$$\left\{\begin{aligned} x_1 + x_2 &= 2 \\ 2x_1 + 2x_2 &= 5 \end{aligned}\right\},$$

where (since the first equation implies that $2x_1 + 2x_2 = 4$) the two equations contradict each other. A system of equations is described as **consistent** if it has one or more solutions, and it is described as **inconsistent** if it has no solution.

A **homogeneous** system of linear equations means a system in which all the right-hand sides are zero: so the system $AX = K$ is homogeneous if and only if the column K is O. Observe that:

27.1 Every homogeneous system of linear equations is consistent.

(For: the arbitrary homogeneous system $AX = O$ has (at least) the particular solution $X = O$ (i.e. $x_1 = 0, x_2 = 0, \ldots, x_n = 0$).)

28. Elementary row operations and systems of linear equations

Consider the arbitrary system of linear equations $AX = K$, where the notational details are as usual. The **augmented matrix** of this system means the $m \times (n+1)$ matrix $[A \quad K]$, i.e. the matrix obtained from the coefficient matrix A by adjoining the extra column K at the right.

The augmented matrix can be used as a representation of the system (or, one might say, as a means of recording what the system is): for, just as the augmented matrix can be written down if we know what the system is, so the system can be written down if we know what the augmented matrix is. Notice that each row of the augmented matrix represents one of the equations in the system. Moreover, the augmented matrix is, for several reasons, a most convenient representation of the system. For a start, it can be more quickly written down than can the actual system with its repeated mentions of the unknowns. Further, changes in the equations in the system can be represented as changes in the rows of the augmented matrix; and, in particular, there is the following correspondence between operations on the system and e.r.o.s on the augmented matrix.

operation on system	e.r.o. on augmented matrix
interchange of ith and jth equations $(i \neq j)$	$R_i \leftrightarrow R_j \quad (i \neq j)$
multiplication of ith equation by nonzero scalar λ	$R_i \to \lambda R_i \quad (\lambda \neq 0)$
addition to the ith equation of λ times the jth $\quad (i \neq j)$	$R_i \to R_i + \lambda R_j \quad (i \neq j)$

The operations in the left-hand column are natural operations to carry out in handling a system of equations (and the information given in the table above does little more than retrace a major element in the motivation for considering e.r.o.s). Notice also that all the operations in the left-hand column, because of their reversibility, will produce a system *equivalent to* (implying and implied by) the system we start with, i.e. a system with exactly the same solution set. By repeated application of this observation, we see that the same thing will be true of any sequence of operations on the system that correspond to e.r.o.s on the augmented matrix. Therefore:

28.1 If the augmented matrix $[A \quad K]$ of the system $AX = K$ is transformed by e.r.o.s to $[B \quad L]$ $(A, B \in F_{m \times n};\ K, L \in F_{m \times 1})$, then the system $BX = L$ is equivalent to the given system $AX = K$.

We know (cf. 24.2) that, in the context described in 28.1, the e.r.o.s can be chosen to make $[B \quad L]$ a reduced echelon matrix. Notice incidentally that when that is the case, B is a reduced echelon matrix row equivalent to A. (It is row equivalent to A by virtue of 23.3, and it is a reduced echelon matrix because a left-hand segment of a reduced echelon matrix is always also a reduced echelon matrix.)

A system of linear equations is said to be in reduced echelon form when its augmented matrix is a reduced echelon matrix. It turns out that it is easy to write down the general solution of a system in reduced echelon form, and hence a strategy for obtaining the general solution of an arbitrary given system $AX = K$ emerges:

(1) Write down the augmented matrix $[A \quad K]$ of the given system and, by e.r.o.s, transform it to a reduced echelon matrix $[B \quad L]$.

(2) Write out the system $BX = L$, which, by 28.1, is equivalent to the given system.

(3) Carry out the easy task of writing down the general solution of the reduced echelon system $BX = L$. (The tactics are to work from the bottom upwards.) As $BX = L$ is equivalent to the given system, this general solution is the general solution of the given system.

As a straightforward example, take the system

$$\begin{Bmatrix} x_1+x_2-x_3-3x_4 = 7 \\ x_1+3x_2+2x_3+4x_4 = -2 \\ 2x_1+x_2+x_3+4x_4 = 5 \end{Bmatrix}.$$

The augmented matrix is

$$\left[\begin{array}{cccc:c} 1 & 1 & -1 & -3 & 7 \\ 1 & 3 & 2 & 4 & -2 \\ 2 & 1 & 1 & 4 & 5 \end{array}\right].$$

By applying to this matrix the usual process for transformation to a reduced echelon matrix, we obtain

$$\left[\begin{array}{cccc:c} 1 & 0 & 0 & 1 & 4 \\ 0 & 1 & 0 & -1 & 0 \\ 0 & 0 & 1 & 3 & -3 \end{array}\right].$$

It will help the student to work through the details of this process and, at each stage, to write down the system of equations corresponding to the matrix produced: this will bring to life all the above remarks about e.r.o.s on the augmented matrix corresponding to operations on the system of equations that are natural and produce systems equivalent to the given one. The outcome of the process is the knowledge (cf. 28.1) that the given system of equations is equivalent to the system

$$\left\{\begin{array}{llll} x_1 & & +\ x_4 & = 4 \\ & x_2 & -\ x_4 & = 0 \\ & & x_3+3x_4 & = -3 \end{array}\right\};$$

and hence (working from the bottom upwards) we obtain the general solution (of this system and of the equivalent given system), namely

$$x_4 = \alpha, \quad x_3 = -3-3\alpha, \quad x_2 = \alpha, \quad x_1 = 4-\alpha \qquad (\alpha \in F),$$

or (the same story arranged in a more natural order)

$$x_1 = 4-\alpha, \quad x_2 = \alpha, \quad x_3 = -3-3\alpha, \quad x_4 = \alpha \qquad (\alpha \in F).$$

Leaving this particular example, we must consider several aspects of the obtaining of the general solution of a system of equations once the system (and its augmented matrix) have been transformed into reduced echelon form.

At the bottom of the reduced echelon augmented matrix, there may be some zero rows. These correspond to tautologous equations

$$0x_1+0x_2+\ldots+0x_n = 0, \qquad \text{i.e. } 0 = 0,$$

which may be ignored.

Consider now the last of the nonzero rows in the reduced echelon augmented matrix. This might be

$$[0 \quad 0\ldots 0 \mid 1],$$

which would correspond to an equation

$$0x_1+0x_2+\ldots+0x_n = 1, \qquad \text{i.e. } 0 = 1.$$

Obviously, no values of $x_1, \ldots, x_n$ satisfy that equation, and so, in this case, the system under discussion has no solutions, i.e. is inconsistent.

In the remaining case, the last nonzero row of the reduced echelon augmented matrix is

$$[0 \quad 0\ldots 0 \quad 1 \quad * \quad *\ldots \mid *],$$

with the leading 1 *not* in the final position. Suppose that the columns of the reduced echelon augmented matrix in which leading 1s occur are (in order) the i_1th, i_2th, …, i_rth columns. Then correspondingly the reduced echelon form of the system of equations has the shape

$$\left.\begin{cases} x_{i_1} + \dots\dots\dots\dots\dots\dots = \lambda_1 \\ \qquad x_{i_2} + \dots\dots\dots\dots = \lambda_2 \\ \qquad\qquad \dots \\ \qquad\qquad\qquad x_{i_r} + \dots = \lambda_r \end{cases}\right\} \qquad (*)$$

("0 = 0" equations being omitted). Here, clearly, $r \leqslant n$ ($x_{i_1}, \dots, x_{i_r}$ being r of the n unknowns $x_1, \dots, x_n$). Because the leading 1s are the only nonzero entries in the columns in which they appear in the reduced echelon matrix, there are, in (*), no appearances of $x_{i_1}, x_{i_2}, \dots, x_{i_r}$ with nonzero coefficients other than those indicated, at the beginnings of equations. Hence it is apparent that the system has, as a particular solution,

$$x_{i_1} = \lambda_1, x_{i_2} = \lambda_2, \dots, x_{i_r} = \lambda_r, \quad \text{all other unknowns} = 0.$$

Therefore, in this second case, the system is consistent.

Putting together the discoveries of the last two paragraphs, we have the following result.

28.2 Let the augmented matrix $[A \;\; K]$ of a given system of equations $AX = K$ be transformed by e.r.o.s to a reduced echelon matrix $[B \;\; L]$ ($A, B \in F_{m \times n}$; $K, L \in F_{m \times 1}$). Then the given system is inconsistent if and only if the last nonzero row of $[B \;\; L]$ is

$$[0 \quad 0 \dots 0 \,\vdots\, 1].$$

So far only one example (and a rather easy one) has been given of the process of writing down the general solution of a consistent system in reduced echelon form. The following example has been designed to illustrate the variety of complications that can arise, and the commentary on the example will explain the general procedure in detail.

Let us consider the following system

$$\begin{cases} x_1 + x_2 \qquad - x_5 \qquad\qquad + 2x_9 = 3 & \dots(1) \\ \qquad x_4 \qquad\qquad\qquad + 2x_9 = 2 & \dots(2) \\ \qquad\qquad x_6 \quad - 3x_8 + x_9 = 1 & \dots(3) \\ \qquad\qquad\quad x_7 - x_8 + x_9 = 2 & \dots(4) \\ \qquad\qquad\qquad\qquad x_{10} = 3 & \dots(5) \end{cases}$$

—a consistent system in reduced echelon form.

As already advised, we work from the bottom of the system upwards; and we give expressions for the unknowns in reverse order (i.e. $x_{10}, x_9, x_8, \dots$).

When we come to a typical equation

$$x_i + \ldots = \ldots,$$

we must first put equal to parameters all unknowns (if any) among $x_{i+1}, x_{i+2}, \ldots$ for which expressions have not yet been given: then (and only then) the equation is used to give an expression for x_i in terms of some (perhaps all, perhaps none) of the parameters so far introduced.

In the above system, equation (5) illustrates an exceptionally straightforward case where an equation gives a definite value for an unknown. We begin the solution by simply transcribing

$$x_{10} = 3.$$

We then turn to equation (4), which reads $x_7 + \ldots = \ldots$. Following the guidelines mapped out in the paragraph before last, we must put equal to parameters the unknowns x_8 and x_9 (i.e. the unknowns higher-numbered than x_7 that have not yet been dealt with); then we use the equation to express x_7 in terms of parameters that have been introduced. This gives as the next part of the solution

$$x_9 = \alpha, \quad x_8 = \beta, \quad x_7 = 2 - \alpha + \beta.$$

Equation (3) reads $x_6 + \ldots = \ldots$. Since expressions have been given already for all the unknowns higher-numbered than x_6, we do not introduce any further parameters at this stage; and so all that there is to do is to use the equation to express x_6 in terms of parameters already introduced. Thus we continue the solution with

$$x_6 = 1 - \alpha + 3\beta.$$

Coming to equation (2), we must avoid the mistake of jumping from x_6 to x_4, forgetting x_5 purely because it doesn't appear explicitly in equation (2). In correct procedure, before we write down an expression for x_4, we must put equal to a parameter every higher-numbered unknown that has not already been dealt with. So the correct continuation of the solution is

$$x_5 = \gamma, \quad x_4 = 2 - 2\alpha.$$

Finally, we come to equation (1). This time there are two hitherto unconsidered unknowns (x_2 and x_3) that must be put equal to parameters (the total non-appearance of x_3 in the system being irrelevant); and then we use the equation to express x_1 in terms of the parameters that have been introduced. So the final instalment of the solution is

$$x_3 = \delta, \quad x_2 = \varepsilon, \quad x_1 = 3 - 2\alpha + \gamma - \varepsilon.$$

Putting all the pieces together (and arranging them in natural order), we obtain what is in fact the general solution of the system, namely

$$x_1 = 3-2\alpha+\gamma-\varepsilon,\quad x_2 = \varepsilon,\quad x_3 = \delta,\quad x_4 = 2-2\alpha,\quad x_5 = \gamma,$$
$$x_6 = 1-\alpha+3\beta,\quad x_7 = 2-\alpha+\beta,\quad x_8 = \beta,\quad x_9 = \alpha,\quad x_{10} = 3$$
$$(\alpha, \beta, \gamma, \delta, \varepsilon \in F).$$

Let us now pause to consider critically why, here and in other examples, our method does produce the *general* solution (of a consistent system in reduced echelon form).

Observe first that, clearly, the method produces statements about the unknowns implied by the equations. E.g. in the specific example above, equation (5) implies $x_{10} = 3$; then x_9 must be something (α, say), x_8 must be something (β, say), and equation (4) implies that $x_7 = 2-\alpha+\beta$; and so on.

It remains to show that the statements in our alleged general solution

$$x_1 = \ldots,\quad x_2 = \ldots,\quad x_3 = \ldots,\quad \ldots \qquad (\alpha, \beta, \gamma, \ldots \in F)$$

do satisfy (i.e. imply the holding of) all the equations, whatever the values given to the parameters $\alpha, \beta, \gamma, \ldots$. Let us refer to our alleged general solution as "AGS". Notice that, because of the reduced echelon form of the system, the unknowns appearing at the beginnings of equations ($x_1, x_4, x_6, x_7, x_{10}$ in our example) do not appear anywhere else with nonzero coefficients and that, in our procedure, it is precisely the other unknowns that are put equal to parameters. There are two types of equation to consider. First an equation may give a specific value for an unknown (e.g. equation (5) in the above example): each such equation is simply transcribed into AGS, and therefore AGS certainly implies that each such equation holds, irrespective of values given to parameters. The other type of equation reads

$$x_i + f(x_j, x_k, \ldots) = \lambda,$$

where $x_j, x_k, \ldots$ are unknowns that are put equal to parameters $\rho, \sigma, \ldots$. Such an equation will be used to get the expression for x_i in AGS, namely

$$x_i = \lambda - f(\rho, \sigma, \ldots);$$

and this, together with the statements $x_j = \rho$, $x_k = \sigma, \ldots$ in AGS, imply the holding of the equation for all values of the parameters.

To sum up, we have now demonstrated that our method will invariably produce a sequence of statements about the unknowns logically equivalent to the system of equations. In other words, the method does produce the general solution (or, as it would be more accurate to say, a form of the general solution). This was worth getting absolutely straight, because a crucial result in our development of linear algebra (28.4 below) depends on it.

In the above discussion we noted that the unknowns that are put equal to parameters are precisely the unknowns that do not occur at the beginnings of equations. The number of unknowns that *do* occur at the beginnings of equations is, of course, equal to the number of equations, and therefore the number of unknowns that are put equal to parameters is

$$(\text{total number of unknowns}) - (\text{number of equations}).$$

This proves the following result.

28.3 Suppose that a consistent system of equations, in reduced echelon form, consists of r equations, "$0 = 0$" equations having been deleted, and that the system involves n unknowns. Then the general solution obtained by the method demonstrated above involves $n - r$ parameters.

The final result, whose crucial importance has already been indicated, is a corollary of 28.3.

28.4 A homogeneous system of linear equations in which there are fewer equations than unknowns has a non-trivial solution (i.e. a solution in which at least one of the unknowns is nonzero).

Proof. Consider a homogeneous system of equations $AX = O$, where $A \in F_{m \times n}$ and $m < n$, so that the number of equations (m) is less than the number of unknowns (n).

On transformation to reduced echelon form, the given system will become, after deletion of "$0 = 0$" equations, a system of, say, r equations, where $r \leqslant m$ ($m - r$ being the number of "$0 = 0$" equations deleted). Since $m < n$, it follows that $r < n$.

The system is consistent (cf. 27.1), and so 28.3 can be applied. It tells us that the general solution of the system obtained by our standard procedure involves $n - r$ parameters. This number of parameters is at least 1, since $r < n$. So the general solution includes

$$\ldots, x_j = \alpha, \ldots \qquad (\alpha \text{ one of the parameters}),$$

and all the particular solutions in which α takes a nonzero value are non-trivial solutions.

This proves the result.

EXERCISES ON CHAPTER THREE

1. Carry out the process of echelon reduction on each of the following matrices:

$$\text{(i)} \begin{bmatrix} 1 & 1 & -1 & 2 \\ -1 & -1 & 4 & 2 \\ 2 & 2 & -1 & 0 \end{bmatrix}, \quad \text{(ii)} \begin{bmatrix} 0 & 1 & 2 & -1 \\ 0 & 1 & -1 & -4 \\ 0 & 2 & 5 & -1 \end{bmatrix},$$

$$\text{(iii)} \begin{bmatrix} 0 & -2 & -1 & 3 \\ -2 & 0 & 1 & 5 \\ -3 & 1 & 0 & 2 \\ 2 & 0 & 1 & -1 \end{bmatrix}, \quad \text{(iv)} \begin{bmatrix} 0 & 2 & 4 & 1 \\ 1 & -1 & 1 & 3 \\ 1 & 2 & 7 & 1 \\ 2 & -1 & 4 & 1 \end{bmatrix}.$$

2. By means of e.r.o.s, transform into a reduced echelon matrix each of the matrices

$$\text{(i)} \begin{bmatrix} 1 & 1 & 0 & 0 \\ 0 & 1 & 1 & 0 \\ 0 & 0 & 1 & 1 \\ 1 & 0 & 0 & 1 \end{bmatrix}, \quad \text{(ii)} \begin{bmatrix} 1 & 1 & 1 & -1 & 1 \\ 1 & 1 & 3 & 0 & 2 \\ 2 & 1 & 3 & -1 & 3 \\ 2 & 1 & 1 & -2 & 4 \end{bmatrix}.$$

Find also a reduced echelon matrix row equivalent to each of the matrices in question 1.

3. In the context of 3-rowed matrices, let $\theta_1, \theta_2, \theta_3$ be the e.r.o.s $R_1 \leftrightarrow R_2$, $R_3 \to R_3 + 4R_1$, $R_3 \to 2R_3$, respectively. Find nonsingular 3×3 matrices Q_1, Q_2 such that premultiplication by Q_1 has the same effect as the composite operation $\theta_3\theta_2\theta_1$ and premultiplication by Q_2 has the same effect as the composite operation $\theta_1\theta_2\theta_3$.

4. Determine which of the following matrices are nonsingular, and find the inverses of those that are nonsingular.

$$\text{(i)} \begin{bmatrix} 1 & 1 & 1 \\ 1 & 2 & 0 \\ -1 & -1 & 1 \end{bmatrix}, \quad \text{(ii)} \begin{bmatrix} 2 & 1 & 1 \\ 3 & -1 & 2 \\ 5 & -5 & 4 \end{bmatrix}, \quad \text{(iii)} \begin{bmatrix} 0 & 2 & -1 \\ 3 & 4 & 1 \\ -1 & 1 & -1 \end{bmatrix},$$

$$\text{(iv)} \begin{bmatrix} 4 & -1 & 0 \\ 1 & 1 & -2 \\ 3 & -1 & 1 \end{bmatrix}, \quad \text{(v)} \begin{bmatrix} 1 & 1 & -1 & 1 \\ -5 & 2 & 1 & -1 \\ 1 & 3 & 3 & -1 \\ 1 & 0 & 2 & -1 \end{bmatrix}.$$

5. The square matrix A is lower triangular (see question 8 on chapter 2), and all the entries on the main diagonal of A are nonzero. Prove that A is nonsingular and that A^{-1} is also lower triangular.

6. In $F_{n \times n}$, the matrices A and B are such that $(AB)^2 = I$. Prove that A and B are nonsingular and that $(BA)^2 = I$.

7. In $F_{n \times n}$, the matrices A and B are such that $A+B = AB$. By considering $(I-A)(I-B)$, prove that A and B commute.

8. In $F_{n \times n}$, the matrices A, B, C are such that $ABC = A+C$. By considering $(I-AB)(I-CB)$, show that $I-AB$ and $I-CB$ are inverses of each other. Deduce, by considering $(I-AB)C$, that $ABC = CBA$.

9. Illustrate 26.7 by expressing the 2×2 matrix $\begin{bmatrix} 1 & 2 \\ 3 & 4 \end{bmatrix}$ as a product of elementary matrices.

10. Prove the converse of 25.4: i.e. prove that if A, B are matrices such that $B = QA$ for some nonsingular matrix Q, then A and B are row equivalent (to each other).

11. The system of equations

$$\left\{\begin{array}{rl} x_1+3x_2+\ x_3 \qquad\quad & = 5 \\ 3x_1+2x_2-4x_3+7x_4 & = k+4 \\ x_1+\ x_2-\ x_3+2x_4 & = k-1 \end{array}\right\}$$

is known to be consistent. Find the value of k and the general solution of the system.

12. Obtain the general solutions of the following systems of equations:

(i) $$\left\{\begin{array}{rl} x_1+\ x_2 \qquad\ +\ x_4+x_5 & = 2 \\ x_1+3x_2-x_3+2x_4 \qquad & = 3 \end{array}\right\};$$

(ii) $$\left\{\begin{array}{rl} x_1 \qquad\qquad -x_4 \qquad\qquad\qquad -\ x_8 & = 1 \\ x_2+2x_3-x_4 \qquad\qquad -x_7 \qquad & = 2 \\ x_5 \qquad\qquad\qquad & = 3 \\ x_6 \qquad +2x_8 & = 4 \end{array}\right\}.$$

13. Find a non-trivial solution of the homogeneous system of equations

$$\left\{\begin{array}{rl} x_1+2x_2+3x_3 & = 0 \\ 2x_1-\ x_2+\ x_3 & = 0 \end{array}\right\}.$$

14.* Suppose that A is a singular matrix in $F_{n \times n}$. Prove that there exists a nonzero column X (in $F_{n \times 1}$) such that $AX = O$.

CHAPTER FOUR

AN INTRODUCTION TO DETERMINANTS

29. Preface to the chapter

In this chapter we are concerned exclusively with square matrices. As in chapters 2 and 3, we work, in the general discussion, over an arbitrary field of scalars F.

A little was said about determinants in chapter 2 (immediately after the proof of 19.4). In particular, it was explained that, for any square matrix A, the determinant of A means a certain scalar calculable from the entries of A.

While it has yet to be said what precisely this "certain scalar" is, the standard basic notation can be introduced at this point. For any square matrix A, the determinant of A will be denoted by det A or, if $A = [\alpha_{ik}]_{n \times n}$, by

$$\begin{vmatrix} \alpha_{11} & \alpha_{12} & \alpha_{13} & \cdots & \alpha_{1n} \\ \alpha_{21} & \alpha_{22} & \alpha_{23} & \cdots & \alpha_{2n} \\ & \cdots & & & \\ \alpha_{n1} & \alpha_{n2} & \alpha_{n3} & \cdots & \alpha_{nn} \end{vmatrix}$$

(the array of entries of A surrounded by vertical lines, instead of brackets).

The determinant of a 1×1 matrix means the sole entry of that matrix, but this case is of virtually no interest. In §19, details were given of the 2×2 case, and these we now recall, using the newly introduced notation:

29.1 If A is the 2×2 matrix $\begin{bmatrix} a & b \\ c & d \end{bmatrix}$, then det $A = ad - bc$. I.e. $\begin{vmatrix} a & b \\ c & d \end{vmatrix} = ad - bc$.

A detailed definition of what det A means when the (square) matrix A is larger than 2×2 will be given in §30.

Every student of linear algebra should know something about determinants. However, a full account, with complete proofs of the properties of determinants, is rather a formidable thing to present to a student at a relatively early stage of a course in linear algebra. For this reason, a compromise is adopted in this book: a clear definition will be given of the determinant of a square matrix of arbitrary size, and it will be indicated that certain properties of determinants apply to square matrices of all sizes; but, in order that the chapter may not be too formidable, proofs given of the properties will, in general, not cover cases where the matrices are larger than 3×3. It is hoped that this compromise treatment will provide the student with a useful introduction to determinants: it will enable him to evaluate determinants of reasonable-sized square matrices, and it should give him some feeling for what the properties of determinants are and why they are important. In due course he may wish to read a more sophisticated treatment that fills the gaps left by this chapter, but that should have a much lower priority in his exploration of linear algebra than the contents of the remaining chapters of this book.

Among the several interesting topics to be touched on in the course of this chapter will be the use of the determinant to ascertain whether a given square matrix is nonsingular (cf. 32.4) and a second method (nothing to do with e.r.o.s and useful in the 3×3 case) for finding the inverse of a nonsingular matrix (see §33).

30. Minors, cofactors, and larger determinants

As a preliminary, we associate a sign (+ or −) with each position (or each entry) in a square matrix. The sign associated with the (i, k)th position is

$$+ \text{ if } i+k \text{ is even}, \qquad - \text{ if } i+k \text{ is odd}.$$

The idea is easily grasped and remembered by means of the "sign chessboard"

$$\begin{bmatrix} + & - & + & - & \cdots \\ - & + & - & + & \cdots \\ + & - & + & - & \cdots \\ & \cdots & & & \end{bmatrix},$$

which displays in each position the sign associated with that position. The pluses and minuses alternate in each row and in each column. All this applies to a square matrix of any size.

Now consider an arbitrary 3×3 matrix

$$A = \begin{bmatrix} \alpha_{11} & \alpha_{12} & \alpha_{13} \\ \alpha_{21} & \alpha_{22} & \alpha_{23} \\ \alpha_{31} & \alpha_{32} & \alpha_{33} \end{bmatrix}.$$

For any relevant i, k, imagine the deletion of the row and column containing the (i, k)th entry of A to leave a 2×2 matrix: the determinant of this 2×2 submatrix is called the **minor** of the (i, k)th entry of A and will be denoted by μ_{ik}.

For example

$$\mu_{23} = \begin{vmatrix} \alpha_{11} & \alpha_{12} \\ \alpha_{31} & \alpha_{32} \end{vmatrix} = \alpha_{11}\alpha_{32} - \alpha_{12}\alpha_{31}. \qquad \text{(Cf. 29.1.)}$$

From this we can define the **cofactor** or **signed minor** σ_{ik} of the (i, k)th entry of A to be $+\mu_{ik}$ if the sign associated with the (i, k)th position is $+$ and $-\mu_{ik}$ if the sign associated with the (i, k)th position is $-$. So, for example, since the sign associated with the $(2, 3)$th position is $-$,

$$\sigma_{23} = -\mu_{23} = -(\alpha_{11}\alpha_{32} - \alpha_{12}\alpha_{31}) = -\alpha_{11}\alpha_{32} + \alpha_{12}\alpha_{31}.$$

We can now define det A (where A is, as before, an arbitrary 3×3 matrix) to be

$$\alpha_{11}\sigma_{11} + \alpha_{12}\sigma_{12} + \alpha_{13}\sigma_{13}$$

(the sum of 3 terms, each the product of an entry of the top row and its cofactor). Thus

$$\begin{aligned} \det A &= \alpha_{11}\mu_{11} - \alpha_{12}\mu_{12} + \alpha_{13}\mu_{13} \\ &= \alpha_{11}\begin{vmatrix} \alpha_{22} & \alpha_{23} \\ \alpha_{32} & \alpha_{33} \end{vmatrix} - \alpha_{12}\begin{vmatrix} \alpha_{21} & \alpha_{23} \\ \alpha_{31} & \alpha_{33} \end{vmatrix} + \alpha_{13}\begin{vmatrix} \alpha_{21} & \alpha_{22} \\ \alpha_{31} & \alpha_{32} \end{vmatrix}. \end{aligned}$$

E.g. $$\begin{vmatrix} 3 & 4 & 5 \\ 1 & -1 & 2 \\ 2 & 1 & 3 \end{vmatrix} = 3\begin{vmatrix} -1 & 2 \\ 1 & 3 \end{vmatrix} - 4\begin{vmatrix} 1 & 2 \\ 2 & 3 \end{vmatrix} + 5\begin{vmatrix} 1 & -1 \\ 2 & 1 \end{vmatrix}$$

$$\begin{aligned} &= 3(-5) - 4(-1) + 5(3) \qquad \text{(by 29.1)} \\ &= 4. \end{aligned}$$

Now that the determinant of a 3×3 matrix has been defined, we can introduce minors and cofactors for an arbitrary 4×4 matrix $A = [\alpha_{ik}]_{4 \times 4}$. For the (i, k)th entry of this matrix, the minor μ_{ik} is defined as the determinant of the 3×3 matrix obtained by deleting the ith row and the kth column; and the cofactor (or signed minor) σ_{ik} of this entry is defined to be $+\mu_{ik}$ or $-\mu_{ik}$, according to the sign associated with the (i, k)th position. Then we can define the determinant of the arbitrary 4×4 matrix A by

$$\det A = \alpha_{11}\sigma_{11} + \alpha_{12}\sigma_{12} + \alpha_{13}\sigma_{13} + \alpha_{14}\sigma_{14}$$

(again the sum of all terms of the form "entry of first row multiplied by its cofactor"); i.e. (cf. sign chessboard)

$$\det A = \alpha_{11}\mu_{11} - \alpha_{12}\mu_{12} + \alpha_{13}\mu_{13} - \alpha_{14}\mu_{14}.$$

The way to progress to larger and larger matrices (5×5, 6×6, etc.) is probably now apparent. At the general step, determinants of $(n-1) \times (n-1)$

matrices will already have been defined, and one wishes now to define the determinant of an arbitrary $n \times n$ matrix $A = [\alpha_{ik}]_{n \times n}$. The procedure is:

(1) for each relevant i, k, the minor μ_{ik} is defined as the determinant of the $(n-1) \times (n-1)$ matrix obtained by deleting the ith row and kth column of A (the determinant of that size of matrix already having a meaning); then

(2) for each relevant i, k, the cofactor (or signed minor) σ_{ik} is defined to be $+\mu_{ik}$ or $-\mu_{ik}$, according to the sign associated with the (i, k)th position; then

(3) det A is defined to be

$$\alpha_{11}\sigma_{11}+\alpha_{12}\sigma_{12}+\alpha_{13}\sigma_{13}+\ldots+\alpha_{1n}\sigma_{1n}$$
$$[=\alpha_{11}\mu_{11}-\alpha_{12}\mu_{12}+\alpha_{13}\mu_{13}-\ldots+(-1)^{n-1}\alpha_{1n}\mu_{1n}].$$

Since the determinants of small matrices have been defined and since the step from $(n-1) \times (n-1)$ matrices up to $n \times n$ matrices can always be made in this way, this procedure gives a meaning to the determinant of a square matrix of arbitrary size.

There is one easy case where the determinant of an $n \times n$ matrix can be quickly found:

30.1 The determinant of a diagonal matrix is the product of the entries on its main diagonal.

Proof. Consider the arbitrary diagonal matrix

$$D = \text{diag}(\alpha_1, \alpha_2, \ldots, \alpha_n).$$

By the definition of the determinant of an $n \times n$ matrix,

$$\begin{aligned}\det D &= \alpha_1 \times (\text{minor of } (1, 1)\text{th entry}) - 0 + 0 - 0 + \ldots \\ &= \alpha_1 \times \det[\text{diag}(\alpha_2, \alpha_3, \ldots, \alpha_n)].\end{aligned}$$

But similarly

$$\begin{aligned}\det[\text{diag}(\alpha_2, \alpha_3, \ldots, \alpha_n)] &= \alpha_2 \times \det[\text{diag}(\alpha_3, \alpha_4, \ldots, \alpha_n)] \\ &= \alpha_2 \times \alpha_3 \times \det[\text{diag}(\alpha_4, \ldots, \alpha_n)],\end{aligned}$$

and so on, till one eventually obtains

$$\begin{aligned}\det[\text{diag}(\alpha_2, \alpha_3, \ldots, \alpha_n)] &= \alpha_2\alpha_3\ldots\alpha_{n-2} \times \det[\text{diag}(\alpha_{n-1}, \alpha_n)] \\ &= \alpha_2\alpha_3\alpha_4\ldots\alpha_{n-1}\alpha_n.\end{aligned}$$

Hence $\det D = \alpha_1\alpha_2\alpha_3\ldots\alpha_{n-1}\alpha_n$, and this establishes the result.

It follows in particular that:

30.2 det $I_n = 1$ (for every n).

31. Basic properties of determinants

The policy outlined in §29 will be followed. The results will be stated for an arbitrary square matrix A, and, in point of fact, they are true for all possible sizes of A (excluding 1×1, where some of them are meaningless); but the proofs given will deal only with the 3×3 case. In effect this lets us regard all cases up to 3×3 as covered, since it is very easy to check the properties in the 2×2 case while in the 1×1 case they are absolute trivialities when meaningful. We shall use the notation introduced in §30: α_{ik} for the (i, k)th entry of A, μ_{ik} and σ_{ik} for the minor and cofactor of this entry.

Our proofs of some of the most important properties in the 3×3 case are obtained by the lowly method of examining the expression for the determinant of a 3×3 matrix A in terms of its entries. The expansion of "$\alpha_{11}\mu_{11} - \alpha_{12}\mu_{12} + \alpha_{13}\mu_{13}$" gives us, for the 3×3 case,

$$\det A = \alpha_{11}\alpha_{22}\alpha_{33} - \alpha_{11}\alpha_{23}\alpha_{32} - \alpha_{12}\alpha_{21}\alpha_{33} + \alpha_{12}\alpha_{23}\alpha_{31} + \alpha_{13}\alpha_{21}\alpha_{32} - \alpha_{13}\alpha_{22}\alpha_{31}.$$

This expression for det A possesses much symmetry of various kinds. For a start, if we go through it making the changes corresponding to the transposing of A (i.e. interchange of α_{12} and α_{21}, of α_{13} and α_{31}, and of α_{23} and α_{32}), we see that the expression as a whole is unchanged. Therefore:

31.1 $\det(A^T) = \det A$.

If, on the other hand, we make the changes corresponding to the e.r.o. $R_1 \leftrightarrow R_2$ (i.e. interchange of α_{1j} and α_{2j} for each j), we find the expression for det A turned into its negative. Thus the matrix obtained from A by interchanging rows 1 and 2 has determinant equal to $-\det A$. The same is found to be true for the interchange of either of the other pairs of rows, and therefore:

31.2 If θ is a type (I) e.r.o., $\det[\theta(A)] = -\det A$.

We come next to a very important fact. We defined det A to be the sum of all the "(entry) $\times$ (cofactor)" products for entries in the first row. However, as it is easy (if a trifle tedious) to verify, the correct expression for det A in the 3×3 case is obtained by taking the sum of the "(entry) $\times$ (cofactor)" products for the entries in any row of A or for the entries in any column of A. The generalization to an $n \times n$ matrix is:

31.3 (i) For each i in the range 1 to n, $\det A = \sum_{j=1}^{n} \alpha_{ij}\sigma_{ij}$ (cf. sum of all "(entry) $\times$ (cofactor)" products for the ith row).

(ii) For each k in the range 1 to n, $\det A = \sum_{j=1}^{n} \alpha_{jk}\sigma_{jk}$ (cf. sum of all "(entry) $\times$ (cofactor)" products for the kth column).

It could justly be claimed that (ii) follows from (i) by 31.1. Use of the equation $\det A = \sum_{j=1}^{n} \alpha_{ij}\sigma_{ij}$ is called expansion of det A by its ith row, and use of the equation $\det A = \sum_{j=1}^{n} \alpha_{jk}\sigma_{jk}$ is called expansion of det A by its kth column.

As an illustration, let us evaluate the determinant used as an example in §30, this time (a) by expanding by the 2nd row, (b) by expanding by the 3rd column.

(a) $$\begin{vmatrix} 3 & 4 & 5 \\ 1 & -1 & 2 \\ 2 & 1 & 3 \end{vmatrix} = -1\begin{vmatrix} 4 & 5 \\ 1 & 3 \end{vmatrix} + (-1)\begin{vmatrix} 3 & 5 \\ 2 & 3 \end{vmatrix} - 2\begin{vmatrix} 3 & 4 \\ 2 & 1 \end{vmatrix}$$

(cf. $-\alpha_{21}\mu_{21}+\alpha_{22}\mu_{22}-\alpha_{23}\mu_{23}$, the "$-\ +\ -$" sign pattern coming from row 2 of the "sign chessboard")

$$= (-1)\times 7 + (-1)\times(-1) - 2\times(-5) = 4.$$

(b) $$\begin{vmatrix} 3 & 4 & 5 \\ 1 & -1 & 2 \\ 2 & 1 & 3 \end{vmatrix} = 5\begin{vmatrix} 1 & -1 \\ 2 & 1 \end{vmatrix} - 2\begin{vmatrix} 3 & 4 \\ 2 & 1 \end{vmatrix} + 3\begin{vmatrix} 3 & 4 \\ 1 & -1 \end{vmatrix}$$

(cf. $\alpha_{13}\mu_{13}-\alpha_{23}\mu_{23}+\alpha_{33}\mu_{33}$)

$$= 5(3)-2(-5)+3(-7) = 4.$$

A cluster of further results can be summed up by saying that:

31.4 det A is a linear function of each row of A.

To make clear what "linear function" means in this context, let $D(R_1, R_2, R_3)$ denote the determinant of the matrix with rows R_1, R_2, R_3. To say that det A is a linear function of the first row of A means that

$$D(S+T, R_2, R_3) = D(S, R_2, R_3) + D(T, R_2, R_3)$$

and $$D(\lambda R_1, R_2, R_3) = \lambda D(R_1, R_2, R_3) \qquad (\lambda \text{ any scalar});$$

i.e. that $$\begin{vmatrix} a+\alpha & b+\beta & c+\gamma \\ p & q & r \\ x & y & z \end{vmatrix} = \begin{vmatrix} a & b & c \\ p & q & r \\ x & y & z \end{vmatrix} + \begin{vmatrix} \alpha & \beta & \gamma \\ p & q & r \\ x & y & z \end{vmatrix}$$

and that $$\begin{vmatrix} \lambda a & \lambda b & \lambda c \\ p & q & r \\ x & y & z \end{vmatrix} = \lambda \begin{vmatrix} a & b & c \\ p & q & r \\ x & y & z \end{vmatrix},$$

both of which are fairly easily seen from the fact that $\det A = \alpha_{11}\mu_{11} - \alpha_{12}\mu_{12} + \alpha_{13}\mu_{13}$.

The proposition 31.4 asserts these things and the corresponding things for each other row. The case of the jth row can be justified through expansion by the jth row or can be deduced from the case of the 1st row by means of 31.2.

Use of the fact that the determinant of the matrix with rows $R_1, \ldots, R_{i-1}$, λR_i, $R_{i+1}, \ldots, R_n$ is λ times the determinant of the matrix with rows $R_1, \ldots, R_{i-1}, R_i, R_{i+1}, \ldots, R_n$ is often termed "taking out the factor λ from the ith row". Another way of putting the fact is:

31.5 If θ is the type (II) e.r.o. $R_i \rightarrow \lambda R_i$ (where $\lambda \neq 0$), then $\det[\theta(A)] = \lambda \det A$.

A corollary is:

31.6 $\det(\lambda A) = \lambda^n \det A \ (A \in F_{n \times n}, \lambda \in F)$.

Proof. This is trivial in the case $\lambda = 0$. In other cases it follows from 31.5 by observing that λA may be obtained from A by the sequence of e.r.o.s $R_1 \rightarrow \lambda R_1$, $R_2 \rightarrow \lambda R_2, \ldots, R_n \rightarrow \lambda R_n$.

From 31.2, we can deduce:

31.7 If the matrix A has two identical rows, then $\det A = 0$.

Proof. Suppose the ith and jth rows of A are the same ($i \neq j$), so that $\theta(A) = A$, where θ is the e.r.o. $R_i \leftrightarrow R_j$. Hence

$$\det A = \det[\theta(A)] = -\det A \qquad \text{(by 31.2)},$$

and therefore $\det A = 0$. This proves the stated result.

(*Technical note*: this proof is valid provided char $F \neq 2$. The result remains true, and can be proved by other means, in the case char $F = 2$.)

Further to 31.2 and 31.5, we are now in a position to prove the useful fact that:

31.8 If θ is any type (III) e.r.o., $\det[\theta(A)] = \det A$.

Proof. Let $D(R_1, R_2, \ldots, R_n)$ denote the determinant of the $n \times n$ matrix with rows $R_1, R_2, \ldots, R_n$.

The truth of the proposition for the e.r.o. $R_1 \rightarrow R_1 + \lambda R_j \ (j \neq 1)$ is seen from the fact that

$$\begin{aligned} D(R_1 + \lambda R_j, R_2, \ldots, R_n) &= D(R_1, R_2, \ldots, R_n) + D(\lambda R_j, R_2, \ldots, R_n) \\ &= D(R_1, R_2, \ldots, R_n) + \lambda D(R_j, R_2, \ldots, R_n) \quad \text{(these two steps by 31.4)} \\ &= D(R_1, R_2, \ldots, R_n) + 0 \quad \text{(by 31.7, the second term being the determinant of a matrix with identical 1st and } j\text{th rows)} \\ &= D(R_1, R_2, \ldots, R_n). \end{aligned}$$

The truth of the proposition for the more general e.r.o. $R_i \rightarrow R_i + \lambda R_j$ $(i \neq j)$ can be proved by the same method (just by adjusting the notation in the previous paragraph, in fact) or can be deduced from the above special case by use of 31.2.

It follows from 31.8 that the value of the determinant is unchanged when a matrix is transformed by any sequence of (1 or more) type (III) e.r.o.s.

This can be used to simplify the evaluation of determinants. Clearly, it is always relatively easy to expand a determinant by a row or column containing just one nonzero entry (since one gets just one nonzero term); and type (III) e.r.o.s can be used to create a column with just one nonzero entry (without changing the value of the determinant). Let us illustrate with a further evaluation of the determinant used in previous illustrations.

$$\begin{vmatrix} 3 & 4 & 5 \\ 1 & -1 & 2 \\ 2 & 1 & 3 \end{vmatrix} = \begin{vmatrix} 0 & 7 & -1 \\ 1 & -1 & 2 \\ 0 & 3 & -1 \end{vmatrix} \quad \begin{array}{l} (R_1 \to R_1 - 3R_2, \text{ then } R_3 \to R_3 - 2R_2) \\ \text{[equality by 31.8]} \end{array}$$

$$= (-1) \times \begin{vmatrix} 7 & -1 \\ 3 & -1 \end{vmatrix} \quad \text{(on expansion by the 1st column)}$$

$$= (-1) \times (-4) = 4.$$

An important observation can now be made in the light of 31.1. Because of the relationship between A and A^T (the relationship roughly expressed by saying that transposing turns rows into columns and vice versa), it can be seen from 31.1 that all the determinantal properties given above that concerned rows will have true counterparts concerning columns. For example, det A is a linear function of each column of A; if the matrix A has two identical columns, then det $A = 0$; etc. To formulate some of these results, we need to introduce (i) the notation $C_1, C_2, C_3, \ldots$ for the 1st, 2nd, 3rd, ... columns of a matrix, and (ii) (the analogue of elementary row operations) *elementary column operations* of three types described by the shorthands $C_i \leftrightarrow C_j \, (i \neq j)$, $C_i \to \lambda C_i \, (\lambda \neq 0)$, $C_i \to C_i + \lambda C_j \, (i \neq j)$. This done, the column analogues of 31.2, 31.5, 31.8 can be written down: e.g. (the case of 31.2) if θ is a type (I) elementary column operation (the interchange of two columns), then $\det[\theta(A)] = -\det A$.

All trace of vagueness is easily eliminated from the deduction of a column result from the corresponding row result. For example, to show that $\det[\theta(A)] = -\det A$ when θ is the column operation $C_i \leftrightarrow C_j$, simply note that θ is the result of the sequence of operations

transpose, then $R_i \leftrightarrow R_j$, then transpose:

the column result now follows by 31.1 and 31.2.

The following piece of work involves the use of elementary column operations and also shows how it may be possible to factorize a determinant with great economy of effort.

$$\begin{vmatrix} 1 & 1 & 1 \\ a & b & c \\ a^2 & b^2 & c^2 \end{vmatrix} = \begin{vmatrix} 1 & 0 & 0 \\ a & b-a & c-a \\ a^2 & b^2-a^2 & c^2-a^2 \end{vmatrix} \quad \begin{array}{l}(C_2 \to C_2 - C_1, \text{ then } C_3 \to C_3 - C_1) \\ \text{[equality by column version of 31.8]}\end{array}$$

$$= \begin{vmatrix} b-a & c-a \\ b^2-a^2 & c^2-a^2 \end{vmatrix} \quad \text{(on expanding by the 1st row)}$$

$$= (b-a)(c-a)\begin{vmatrix} 1 & 1 \\ b+a & c+a \end{vmatrix} \quad \text{(taking out factors from the columns)}$$

$$= (b-a)(c-a)(c-b) \quad \text{(just by 29.1)}$$
$$= (a-b)(b-c)(c-a).$$

A second factorization, building on the above, now gives further illustrations of the use of results in this section. Let Δ denote the value of the determinant discussed above. Then:

$$\begin{vmatrix} 1+a^3 & 1+b^3 & 1+c^3 \\ a & b & c \\ a^2 & b^2 & c^2 \end{vmatrix} = \Delta + \begin{vmatrix} a^3 & b^3 & c^3 \\ a & b & c \\ a^2 & b^2 & c^2 \end{vmatrix}$$

(by the fact that the determinant is a linear function of the first row)

$$= \Delta + abc\begin{vmatrix} a^2 & b^2 & c^2 \\ 1 & 1 & 1 \\ a & b & c \end{vmatrix} \quad \text{(on taking factors out of the columns)}$$

$$= \Delta + abc(-1)^2\begin{vmatrix} 1 & 1 & 1 \\ a & b & c \\ a^2 & b^2 & c^2 \end{vmatrix} \quad (R_1 \leftrightarrow R_2, \text{ then } R_2 \leftrightarrow R_3 \text{ [cf. 31.2]})$$

$$= \Delta + abc\Delta$$
$$= (a-b)(b-c)(c-a)(abc+1).$$

32. The multiplicative property of determinants

(In this section, further results about determinants will be deduced from those in §31. On the face of it, there will be no limitation on the size of the matrices discussed in this section; but, of course, since we are building on §31, we can claim to be proving results only in the cases in which the results of §31 have been proved—i.e. only for matrices no bigger than 3 × 3.)

We begin with a table giving information, culled from §31, about the effect of every possible e.r.o. θ on an arbitrary $n \times n$ matrix B and, in each case, about the $n \times n$ elementary matrix P corresponding to θ.

e.r.o. θ	$\det[\theta(B)]$	$\det P$
$R_i \leftrightarrow R_j \quad (i \neq j)$	$-\det B$	-1
$R_i \to \lambda R_i \quad (\lambda \neq 0)$	$\lambda \det B$	λ
$R_i \to R_i + \lambda R_j \quad (i \neq j)$	$\det B$	1

The contents of the third column are deduced from those of the second by using the facts that (by definition) $P = \theta(I_n)$ and that (cf. 30.2) $\det I_n = 1$. It can be seen that, in every case,

$$\det[\theta(B)] = (\det P) \times (\det B),$$

i.e. (in view of 25.1) $\quad \det(PB) = (\det P) \times (\det B).$

Thus:

32.1 If $B \in F_{n \times n}$ and P is an $n \times n$ elementary matrix, then $\det(PB) = (\det P) \times (\det B)$.

It is also apparent from the above table that:

32.2 Every elementary matrix has nonzero determinant.

By repeated use of 32.1, we discover that if $B \in F_{n \times n}$ and $P_1, P_2, \ldots, P_s$ are $n \times n$ elementary matrices, then

$$\det(P_2 P_1 B) = (\det P_2)(\det(P_1 B)) = (\det P_2)(\det P_1)(\det B),$$
$$\det(P_3 P_2 P_1 B) = (\det P_3)(\det(P_2 P_1 B)) = (\det P_3)(\det P_2)(\det P_1)(\det B),$$

etc., and eventually

$$\det(P_s P_{s-1} \ldots P_2 P_1 B) = (\det P_s)(\det P_{s-1}) \ldots (\det P_2)(\det P_1)(\det B).$$

This prepares the way for the proof of:

32.3 If $A, B \in F_{n \times n}$ and A is nonsingular, then

$$\det(AB) = (\det A)(\det B).$$

Proof. Suppose that $A, B \in F_{n \times n}$ and that A is nonsingular. By 26.8, A can be expressed as the product $P_s P_{s-1} \ldots P_2 P_1$ of some elementary matrices $P_1, P_2, \ldots, P_s$. Therefore

$$\begin{aligned}\det(AB) &= \det(P_s P_{s-1} \ldots P_2 P_1 B)\\ &= (\det P_s)(\det P_{s-1}) \ldots (\det P_2)(\det P_1)(\det B),\end{aligned}$$

by the preamble. But from the case $B = I_n$ we have

$$\begin{aligned}\det A &= (\det P_s)(\det P_{s-1})\dots(\det P_2)(\det P_1)(\det I_n)\\ &= (\det P_s)(\det P_{s-1})\dots(\det P_2)(\det P_1) \qquad \text{(by 30.2)}.\end{aligned}$$

Hence $\det(AB) = (\det A)(\det B)$, which proves the result.

Shortly (in 32.6) we shall extend 32.3 to cover also the case where A is singular. First, without assuming more than 32.3 actually tells us, we can now establish one of the most significant facts about determinants—the determinantal criterion for a square matrix to be nonsingular.

32.4 Let A be a square matrix. Then A is nonsingular if and only if $\det A \neq 0$.

Proof. Consider first the case A nonsingular. In this case, A^{-1} exists, and we have

$$\begin{aligned}(\det A)(\det A^{-1}) &= \det(AA^{-1}) \qquad \text{(by 32.3, } A \text{ being nonsingular)}\\ &= \det I = 1 \qquad \text{(cf. 30.2)}.\end{aligned}$$

Therefore in this case $\det A \neq 0$.

Now consider the case A singular. We know from chapter 3 that A is row equivalent to a reduced echelon matrix E; and, A being singular, the reduced echelon matrix E must have (by 26.2(ii)) a zero row: indeed its bottom row must be zero. From expansion of $\det E$ by the bottom row, it is clear that $\det E = 0$. But, since A and E are row equivalent to each other, $A = QE$ for some nonsingular matrix Q (cf. 25.4). Hence

$$\begin{aligned}\det A &= \det(QE)\\ &= (\det Q)(\det E) \qquad \text{(by 32.3 since } Q \text{ is nonsingular)}\\ &= (\det Q) \times 0 = 0.\end{aligned}$$

We have now shown that $\det A$ is nonzero if A is nonsingular and zero if A is singular, and thus the stated result is proved.

In discussing the case A nonsingular in the above proof, we obtained the equation $(\det A)(\det A^{-1}) = 1$. Hence it is evident that:

32.5 If A is a nonsingular matrix, $\det(A^{-1}) = 1/(\det A)$.

Now that 32.4 has been proved, it is easy to obtain the promised improved version of 32.3, namely:

32.6 For *all* $A, B \in F_{n\times n}$, $\det(AB) = (\det A)(\det B)$.

(This general result is called the *multiplicative property* of determinants.)

Proof. Let $A, B \in F_{n\times n}$. If A is singular, then (by 26.7) AB is also singular, so that (by 32.4) $\det A = \det(AB) = 0$. Hence the equation $\det(AB) = (\det A)(\det B)$ holds in the A singular case and therefore (in view of 32.3) in all cases.

33. Another method for inverting a nonsingular matrix

Throughout this section A will denote an $n \times n$ matrix ($n \geqslant 2$) with (i, k)th entry α_{ik}, and, as in earlier work, we use σ_{ik} for the cofactor of the (i, k)th entry of A. As in §32, the results here are based on those in §31, and so we can claim to have proved them only for cases where n is 2 or 3.

We begin with an apparently technical result.

33.1 Let i, k be unequal integers in the range 1 to n. Then

$$\text{(i)} \sum_{j=1}^{n} \alpha_{ij}\sigma_{kj} = 0, \qquad \text{and} \qquad \text{(ii)} \sum_{j=1}^{n} \alpha_{jk}\sigma_{ji} = 0.$$

(In the left-hand side of (i) we have the entries $\alpha_{i1}, \alpha_{i2}, \ldots$ of the ith row married with the cofactors $\sigma_{k1}, \sigma_{k2}, \ldots$ of the entries of a different row (the kth row). Similarly in the left-hand side of (ii) we have the entries of one column married with the cofactors of the entries in a different column.)

Proof. Consider an $n \times n$ matrix B identical with A in all rows except the kth, but having as its kth row another copy of the ith row of A. Observe that, for the matrix B, the cofactors of the entries in the kth row, being (by definition) calculable from entries in rows other than the kth, coincide with the cofactors $\sigma_{k1}, \sigma_{k2}, \ldots$ of the entries in the kth row of A. Hence, on expanding det B by the kth row of B (which is $[\alpha_{i1}\ \alpha_{i2} \ldots \alpha_{in}]$), we obtain

$$\det B = \sum_{j=1}^{n} \alpha_{ij}\sigma_{kj}.$$

But, by 31.7, det $B = 0$, since the ith and kth rows of B are both the same as the ith row of A. Therefore

$$\sum_{j=1}^{n} \alpha_{ij}\sigma_{kj} = 0,$$

and this is the result (i). Part (ii) is similarly proved: it involves consideration of columns rather than rows.

Using the Kronecker delta, we can combine this last result with 31.3 into a single result, namely:

33.2 For all relevant i, k,

$$\sum_{j=1}^{n} \alpha_{ij}\sigma_{kj} = \delta_{ik}(\det A) \qquad \text{and} \qquad \sum_{j=1}^{n} \alpha_{jk}\sigma_{ji} = \delta_{ik}(\det A).$$

The significance of this begins to become apparent when we introduce a matrix called the **adjugate** of A (and denoted by adj A), defined to be the $n \times n$

matrix whose (i, k)th entry is σ_{ki}. (So adj A is the matrix formed from A by replacing each entry by its cofactor and then transposing.)

In the $n \times n$ matrix $A(\text{adj } A)$, for all relevant i, k,

$$\begin{aligned}(i,k)\text{th entry} &= \sum_{j=1}^{n} \alpha_{ij} \times (j,k)\text{th entry of adj } A \\ &= \sum_{j=1}^{n} \alpha_{ij}\sigma_{kj} \qquad \text{(cf. definition of adj } A) \\ &= \delta_{ik}(\det A) \qquad \text{(by 33.2).}\end{aligned}$$

As the (i, k)th entry of $(\text{adj } A)A$ also works out to be (for all relevant i, k) $\delta_{ik}(\det A)$, it is apparent that:

33.3 $A(\text{adj } A) = (\text{adj } A)A = (\det A)I$.

From this fact and 32.4 it follows immediately that:

33.4 If A is nonsingular, then $A^{-1} = (1/\det A) \times (\text{adj } A)$.

Observe that this is a generalization of the recipe for the inverse of a 2×2 matrix given in 19.4.

For an illustration of the application of 33.4 to a 3×3 matrix, consider the matrix

$$A = \begin{bmatrix} 3 & 4 & 5 \\ 1 & -1 & 2 \\ 2 & 1 & 3 \end{bmatrix}.$$

In the course of this chapter, we have found (by four different methods) that $\det A = 4$. By 32.4, therefore, A is nonsingular. By working out the cofactors of the entries, we obtain

$$\text{adj } A = \begin{bmatrix} -5 & 1 & 3 \\ -7 & -1 & 5 \\ 13 & -1 & -7 \end{bmatrix}^T = \begin{bmatrix} -5 & -7 & 13 \\ 1 & -1 & -1 \\ 3 & 5 & -7 \end{bmatrix}.$$

Hence (by 33.4)

$$A^{-1} = \tfrac{1}{4}\begin{bmatrix} -5 & -7 & 13 \\ 1 & -1 & -1 \\ 3 & 5 & -7 \end{bmatrix}.$$

The same matrix was inverted by the use of e.r.o.s in the worked example in §26. On comparing the amounts of work entailed in the two methods, the student may well come to the conclusion that there is little to choose between them. While that may be so for the inversion of a 3×3 matrix, it is not true for larger matrices. For larger matrices, the method that 33.4 provides for finding the inverse is inferior because of the great amount of work needed to compute the values of the cofactors of all the entries.

EXERCISES ON CHAPTER FOUR

1. Evaluate the following determinants. Where appropriate, simplify your work by using row and/or column operations.

$$\text{(i)} \begin{vmatrix} 2 & 5 \\ -1 & 3 \end{vmatrix}, \quad \text{(ii)} \begin{vmatrix} 2 & 4 & 5 \\ 1 & 3 & -1 \\ 3 & -2 & 3 \end{vmatrix}, \quad \text{(iii)} \begin{vmatrix} 3 & 5 & 3 \\ 3 & 3 & 5 \\ 5 & 3 & 3 \end{vmatrix},$$

$$\text{(iv)} \begin{vmatrix} 1 & 1 & 2 \\ -1 & 1 & 3 \\ -2 & -3 & 1 \end{vmatrix}, \quad \text{(v)} \begin{vmatrix} 1 & 3 & -2 & 4 \\ -1 & 1 & 5 & 2 \\ 2 & 2 & 1 & -2 \\ 1 & 4 & -2 & 3 \end{vmatrix}.$$

2. Show that $\begin{vmatrix} 1 & 1 & 1 \\ a & b & c \\ a^3 & b^3 & c^3 \end{vmatrix} = (a-b)(b-c)(c-a)(a+b+c).$

(Recall that $x^3-y^3 = (x-y)(x^2+xy+y^2)$.) Obtain also a factorization of the determinant

$$\begin{vmatrix} 1-a^2 & 1-b^2 & 1-c^2 \\ a & b & c \\ a^3 & b^3 & c^3 \end{vmatrix}.$$

3. Show that

$$\begin{vmatrix} a & b & c \\ a(b+c) & b(c+a) & c(a+b) \\ (b+c)^2 & (c+a)^2 & (a+b)^2 \end{vmatrix} = -(a-b)(b-c)(c-a)(a+b+c)^2.$$

4. Show that if n is odd, every real $n \times n$ skew-symmetric matrix has zero determinant.

5. Let A, B, P be matrices in $F_{n\times n}$ such that P is nonsingular and $B = P^{-1}AP$. Using 32.6 and 32.5, show that $\det A = \det B$. Show also that, for every scalar t, $\det(tI_n - A) = \det(tI_n - B)$.

6. (i) Show that every real orthogonal matrix has determinant equal to 1 or -1.

(ii)* Let A and B be real orthogonal matrices such that $\det A = 1$ and $\det B = -1$. By considering $A^T(A+B)B^T$, show that $A+B$ is singular.

(iii) Give an example of two non-orthogonal real 2×2 matrices A, B such that $\det A = 1$, $\det B = -1$ and $A+B$ is nonsingular.

7.* Let J be the $n \times n$ matrix with every entry equal to 1, and let $A = \alpha I_n + \beta J$, where α, β are scalars. Show that $\det A = \alpha^{n-1}(\alpha + n\beta)$. (*Hint*: first obtain the factor $\alpha + n\beta$ of $\det A$ by performing the row operation $R_1 \to R_1 + R_2 + \ldots + R_n$, which is the resultant of the sequence of e.r.o.s $R_1 \to R_1 + R_2$, $R_1 \to R_1 + R_3, \ldots, R_1 \to R_1 + R_n$.) Deduce that if $\alpha \neq 0$, A is singular if and only if $\alpha = -n\beta$. (Cf. exercise 16 on chapter 2.)

8. Let $A \in F_{m\times m}$, $B \in F_{n\times n}$, and let M be an $(m+n)\times(m+n)$ matrix of the form $\begin{bmatrix} A & O \\ X & B \end{bmatrix}$ $(X \in F_{n\times m})$. Suppose that B is nonsingular. By postmultiplying M by

$N = \begin{bmatrix} I_m & O \\ -B^{-1}X & I_n \end{bmatrix}$, show that det M = (det A) $\times$ (det B). Deduce that every matrix of the form $\begin{bmatrix} A & Y \\ O & B \end{bmatrix}$ $(Y \in F_{m \times n})$ also has determinant equal to (det A) $\times$ (det B). (*Remark*: insights from later chapters make it clear that these results also hold in the case B singular.)

9. Use the method provided by 33.4 to find the inverses of the nonsingular 3×3 matrices in exercise 4 on chapter 3.

10. Let A be a nonsingular $n \times n$ matrix. Show that $\det(\operatorname{adj} A) = (\det A)^{n-1}$. Show also that $\operatorname{adj}(\operatorname{adj} A) = (\det A)^{n-2}A$.

11.* Consider (with the usual notation) a system of equations $AX = K$, and suppose that $m = n$ and that A is nonsingular. Show that the unique solution $(X = A^{-1}K)$ of the system can be given by $x_1 = \Delta_1/\Delta, x_2 = \Delta_2/\Delta, \ldots, x_n = \Delta_n/\Delta$, where $\Delta = \det A$ and, for each relevant j, Δ_j is the determinant of the matrix obtained by replacing the jth column of A by K. (This result is known as Cramer's rule.)

CHAPTER FIVE

VECTOR SPACES

34. Introduction

In accordance with our standing convention, the symbol F will, throughout this chapter, stand for an arbitrary field.

It is often true in mathematics (and elsewhere) that when we look only at one or two examples of a phenomenon, superficial features that belong only to these few examples may dominate our attention and hinder us from understanding, or even from noticing, the most important facts about the phenomenon. On the other hand, if we are willing to embark on the tougher intellectual task of studying the phenomenon in full generality, we may expect to be rewarded by deeper and clearer insight into the heart of the subject-matter.

So it is with the study of systems of vectors. The intended meaning of "system of vectors" here is any system which is, algebraically, like E_3 in two basic respects: (1) the member objects of the system can be added and multiplied by scalars; and (2) these operations obey all natural-looking laws. It was pointed out in §14 that $F_{m\times n}$ is like E_3 in these respects; and thus we have seen two examples of a phenomenon ("a system of vectors") which one can (very advantageously, as it turns out) discuss in greater generality. This chapter is devoted to precisely such a more general discussion, and the discussion is formally begun in §35 with the introduction of the concept of "a vector space over F". Roughly speaking, "a vector space" is the accepted proper term for "a system of vectors" in the sense outlined above, while the words "over F" indicate that F is the set of scalars by which one can multiply the member objects of the system. As well as giving the precise definition of "a vector space over F", §35 will mention further examples of vector spaces; and then §36 will develop the most elementary consequences of the definition.

Further to earlier remarks, the student should understand that this chapter is not intended as a sterile exercise in generalization for generalization's sake. The purpose of the chapter, rather, is to introduce the ideas that are of fundamental importance wherever systems of vectors occur—and to do so with the clarity and efficiency that are possible only in the context of an appropriately general discussion. Accordingly, §§37, 38, 39 introduce three cardinally important concepts—the idea of a *subspace*, the idea of a *spanning sequence*, and the idea of *linear dependence*. It is imperative that the student should achieve a firm and confident grasp of these concepts: they underlie all the further exploration of linear algebra in this book. Many fundamentally important theorems about these concepts are given in §§40, 41; and, in particular, some of these enable the notion of the *dimension* of a vector space to be introduced in very many cases. The understanding of all this material is a decisive advance in the exploration of linear algebra: having understood the ideas and the theorems, the student will be enormously better equipped to tackle with insight the naturally arising problems of linear algebra (problems that come from matrices, systems of linear equations, etc.).

This chapter ends with two sections (§§42 and 43) on the technical matter of sums of subspaces.

35. The definition of a vector space, and examples

The definition involves an arbitrary field (denoted as always by F). In this context, as elsewhere, we refer to the elements of F as "scalars".

Let V be a set. We describe V as a **vector space** over F if:

(a) addition and multiplication by a scalar are defined on V (i.e. $\mathbf{x}+\mathbf{y}$ has a meaning for all $\mathbf{x}, \mathbf{y} \in V$, and $\lambda\mathbf{x}$ has a meaning for all $\mathbf{x} \in V$ and for all $\lambda \in F$); and

(b) all the following ten postulates (A0), ..., (M4) are satisfied.

(A0) V is closed under addition; i.e. (cf. remark (*a*) in §4) for all $\mathbf{x}, \mathbf{y} \in V$, $\mathbf{x}+\mathbf{y}$ is a member of V.

(A1) Addition is associative on V; i.e. for all $\mathbf{x}, \mathbf{y}, \mathbf{z} \in V$,

$$(\mathbf{x}+\mathbf{y})+\mathbf{z} = \mathbf{x}+(\mathbf{y}+\mathbf{z}).$$

(A2) Addition is commutative on V; i.e. for all $\mathbf{x}, \mathbf{y} \in V$,

$$\mathbf{x}+\mathbf{y} = \mathbf{y}+\mathbf{x}.$$

(A3) There is a zero object, $\mathbf{0}$, in V satisfying

$$\mathbf{x}+\mathbf{0} = \mathbf{x} \qquad \text{for all } \mathbf{x} \in V.$$

(A4) Each object in V has a negative in V; i.e. for each $\mathbf{x} \in V$ there is a corresponding object $\mathbf{y}$, also a member of V, satisfying $\mathbf{x}+\mathbf{y} = \mathbf{0}$.

(M0) V is closed under the scalar multiplication operation; i.e. for all $\mathbf{x} \in V$ and for all $\lambda \in F$, $\lambda\mathbf{x}$ is a member of V.

(M1) For all $\mathbf{x}, \mathbf{y} \in V$ and for all $\lambda \in F$, $\lambda(\mathbf{x}+\mathbf{y}) = \lambda\mathbf{x}+\lambda\mathbf{y}$.

(M2) For all $\mathbf{x} \in V$ and for all $\lambda, \mu \in F$, $(\lambda+\mu)\mathbf{x} = \lambda\mathbf{x}+\mu\mathbf{x}$.

(M3) For all $\mathbf{x} \in V$ and for all $\lambda, \mu \in F$, $\lambda(\mu\mathbf{x}) = (\lambda\mu)\mathbf{x}$.
(M4) For all $\mathbf{x} \in V$, $1\mathbf{x} = \mathbf{x}$.

Remarks. (*a*) When V is such a vector space, we usually refer to its member objects as "vectors" and, if practical, denote them by bold-type symbols.

(*b*) The ten postulates (A0),...,(M4) are called the *vector space axioms.* It must not be imagined that the human race received them, engraved on tablets of stone, at some moment in past history! On the contrary, they were evolved by the mathematical community in the attempt to write down a concise set of properties satisfying two criteria: (1) the properties must be possessed by every system that one would naturally be inclined to call "a system of vectors"; and (2) they must imply as consequences all other properties that one would expect every system of vectors to possess. Clearly the vector space axioms measure up well to the first of these criteria, and the results in §36 will reveal that they also measure up well to the second.

(*c*) Note that a vector space must be a non-empty set because (cf. axiom (A3)) it must contain a zero object.

(*d*) There follows a list of examples of vector spaces. The varied nature of these examples shows the versatility of the concept of a vector space. Proofs of the assertions that the systems mentioned actually are vector spaces will be omitted. In every case, the proof would consist of a lengthy but essentially trivial verification that all ten vector space axioms hold for the system in question. Students may, however, gain some sense of perspective in the subtler cases by working through parts of these proofs.

(1) The system E_3 of vectors discussed in chapter 1 is a vector space over $\mathbb{R}$. So is the restricted system E_2 described in remark (*a*) of §10.

(2) For any given positive integers m, n, the set $F_{m \times n}$ is a vector space over F.

(3) For each positive integer n, we denote by F^n the set of all n-tuples of elements of F (i.e. objects of the form $(\alpha_1, \alpha_2, \ldots, \alpha_n)$ with each $\alpha_j \in F$). The set F^n becomes a vector space over F when one defines addition of n-tuples and multiplication of an n-tuple by a scalar "componentwise", i.e. by the rules

$$(\alpha_1, \alpha_2, \ldots, \alpha_n) + (\beta_1, \beta_2, \ldots, \beta_n) = (\alpha_1 + \beta_1, \alpha_2 + \beta_2, \ldots, \alpha_n + \beta_n),$$
$$\lambda(\alpha_1, \alpha_2, \ldots, \alpha_n) = (\lambda\alpha_1, \lambda\alpha_2, \ldots, \lambda\alpha_n) \qquad (\lambda \in F).$$

It is apparent that elements of F^n are to all intents and purposes just row matrices—both in form and in the way we add them and multiply them by scalars. Indeed, the three vector spaces F^n, $F_{1 \times n}$, $F_{n \times 1}$ differ only in the manner in which we write their member objects on paper! It is sometimes essential to respect these differences (e.g. to avoid confusion between rows and columns in matrix theory). But there are other occasions where the differences can helpfully be blurred or ignored: e.g. given a finite sequence of members of F^n, we may choose to interpret them as row matrices and consider the many-rowed matrix having the given vectors as rows.

We do not normally observe the distinction between the 1-tuple (α) and the scalar α. Accordingly we identify F^1 with the field F itself, and this gives us F $(=F^1)$ as a vector space over itself.

(4) We can deal similarly with F^∞, the set of all infinite sequences of elements of F. This set too becomes a vector space over F when operations are defined componentwise, i.e. by

$$(\alpha_1,\alpha_2,\alpha_3,\ldots)+(\beta_1,\beta_2,\beta_3,\ldots)=(\alpha_1+\beta_1,\alpha_2+\beta_2,\alpha_3+\beta_3,\ldots),$$
$$\lambda(\alpha_1,\alpha_2,\alpha_3,\ldots)=(\lambda\alpha_1,\lambda\alpha_2,\lambda\alpha_3,\ldots) \qquad (\lambda\in F).$$

(5) The set of all formal polynomials in an indeterminate t with coefficients in F is a vector space over F (addition of polynomials and multiplication of a polynomial by an element of F having their natural meanings).

Let n be any nonnegative integer. A further vector space over F is obtained by taking, instead, just the set of polynomials with degree not exceeding n.

(6) Let S be a non-empty set. We denote by F^S the set of all mappings from S to F. This becomes a vector space over F when we define operations "pointwise": i.e., for $f,g\in F^S$, we define $f+g$ and $\lambda f\,(\lambda\in F)$ by

$$(f+g)(x)=f(x)+g(x) \quad \text{and} \quad (\lambda f)(x)=\lambda(f(x)) \qquad (x\in S).$$

For example if $F=\mathbb{R}$ and S is the interval $[0,1]$, then F^S is a vector space whose "vectors" are the real-valued functions with domain $[0,1]$. (*Note*: the appendix explains certain things about mappings, including such basic items of nomenclature as "domain".)

(7) The complex field $\mathbb{C}$ is a vector space over $\mathbb{R}$. (And, similarly, whenever a field F is a subfield of a bigger field K, K is a vector space over F.)

36. Elementary consequences of the vector space axioms

Throughout the results and remarks of this section, V denotes an arbitrary vector space over the field F.

Attention is drawn first to the general remark (*b*) in §4: this applies to sums of vectors in V because of the vector space axioms (A1) and (A2), and so henceforth, in dealing with such sums, we shall omit brackets and change the order of terms without further comment. In the same vein, notice that it is feasible to use the standard Σ-notation shorthand for long sums of vectors in V.

The first two propositions enable some items of nomenclature and notation to be cleared up.

36.1 There is only one zero object in V.

Proof. Suppose that $\mathbf{0}$ and $\mathbf{o}$ are both zero objects in V. The result will follow if we can prove that $\mathbf{0}$ and $\mathbf{o}$ coincide; and this is so because

$$\begin{aligned} \mathbf{0} &= \mathbf{o}+\mathbf{0} \qquad \text{(since } \mathbf{o} \text{ is a zero in } V) \\ &= \mathbf{o} \qquad \text{(since } \mathbf{0} \text{ is a zero in } V). \end{aligned}$$

Because of 36.1 (and (A3)) we may speak of *the* zero vector in V. Whenever practical, we use the symbol $\mathbf{0}$ for the zero vector in a vector space.

36.2 Each vector in V has just one negative.

Proof. Let $\mathbf{x}$ be a vector in V. Suppose that $\mathbf{y}$ and $\mathbf{z}$ are both negatives of $\mathbf{x}$. The result will follow if we can prove that $\mathbf{y}$ and $\mathbf{z}$ coincide; and this is so because

$$\begin{aligned} \mathbf{y} &= \mathbf{0}+\mathbf{v} = \mathbf{z}+\mathbf{x}+\mathbf{y} \qquad \text{(since } \mathbf{z} \text{ is a negative of } \mathbf{x}) \\ &= \mathbf{z}+\mathbf{0} \qquad \text{(since } \mathbf{y} \text{ is a negative of } \mathbf{x}) \\ &= \mathbf{z}. \end{aligned}$$

Because of 36.2 (and (A4)) we may speak of *the* negative of a vector $\mathbf{x}$ in V, and we shall henceforth denote the negative of $\mathbf{x}$ by $-\mathbf{x}$. Moreover, we may now define subtraction on V in accordance with remark (*c*) in §4: so $\mathbf{x}-\mathbf{y}$ shall mean $\mathbf{x}+(-\mathbf{y})$ $(\mathbf{x},\mathbf{y} \in V)$.

Here now are five further basic general results.

36.3 For $\mathbf{x},\mathbf{y},\mathbf{z} \in V$, $\mathbf{x}+\mathbf{y} = \mathbf{z}$ is equivalent to $\mathbf{x} = \mathbf{z}-\mathbf{y}$.

36.4 For all $\mathbf{x} \in V$, $0\mathbf{x} = \mathbf{0}$.

36.5 For all $\lambda \in F$, $\lambda\mathbf{0} = \mathbf{0}$.

36.6 For $\lambda \in F$ and $\mathbf{x} \in V$, $\lambda\mathbf{x} = \mathbf{0} \Rightarrow \lambda = 0$ or $\mathbf{x} = \mathbf{0}$.

36.7 For all $\lambda \in F$ and all $\mathbf{x} \in V$, $(-\lambda)\mathbf{x} = \lambda(-\mathbf{x}) = -(\lambda\mathbf{x})$; and, in particular, $(-1)\mathbf{x} = -\mathbf{x}$ for all $\mathbf{x} \in V$.

There follow proofs of 36.4, 36.6 and 36.7. Proofs of 36.3 and 36.5 are left as exercises. In our proof of 36.6 we shall use 36.5, assuming that the student has already carried out the task of proving it.

Proof of 36.4. Let $\mathbf{x}$ be an arbitrary vector in V. We have

$$0\mathbf{x} = (0+0)\mathbf{x},$$

$$\text{i.e. (by (M2))} \qquad 0\mathbf{x} = 0\mathbf{x}+0\mathbf{x}.$$

Adding $-(0\mathbf{x})$ to both sides, we deduce

$$\mathbf{0} = 0\mathbf{x}+\mathbf{0}, \qquad \text{i.e. } \mathbf{0} = 0\mathbf{x}.$$

This proves the result.

Proof of 36.6. Suppose that $\lambda\mathbf{x} = \mathbf{0}$, where $\mathbf{x} \in V$ and $\lambda \in F$.
If $\lambda \neq 0$, then λ^{-1} exists in the field F, and we have

$$\lambda^{-1}(\lambda\mathbf{x}) = \lambda^{-1}\mathbf{0},$$

i.e. $(\lambda^{-1}\lambda)\mathbf{x} = \mathbf{0}$ (by (M3) and 36.5, which we suppose already proved),
i.e. $1\mathbf{x} = \mathbf{0}$; and hence (by (M4)) $\mathbf{x} = \mathbf{0}$.

The only remaining case is $\lambda = 0$, and so the result follows.

Proof of 36.7. Let $\lambda \in F$ and $\mathbf{x} \in V$.
We have

$$\begin{aligned}(-\lambda)\mathbf{x} + \lambda\mathbf{x} &= ((-\lambda)+\lambda)\mathbf{x} \quad \text{(by (M2))}\\ &= 0\mathbf{x} = \mathbf{0} \quad \text{(by 36.4).}\end{aligned}$$

This shows that $(-\lambda)\mathbf{x}$ is a negative, and therefore (cf. 36.2) the negative, of $\lambda\mathbf{x}$. Thus $(-\lambda)\mathbf{x} = -(\lambda\mathbf{x})$.

It follows in particular (the case $\lambda = 1$) that

$$(-1)\mathbf{x} = -(1\mathbf{x}) = -\mathbf{x} \quad \text{(cf. (M4)).}$$

Further, we now deduce

$$\begin{aligned}\lambda(-\mathbf{x}) &= \lambda((-1)\mathbf{x}) \quad \text{(by the particular case just noted)}\\ &= (\lambda(-1))\mathbf{x} \quad \text{(by (M3))}\\ &= (-\lambda)\mathbf{x}\\ &= -(\lambda\mathbf{x}) \quad \text{(by the first part of the proof).}\end{aligned}$$

All parts of 36.7 are now proved.

The evidence of the results in this section may lead the student to surmise that the operations in an arbitrary vector space are "no-catch" operations in the sense explained in remark (*d*) in §4, and an assurance can be given that this is indeed the case.

This final remark may be ignored by students with no knowledge of the theory of groups. But those who do know something about group theory will recognize that the vector space axioms (A0), (A1), ..., (A4) state that a vector space is an abelian group with respect to addition, and they will regard the propositions 36.1, 36.2 and 36.3 as simple consequences of that observation.

37. Subspaces

Let V denote a vector space over F.

Consider a non-empty subset S of V. Because addition and multiplication by scalars are defined on V, we can add together vectors belonging to S and we can multiply them by scalars. Moreover, certain laws will hold for all vectors in

S because they hold for all vectors in V: e.g. it must be true that

$$\lambda(\mathbf{x}+\mathbf{y}) = \lambda\mathbf{x}+\lambda\mathbf{y} \qquad \text{for all } \mathbf{x}, \mathbf{y} \in S \text{ and all } \lambda \in F.$$

Nevertheless, a randomly chosen subset S of V is unlikely to be a vector space: e.g. there may be no zero vector in S, or S may not be closed under addition or under multiplication by scalars. On the other hand, it is often useful to recognize cases where S *is* a vector space: in any such case, naturally, we describe S as a subspace of V.

Thus (to provide a less discursive definition) a **subspace** of a given vector space V means a subset of V that is, in its own right, a vector space with respect to the operations of addition and multiplication by scalars defined on V (and therefore defined on the subset). (*Note*: a subspace must be a non-empty set, because a subspace is a vector space and every vector space is a non-empty set [cf. remark (*c*) in §35].)

Illustrations. (1) The vector space E_2 (cf. remark (*a*) in §10) is a subspace of E_3. Also a subspace of E_3 is the set of vectors corresponding to the points in any plane through the origin (i.e. the set of vectors that are position vectors of points in the plane). The same goes for the set of vectors corresponding to the points in any line through the origin.

(2) In the vector space of all formal polynomials with coefficients in F, the polynomials with degrees not exceeding any given integer n form a subspace.

(3) Trivially, for every vector space V, V is a subspace of itself.

The following result gives a useful simple test by which one can prove that certain subsets of vector spaces are subspaces.

37.1 Suppose that S is a subset of the vector space V. Then S is a subspace of V if the following two conditions both hold:

(1) $S \neq \varnothing$;

(2) $\mathbf{x}, \mathbf{y} \in S$ and $\lambda, \mu \in F \Rightarrow \lambda\mathbf{x}+\mu\mathbf{y} \in S$.

Proof. Suppose that the two conditions hold. We show that the vector space axioms all hold for S. Among these, (A1), (A2), (M1), (M2), (M3), (M4) obviously hold for S: for they concern properties of vectors that are true for all vectors in the whole space V *and therefore* are true for all vectors in the subset S. It remains to show that (A0), (A3), (A4), (M0) hold for S, and we take these in turn.

(A0) $\mathbf{x}, \mathbf{y} \in S \Rightarrow 1\mathbf{x}+1\mathbf{y} \in S$ (by condition (2))
$\Rightarrow \mathbf{x}+\mathbf{y} \in S$ (by vector space axiom (M4)).

Thus S is closed under addition: i.e. (A0) holds for S.

(A3) Because (by condition (1)) $S \neq \varnothing$, we can introduce a vector $\mathbf{x} \in S$. By condition (2), $0\mathbf{x}+0\mathbf{x} \in S$, i.e. $\mathbf{0} \in S$ (cf. 36.4). Hence (A3) holds for S.

(A4) $\mathbf{x} \in S \Rightarrow (-1)\mathbf{x} + 0\mathbf{x} \in S$ (by condition (2))
$\Rightarrow -\mathbf{x} \in S$ (by 36.7 and 36.4).

Hence (A4) holds for S.

(M0) $\mathbf{x} \in S$ and $\lambda \in F \Rightarrow 0\mathbf{x} + \lambda\mathbf{x} \in S$ (by condition (2))
$\Rightarrow \lambda\mathbf{x} \in S$ (by 36.4).

Thus S is closed under multiplication by scalars: i.e. (M0) holds for S.

It now follows that S is a vector space (with respect to the operations defined on V) and is thus a subspace of V.

Condition (2) in the above result may be informally restated as "S is closed under the taking of linear combinations". This prompts the remark that obviously, because of the basic closure properties of a vector space (axioms (A0) and (M0)):

37.2 Any linear combination of any vectors in a subspace must also belong to the subspace.

Further to this, it should be realized that the conditions (1) and (2) in 37.1 are certainly both necessary conditions for S to be a subspace of V: 37.1 tells us the more interesting fact that together they constitute sufficient conditions for S to be a subspace.

There is a small point that deserves to be brought out explicitly at this stage. Suppose that S is a subspace of the vector space V. There is a zero vector $\mathbf{0}$ in the vector space V and there is a zero vector $\mathbf{o}$ in the vector space S; but need these coincide? The answer is yes, as the following simple argument shows.

By (M0) as it applies to S, $0\mathbf{o} \in S$. But, by 36.4, $0\mathbf{x} = \mathbf{0}$ for all $\mathbf{x} \in V$ and so, in particular, $0\mathbf{o} = \mathbf{0}$. Hence $\mathbf{0} \in S$. Clearly, therefore, $\mathbf{0}$ acts as a zero in S and hence (cf. 36.1) is *the* zero in S. So $\mathbf{0}$ and $\mathbf{o}$ do coincide.

Notice in particular that:

37.3 In a vector space V, the zero vector of V belongs to every subspace.

A trivial application of 37.1 shows that:

37.4 In every vector space $\{\mathbf{0}\}$ is a subspace.

A vector space (or subspace) that consists of the zero vector alone is called a zero vector space (or subspace).

Less trivial examples of the use of 37.1 are provided by the proofs of the following two theorems.

37.5 Let $A \in F_{m \times n}$. Then the set $\{X \in F_{n \times 1} : AX = O\}$ is a subspace of $F_{n \times 1}$.

Note: "$AX = O$" can be interpreted as an arbitrary system of homogeneous linear equations (cf. §27); and so the theorem tells us that the solution set of the homogeneous system $AX = O$ (where $A \in F_{m \times n}$) is a subspace of $F_{n \times 1}$.

Proof. Let S be the subset $\{X \in F_{n\times 1} : AX = O\}$ of $F_{n\times 1}$.

(1) Clearly the zero $n \times 1$ column belongs to S. So $S \neq \emptyset$.

(2) Let $X, Y \in S$ and let $\lambda, \mu \in F$. Then $AX = AY = O$, and hence (by elementary properties of matrices)

$$A(\lambda X + \mu Y) = \lambda(AX) + \mu(AY) = \lambda O + \mu O = O.$$

Thus $\lambda X + \mu Y \in S$.

This proves that: $X, Y \in S$ and $\lambda, \mu \in F \Rightarrow \lambda X + \mu Y \in S$.

It now follows (by 37.1) that S is a subspace of $F_{n\times 1}$.

37.6 The intersection of two subspaces of a vector space V is also a subspace of V.

Proof. Let S, T be subspaces of V. Certainly, $S \cap T \subseteq V$.

(1) By 37.3, $\mathbf{0} \in S$ and $\mathbf{0} \in T$. So $\mathbf{0} \in S \cap T$. Hence $S \cap T \neq \emptyset$.

(2) Let $\mathbf{x}, \mathbf{y} \in S \cap T$, and let λ, μ be scalars. Since $\mathbf{x}, \mathbf{y} \in S$ and S is a subspace, $\lambda\mathbf{x} + \mu\mathbf{y} \in S$ (cf. 37.2). Similarly, $\lambda\mathbf{x} + \mu\mathbf{y} \in T$. Hence $\lambda\mathbf{x} + \mu\mathbf{y} \in S \cap T$. This has shown that $S \cap T$ is "closed under the taking of linear combinations".

It now follows (by 37.1) that $S \cap T$ is a subspace of V.

The above argument is easily generalized to prove the fact that the intersection of any collection of subspaces of a vector space V is also a subspace of V.

38. Spanning sequences

Throughout the general discussion in this section, V will denote an arbitrary vector space over the field F.

The ideas to be developed in this section stem from the following fairly easy theorem.

38.1 Let $(\mathbf{x}_1, \mathbf{x}_2, \ldots, \mathbf{x}_n)$ be a (finite) sequence of vectors in V, and let S be the set of all linear combinations of $\mathbf{x}_1, \mathbf{x}_2, \ldots, \mathbf{x}_n$. Then:

(i) S is a subspace of V;

(ii) each of $\mathbf{x}_1, \mathbf{x}_2, \ldots, \mathbf{x}_n$ belongs to S;

(iii) S is the "smallest" subspace of V that contains $\mathbf{x}_1, \mathbf{x}_2, \ldots, \mathbf{x}_n$—"smallest" in the sense that if T is any subspace containing $\mathbf{x}_1, \mathbf{x}_2, \ldots, \mathbf{x}_n$, then $S \subseteq T$.

Proof. (i) It is clear that $S \subseteq V$ (cf. 37.2). This noted, one can see mentally (by 37.1) that S is a subspace of V.

(ii) is clear from the fact that, for each i, $\mathbf{x}_i = \sum_{j=1}^{n} \delta_{ij}\mathbf{x}_j$. (Cf. 17.1).

(iii) Now suppose that T is any subspace (of V) that contains $\mathbf{x}_1, \mathbf{x}_2, \ldots, \mathbf{x}_n$. Then, by 37.2, T contains every linear combination of $\mathbf{x}_1, \mathbf{x}_2, \ldots, \mathbf{x}_n$; i.e. T

contains every member of S; i.e. $S \subseteq T$. This shows that S is the "smallest" subspace of V that contains $\mathbf{x}_1, \ldots, \mathbf{x}_n$.

Remarks. (*a*) The subspace S described in 38.1 is called the *subspace spanned by* $(\mathbf{x}_1, \mathbf{x}_2, \ldots, \mathbf{x}_n)$. We denote this subspace by

$$\mathrm{sp}\,(\mathbf{x}_1, \mathbf{x}_2, \ldots, \mathbf{x}_n).$$

Notice at the outset that this notation stands neither for a single vector nor for a finite sequence of vectors, but rather for a subspace as detailed in 38.1.

(*b*) As a simple example, consider, in the context of E_3, $\mathrm{sp}\,(\mathbf{i}, \mathbf{j})$. This means the set of all linear combinations of $\mathbf{i}$ and $\mathbf{j}$, and it is, therefore, the subspace that we have labelled E_2. In accordance with 38.1(iii), this subspace can also be characterized as the "smallest" subspace of E_3 containing $\mathbf{i}$ and $\mathbf{j}$.

For a second simple example, observe that, in F^4,

$$\begin{aligned}\mathrm{sp}\,((1,0,0,0),(0,1,0,0),(0,0,0,1)) &= \{(\alpha, \beta, 0, \gamma) : \alpha, \beta, \gamma \in F\} \\ &= \{(x_1, x_2, x_3, x_4) : x_3 = 0\}.\end{aligned}$$

(*c*) In the rather trivial case of a sequence $(\mathbf{x})$ of length 1, $\mathrm{sp}\,(\mathbf{x})$ simply consists of all the scalar multiples of the vector $\mathbf{x}$.

(*d*) It is convenient to give a meaning to the subspace spanned by the "empty sequence"—i.e. sequence of length zero, which, with some abuse of notation, we shall denote by the empty set symbol $\varnothing$. We define $\mathrm{sp}\,(\varnothing)$ to be $\{\mathbf{0}\}$, the zero subspace. This turns out to be a sensible definition to the extent that it enables several future results to be concisely stated, without the need for an extra clause covering or excluding the case of the zero (sub)space. Notice that the zero subspace is also equal to $\mathrm{sp}\,(\mathbf{0})$.

(*e*) It is clear that if we change the order of the vectors in the sequence $(\mathbf{x}_1, \mathbf{x}_2, \ldots, \mathbf{x}_n)$, this makes no difference to the subspace spanned by the sequence. Indeed, $\mathrm{sp}\,(\mathbf{x}_1, \mathbf{x}_2, \ldots, \mathbf{x}_n)$ depends only on the *set* of vectors appearing in the sequence $(\mathbf{x}_1, \mathbf{x}_2, \ldots, \mathbf{x}_n)$, and does not depend on the order in which they appear or on how often any vector is repeated in the sequence. For example, in E_3, $\mathrm{sp}\,(\mathbf{j}, \mathbf{i}, \mathbf{i}) = \mathrm{sp}\,(\mathbf{i}, \mathbf{j})$. (One could accordingly begin the whole discussion by considering the subspace spanned by a set of vectors rather than a sequence; but the sequence approach fits in much better with other ideas to be introduced later in the chapter.)

(*f*) Generalization of the idea to the subspace spanned by an infinite sequence (or set) of vectors is possible; but that generalization lies outside the scope of this book.

The characterization of $\mathrm{sp}\,(\mathbf{x}_1, \ldots, \mathbf{x}_n)$ given by 38.1(iii) enables us to prove efficiently the three useful technical lemmas which now follow.

38.2 Let $(\mathbf{x}_1, \mathbf{x}_2, \ldots, \mathbf{x}_n)$ and $(\mathbf{y}_1, \mathbf{y}_2, \ldots, \mathbf{y}_m)$ be two sequences of vectors in V with the property that each $\mathbf{x}_i$ is expressible as a linear combination of the $\mathbf{y}$s. Then $\mathrm{sp}\,(\mathbf{x}_1, \ldots, \mathbf{x}_n) \subseteq \mathrm{sp}\,(\mathbf{y}_1, \ldots, \mathbf{y}_m)$.

Proof. The expressibility of each $\mathbf{x}_i$ as a linear combination of the ys tells us that $\mathrm{sp}(\mathbf{y}_1,\ldots,\mathbf{y}_m)$ is *a* subspace of V that contains $\mathbf{x}_1,\ldots,\mathbf{x}_n$, while (by 38.1(iii)) $\mathrm{sp}(\mathbf{x}_1,\ldots,\mathbf{x}_n)$ is the "smallest" such subspace. Therefore, $\mathrm{sp}(\mathbf{x}_1,\ldots,\mathbf{x}_n) \subseteq \mathrm{sp}(\mathbf{y}_1,\ldots,\mathbf{y}_m)$.

38.3 Let $(\mathbf{x}_1,\mathbf{x}_2,\ldots,\mathbf{x}_n)$ and $(\mathbf{y}_1,\mathbf{y}_2,\ldots,\mathbf{y}_m)$ be two sequences of vectors in V with the property that each $\mathbf{x}_i$ is expressible as a linear combination of the ys *and* each $\mathbf{y}_j$ is expressible as a linear combination of the xs. Then $\mathrm{sp}(\mathbf{x}_1,\ldots,\mathbf{x}_n) = \mathrm{sp}(\mathbf{y}_1,\ldots,\mathbf{y}_m)$.

Proof. This is an immediate consequence of 38.2, which tells us that $\mathrm{sp}(\mathbf{x}_1,\ldots,\mathbf{x}_n) \subseteq \mathrm{sp}(\mathbf{y}_1,\ldots,\mathbf{y}_m)$ *and* $\mathrm{sp}(\mathbf{y}_1,\ldots,\mathbf{y}_m) \subseteq \mathrm{sp}(\mathbf{x}_1,\ldots,\mathbf{x}_n)$.

38.4 Suppose that $\mathbf{x}_1,\mathbf{x}_2,\ldots,\mathbf{x}_n$ are vectors in V with the property that (for a certain value of j) $\mathbf{x}_j$ is expressible as a linear combination of the other xs (i.e. of $\mathbf{x}_1,\ldots,\mathbf{x}_{j-1},\mathbf{x}_{j+1},\ldots,\mathbf{x}_n$). Then $\mathrm{sp}(\mathbf{x}_1,\ldots,\mathbf{x}_n) = \mathrm{sp}(\mathbf{x}_1,\ldots,\mathbf{x}_{j-1},\mathbf{x}_{j+1},\ldots,\mathbf{x}_n)$.

(This, it will be noted, says that deletion, from a sequence, of a vector expressible as a linear combination of the others makes no difference to the subspace spanned.)

Proof. A little reflection reveals that if we take $(\mathbf{y}_1,\ldots,\mathbf{y}_m)$ to be $(\mathbf{x}_1,\ldots,\mathbf{x}_{j-1},\mathbf{x}_{j+1},\ldots,\mathbf{x}_n)$, then the hypotheses of 38.3 are satisfied. Thus the present result follows immediately from 38.3.

The ideas of this section lead to useful lines of thought about matrices, via the following definitions.

Let $A \in F_{m\times n}$. (1) The **row space** of A is defined to be the subspace of $F_{1\times n}$ spanned by the rows of A (i.e. the set of all rows expressible as linear combinations of the rows of A). And similarly (2) the **column space** of A is defined to be the subspace of $F_{m\times 1}$ spanned by the columns of A.

We are already in a position to prove that:

38.5 Two matrices that are row equivalent to each other have the same row space.

Proof. First suppose that $B = \theta(A)$, where θ is an e.r.o. $(A, B \in F_{m\times n})$. By considering in turn each of the three types of e.r.o. ($R_i \leftrightarrow R_j$ $(i \neq j)$, $R_i \to \lambda R_i$ $(\lambda \neq 0)$, $R_i \to R_i + \lambda R_j$ $(i \neq j)$), one soon sees that, whatever θ may be, every row of B is expressible as a linear combination of the rows of A. But equally, since $A = \theta^{-1}(B)$ (cf. 23.1), every row of A is expressible as a linear combination of the rows of B. Hence, by 38.3, the subspace of $F_{1\times n}$ spanned by the rows of A coincides with the subspace spanned by the rows of B; i.e. A and B have the same row space.

The conclusion so far is that the performing of an e.r.o. produces a matrix with the same row space as the original matrix. It follows that the same must be true for any sequence of e.r.o.s, and this is the stated result.

Our approach hitherto in this section has been to start with a sequence of vectors and then to consider the subspace spanned by the given sequence. Often, though, we shall want to take the reverse approach: i.e. we shall start with a given vector space or subspace S and look for a sequence of vectors that spans S—i.e. a sequence $(\mathbf{x}_1, \mathbf{x}_2, \ldots, \mathbf{x}_n)$ of vectors in S such that $\text{sp}(\mathbf{x}_1, \mathbf{x}_2, \ldots, \mathbf{x}_n)$ is the whole of S. The advantages of having such a sequence are considerable: in most cases our given space (or subspace) S will be an infinite set of vectors, but if the sequence $(\mathbf{x}_1, \mathbf{x}_2, \ldots, \mathbf{x}_n)$ spans S, then all the vectors in S can be expressed as linear combinations of $\mathbf{x}_1, \mathbf{x}_2, \ldots, \mathbf{x}_n$ and thus all calculations within S can be conducted in terms of finitely many vectors, viz. $\mathbf{x}_1, \mathbf{x}_2, \ldots, \mathbf{x}_n$.

To give a very simple illustration, suppose we start with the vector space E_3 and look for a sequence of vectors that spans E_3. That is a very easy task: clearly, the sequence $(\mathbf{i}, \mathbf{j}, \mathbf{k})$ spans E_3. (And correspondingly all calculations within E_3 can be accomplished through the handling of linear combinations of the three vectors $\mathbf{i}, \mathbf{j}, \mathbf{k}$.)

Let us now work through an instructive, and less trivial, example—to find a sequence of vectors that spans

$$S = \{(x_1, x_2, x_3, x_4) \in \mathbb{R}^4 : x_1 + 2x_2 - x_3 + 3x_4 = 0\}.$$

That this set S is a subspace of $\mathbb{R}^4$ can be seen from 37.5 by regarding "$x_1 + 2x_2 - x_3 + 3x_4 = 0$" as a homogeneous system of 1 equation.

(*Note*: strictly speaking, in terms of our previous work the solution set of this equation is a set of columns, but we can identify this set with S—in accordance with the remark in §35 about the convenience of blurring or ignoring the difference between F^n and $F_{n \times 1}$.)

From the work of chapter 3, we know that the general solution of the equation $x_1 + 2x_2 - x_3 + 3x_4 = 0$ is

$$x_1 = -3\alpha + \beta - 2\gamma, \quad x_2 = \gamma, \quad x_3 = \beta, \quad x_4 = \alpha \qquad (\alpha, \beta, \gamma \in \mathbb{R}).$$

Therefore S is the set of all vectors of the form

$$(-3\alpha + \beta - 2\gamma, \gamma, \beta, \alpha) \qquad \text{with } \alpha, \beta, \gamma \in \mathbb{R}.$$

To obtain an answer to the problem, we must express the above general member of S in the form

$$\alpha(\ ,\ ,\ ,\ ) + \beta(\ ,\ ,\ ,\ ) + \gamma(\ ,\ ,\ ,\ ).$$

It is easily seen that the one and only correct way to fill the blanks is to use the vectors $(-3, 0, 0, 1)$, $(1, 0, 1, 0)$, $(-2, 1, 0, 0)$, each of which is a member of S. Thus every member of S is a linear combination of the three member vectors just listed, and so we are able to conclude that the sequence

$$((-3,0,0,1),(1,0,1,0),(-2,1,0,0))$$

spans S.

It must not be supposed that, no matter what vector space V we consider, there will exist a finite sequence of vectors that spans V. In fact, there are many vector spaces in which no such finite spanning sequence exists. As an exercise at the end of the chapter points out, the vector space of all formal polynomials with coefficients in F falls into this category.

We describe a vector space V as **finite-dimensional** if there does exist a finite sequence of vectors that spans V. Otherwise, we describe V as infinite-dimensional. We use the abbreviation f.d. for "finite-dimensional".

39. Linear dependence and independence

Throughout the general discussion in this section, V will again denote an arbitrary vector space over the field F.

One doesn't have to go very far into the subject-matter of linear algebra to find oneself remarking on situations where one object is a linear combination of the others under consideration. There was a result (38.4) about such a situation in the previous section; and, at a more rudimentary level, one may notice that, in a system of linear equations, one of the equations is redundant because it is a linear combination of (and therefore a consequence of) others in the system.

These thoughts lead naturally to certain discussions in the context of the arbitrary vector space V—discussions whose outcome is of the greatest importance in vector space theory and in linear algebra generally. The following lemma points the way to the definitions on which these important discussions will be based.

39.1 Let L stand for the sequence $(\mathbf{x}_1, \mathbf{x}_2, \ldots, \mathbf{x}_n)$ of vectors in V, and suppose that the length n of this sequence is at least 2. Then the following two conditions are equivalent to each other:

(D1) one of the vectors in L can be expressed as a linear combination of the others;

(D2) there is a non-trivial way to express $\mathbf{0}$ as a linear combination of the vectors in L; i.e. there exist scalars $\lambda_1, \lambda_2, \ldots, \lambda_n$, not all zero, such that $\sum_{i=1}^{n} \lambda_i \mathbf{x}_i = \mathbf{0}$.

Before the proof is given, there are two things to say about the condition (D2).

(i) It should be noted once and for all that there is a big difference in meaning between the phrases "not all zero" and "all nonzero". It is the former that appears in (D2) and elsewhere in this section.

(ii) To appreciate the force of the condition (D2), observe that there is, in any event, the trivial way to express $\mathbf{0}$ as a linear combination of the vectors in L—viz. to take all the coefficients equal to 0. The condition (D2) holds if and only if there is *also* a non-trivial way to express $\mathbf{0}$ in the form $\sum_{i=1}^{n} \lambda_i \mathbf{x}_i$.

Proof of 39.1. (*A*) Suppose that (D1) holds. Then, for some j, we have

$$\mathbf{x}_j = \alpha_1 \mathbf{x}_1 + \ldots + \alpha_{j-1}\mathbf{x}_{j-1} + \alpha_{j+1}\mathbf{x}_{j+1} + \ldots + \alpha_n \mathbf{x}_n,$$

for some scalars $\alpha_1, \ldots, \alpha_{j-1}, \alpha_{j+1}, \ldots, \alpha_n$. Hence

$$\alpha_1 \mathbf{x}_1 + \ldots + \alpha_{j-1}\mathbf{x}_{j-1} + (-1)\mathbf{x}_j + \alpha_{j+1}\mathbf{x}_{j+1} + \ldots + \alpha_n \mathbf{x}_n = \mathbf{0}.$$

The coefficient of $\mathbf{x}_j$ being nonzero, this equation exhibits $\mathbf{0}$ expressed as a linear combination of the vectors in L in a non-trivial way. Thus (D2) holds.

(*B*) Suppose conversely that (D2) holds, so that $\sum_{i=1}^{n} \lambda_i \mathbf{x}_i = \mathbf{0}$ for some scalars $\lambda_1, \lambda_2, \ldots, \lambda_n$ that are not all zero. Suppose that λ_j is nonzero. Then, starting with the fact that

$$\lambda_j \mathbf{x}_j = \sum_{\substack{i=1 \\ i \neq j}}^{n} (-\lambda_i \mathbf{x}_i),$$

we obtain, on multiplying through by $1/\lambda_j$,

$$\mathbf{x}_j = \sum_{\substack{i=1 \\ i \neq j}}^{n} \beta_i \mathbf{x}_i,$$

where $\beta_i = -\lambda_i/\lambda_j$. So $\mathbf{x}_j$ is expressible as a linear combination of the other vectors in L, and thus (D1) holds.

The paragraphs (*A*) and (*B*) have shown, respectively, that (D1) $\Rightarrow$ (D2) and that (D2) $\Rightarrow$ (D1). It follows that (D1) and (D2) are equivalent, as asserted.

It is natural to regard the vectors $\mathbf{x}_1, \mathbf{x}_2, \ldots, \mathbf{x}_n$ as in some sense mutually dependent if the equivalent conditions (D1) and (D2) hold (and independent if they do not hold). Before we formalize that idea, notice that, while the original motivation for our discussion shines clearly in (D1), (D2) is nevertheless a more satisfactory condition to consider, for at least two reasons. First (D2) involves all the vectors $\mathbf{x}_1, \mathbf{x}_2, \ldots, \mathbf{x}_n$ in a symmetrical way (whereas (D1) is about the relationship of one of them to the rest); and secondly (D2), unlike (D1), makes sense in the so far neglected case $n = 1$.

The way has now been fully prepared for the following definitions.

Let $(\mathbf{x}_1, \mathbf{x}_2, \ldots, \mathbf{x}_n)$ be a sequence of vectors in V $(n \geqslant 1)$.

(1) We say that the sequence $(\mathbf{x}_1, \mathbf{x}_2, \ldots, \mathbf{x}_n)$ is **linearly dependent** (abbreviation L.D.) if the sequence satisfies the condition (D2), i.e. if $\mathbf{0}$ can be expressed in a non-trivial way as a linear combination of $\mathbf{x}_1, \mathbf{x}_2, \ldots, \mathbf{x}_n$, i.e. if there exist scalars $\lambda_1, \lambda_2, \ldots, \lambda_n$, *not all zero*, such that $\sum_{i=1}^{n} \lambda_i \mathbf{x}_i = \mathbf{0}$.

(2) We say that the sequence $(\mathbf{x}_1, \mathbf{x}_2, \ldots, \mathbf{x}_n)$ is **linearly independent** (abbreviation L.I.) if it is not L.D., i.e. if the *only* way to express $\mathbf{0}$ as a linear combination of $\mathbf{x}_1, \mathbf{x}_2, \ldots, \mathbf{x}_n$ is the trivial way with all the coefficients equal to zero.

Thus, to rephrase the latter definition in a form which will prove helpful:

39.2 The sequence $(\mathbf{x}_1, \mathbf{x}_2, \ldots, \mathbf{x}_n)$ of vectors in V (where $n \geqslant 1$) is L.I. if and only if

$$\sum_{i=1}^{n} \lambda_i \mathbf{x}_i = \mathbf{0} \Rightarrow \lambda_1 = \lambda_2 = \ldots = \lambda_n = 0 \qquad (\lambda_1, \ldots, \lambda_n \in F).$$

It is also immediately apparent (from the equivalence of (D1) and (D2) in the case $n \geqslant 2$) that:

39.3 A sequence of vectors $(\mathbf{x}_1, \mathbf{x}_2, \ldots, \mathbf{x}_n)$ whose length n is $\geqslant 2$ is L.D. if and only if one of the vectors in the sequence can be expressed as a linear combination of the others.

Remarks. (*a*) By convention, we take the empty sequence of vectors, $\varnothing$, to be L.I.

(*b*) In connection with 39.3, notice that when a sequence of vectors $(\mathbf{x}_1, \ldots, \mathbf{x}_n)$, with $n \geqslant 2$, is L.D., it *may not* be true that *each* of the vectors in the sequence is expressible as a linear combination of the others. As an instructive example, consider in $\mathbb{R}^2$ the L.D. sequence $(\mathbf{x}, \mathbf{y}, \mathbf{z})$, where $\mathbf{x} = (1, 0)$, $\mathbf{y} = (0, 1)$, $\mathbf{z} = (0, 2)$. Here $\mathbf{z}$ is expressible as a linear combination of the others (since $\mathbf{z} = 0\mathbf{x} + 2\mathbf{y}$); but $\mathbf{x}$ is not expressible as a linear combination of $\mathbf{y}$ and $\mathbf{z}$ (since, obviously, all linear combinations of $\mathbf{y}$ and $\mathbf{z}$ are of the form $(0, \alpha)$).

(*c*) Subtle and indirect methods of deciding whether a given sequence of vectors is L.D. or L.I. will become available to us as we develop the theory further (specifically in §40). Already, however, simple procedures are open to us by which we can prove a given sequence L.D. or L.I. as the case may be.

To prove that a given sequence of vectors is L.D., the snappiest method is to find (through suitable exploratory calculations) a non-trivial way of expressing $\mathbf{0}$ as a linear combination of the given vectors. For example, if one discovered (by any means) that $7\mathbf{w} + 4\mathbf{x} - 2\mathbf{y} + \mathbf{z} = \mathbf{0}$, that equation by itself would prove the linear dependence of the sequence $(\mathbf{w}, \mathbf{x}, \mathbf{y}, \mathbf{z})$.

On the other hand, to prove the linear independence of a given sequence of vectors $(\mathbf{x}_1, \ldots, \mathbf{x}_n)$, we often use 39.2. We begin by supposing that $\sum_{i=1}^{n} \lambda_i \mathbf{x}_i = \mathbf{0}$

($\lambda_1, \ldots, \lambda_n$ being scalars), and we prove it to be a consequence of this supposition that $\lambda_1, \ldots, \lambda_n$ must all be zero. By that procedure we demonstrate that

$$\sum_{i=1}^{n} \lambda_i \mathbf{x}_i = \mathbf{0} \Rightarrow \lambda_1 = \lambda_2 = \ldots = \lambda_n = 0 \qquad (\lambda_1, \ldots, \lambda_n \in F),$$

which (cf. 39.2) establishes that the sequence $(\mathbf{x}_1, \ldots, \mathbf{x}_n)$ is L.I.

As a simple illustration, let us prove the linear independence of the sequence $(\mathbf{i}, \mathbf{j}, \mathbf{k})$ in E_3. We begin by supposing that

$$\lambda\mathbf{i} + \mu\mathbf{j} + \nu\mathbf{k} = \mathbf{0} \qquad (\lambda, \mu, \nu \in \mathbb{R}).$$

The zero vector, then, is equal both to $0\mathbf{i} + 0\mathbf{j} + 0\mathbf{k}$ and to $\lambda\mathbf{i} + \mu\mathbf{j} + \nu\mathbf{k}$. So, by 8.3, $\lambda = 0$, $\mu = 0$ and $\nu = 0$.

This (the demonstration that $\lambda\mathbf{i} + \mu\mathbf{j} + \nu\mathbf{k} = \mathbf{0} \Rightarrow \lambda = \mu = \nu = 0$) proves that the sequence $(\mathbf{i}, \mathbf{j}, \mathbf{k})$ is L.I.

In the proofs of theorems (39.11, for instance) the student will see many further examples of this approach to the task of proving a sequence of vectors to be L.I.

Our next task is to take note of the simpler properties of linear dependence and independence.

39.4 Changing the order of the vectors in a sequence does not affect whether the sequence is L.D. or L.I.

(This is obvious from the definitions.)

39.5 Any sequence containing a repetition is L.D.

(That $(\mathbf{x}_1, \mathbf{x}_1, \mathbf{x}_3, \ldots, \mathbf{x}_n)$ is L.D. is shown by the equation

$$1\mathbf{x}_1 + (-1)\mathbf{x}_1 + 0\mathbf{x}_3 + \ldots + 0\mathbf{x}_n = \mathbf{0};$$

and the general result follows by 39.4.)

In the light of this result, notice that the linear dependence or independence of a sequence of vectors is *not* determined merely by the *set* of vectors appearing in the sequence. For example the same set of vectors (namely $\{\mathbf{i}, \mathbf{j}, \mathbf{k}\}$) appears in both of the sequences $(\mathbf{i}, \mathbf{j}, \mathbf{k})$ and $(\mathbf{i}, \mathbf{i}, \mathbf{j}, \mathbf{k})$; but, while the former is (as we have shown) L.I., the latter is clearly L.D. because of the repetition of $\mathbf{i}$.

39.6 Any sequence in which $\mathbf{0}$ appears is L.D.

(That $(\mathbf{0}, \mathbf{x}_2, \mathbf{x}_3, \ldots, \mathbf{x}_n)$ is L.D. is shown by the equation

$$1\mathbf{0} + 0\mathbf{x}_2 + 0\mathbf{x}_3 + \ldots + 0\mathbf{x}_n = \mathbf{0};$$

and the general result follows by 39.4.)

39.7 Every subsequence of a L.I. sequence is also L.I.; and every extension of a L.D. sequence is also L.D.

Here the words "subsequence" and "extension" are used with the obvious meanings: e.g. if L is the sequence $(\mathbf{x}_1, \mathbf{x}_2, \mathbf{x}_3, \mathbf{x}_4, \mathbf{x}_5)$, then the sequence $(\mathbf{x}_1, \mathbf{x}_3, \mathbf{x}_4)$ is a subsequence of L, while the sequence $(\mathbf{x}_0, \mathbf{x}_1, \mathbf{x}_2, \mathbf{x}_3, \mathbf{x}_4, \mathbf{x}_5, \mathbf{x}_6, \mathbf{x}_7)$ is an extension of L. The empty sequence $\varnothing$ counts as a subsequence of any given sequence, and 39.7 covers this case in view of the convention that $\varnothing$ is L.I.

It is not difficult to justify 39.7, and the details are left for the student to think through. He should begin with the second clause and note that (a separate remark having taken care of the case of the empty subsequence) the first clause follows from the second.

39.8 A sequence $(\mathbf{x})$ of length 1 is L.D. if and only if $\mathbf{x} = \mathbf{0}$.

Proof. If $\mathbf{x} = \mathbf{0}$, then the fact that $1\mathbf{0} = \mathbf{0}$ shows that $(\mathbf{x})$ is L.D. (This is a degenerate, but valid, case of 39.6.) If, on the other hand, $\mathbf{x} \neq \mathbf{0}$, then, by 36.6,

$$\lambda\mathbf{x} = \mathbf{0} \Rightarrow \lambda = 0 \qquad (\lambda \in F),$$

which (cf. 39.2) shows that $(\mathbf{x})$ is L.I. The stated result now follows.

39.9 Suppose that $\mathbf{x}, \mathbf{y}$ are both nonzero vectors in V. Then the sequence $(\mathbf{x}, \mathbf{y})$ is L.D. if and only if $\mathbf{x}$ and $\mathbf{y}$ are scalar multiples of each other.

Proof. If $\mathbf{x}$ and $\mathbf{y}$ are scalar multiples of each other, then $(\mathbf{x}, \mathbf{y})$ is a L.D. sequence, by a trivial case of 39.3.

To prove the converse, suppose that $(\mathbf{x}, \mathbf{y})$ is L.D. Then, for some scalars α, β which are not both zero,

$$\alpha\mathbf{x} + \beta\mathbf{y} = \mathbf{0}. \tag{1}$$

Here α cannot be 0: for if α were 0, (1) would reduce to $\beta\mathbf{y} = \mathbf{0}$, and hence (by 36.6 and the fact that $\mathbf{y} \neq \mathbf{0}$) β would also have to be 0—a contradiction (since α, β are not both zero). Similarly $\beta \neq 0$. Hence from (1) we deduce $\mathbf{x} = (-\beta/\alpha)\mathbf{y}$ and $\mathbf{y} = (-\alpha/\beta)\mathbf{x}$; and thus $\mathbf{x}$ and $\mathbf{y}$ are scalar multiples of each other.

The result is now established.

There is a nice geometrical interpretation of the linear dependence of a sequence of three vectors in E_3: the sequence $(\mathbf{x}, \mathbf{y}, \mathbf{z})$ of vectors in E_3 is (one can show) L.D. if and only if $\mathbf{x}, \mathbf{y}, \mathbf{z}$ are coplanar. (See §10, remark (*b*), for the precise meaning of "coplanar" in this context.)

One of the most significant properties of a L.I. sequence is given by the following theorem.

39.10 Suppose that the sequence $(\mathbf{x}_1, \mathbf{x}_2, \ldots, \mathbf{x}_n)$ of vectors in V is L.I. $(n \geqslant 1)$ and that $\mathbf{y} \in \mathrm{sp}(\mathbf{x}_1, \mathbf{x}_2, \ldots, \mathbf{x}_n)$. Then there is *only one* way of expressing $\mathbf{y}$ as a linear combination of $\mathbf{x}_1, \mathbf{x}_2, \ldots, \mathbf{x}_n$.

Proof. Suppose that

$$\mathbf{y} = \sum_{i=1}^{n} \lambda_i \mathbf{x}_i = \sum_{i=1}^{n} \mu_i \mathbf{x}_i \qquad \text{(where each } \lambda_i \in F \text{ and each } \mu_i \in F\text{)}.$$

We shall show that $\lambda_i = \mu_i$ for every i (in the range 1 to n).

By subtracting the second expression for $\mathbf{y}$ from the first, we obtain

$$\mathbf{0} = \mathbf{y} - \mathbf{y} = \sum_{i=1}^{n} (\lambda_i - \mu_i)\mathbf{x}_i.$$

Hence, since the sequence $(\mathbf{x}_1, \ldots, \mathbf{x}_n)$ is L.I., $\lambda_i - \mu_i = 0$ for every i; i.e. $\lambda_i = \mu_i$ for every i.

The stated result follows.

We conclude this section with an important fact about echelon matrices:

39.11 The nonzero rows in an echelon matrix form a L.I. sequence.

Proof. There is a trivial case where the echelon matrix is O and the sequence of nonzero rows is therefore empty and so L.I.

Henceforth consider the remaining case where there is at least one nonzero row in the echelon matrix. Let its nonzero rows be (in order from the top downwards) $R_1, R_2, \ldots, R_l$.

Suppose that $\lambda_1 R_1 + \lambda_2 R_2 + \ldots + \lambda_l R_l = O$ ($\lambda_1, \ldots, \lambda_l$ being scalars).

Each side of this equation is a row matrix. We pick out from each side the entry in the position where R_1 has its leading entry and where, therefore, the lower rows $R_2, \ldots, R_l$ of the echelon matrix have zeros. Hence we deduce

$$\lambda_1 \times (\text{leading entry of } R_1) = 0,$$

and thus $\lambda_1 = 0$. It follows that

$$\lambda_2 R_2 + \lambda_3 R_3 + \ldots + \lambda_l R_l = O.$$

We now focus on the position where R_2 has its leading entry (and therefore $R_3, \ldots, R_l$ have zeros). Hence we deduce that $\lambda_2 = 0$. Proceeding in this way, we successively deduce $\lambda_3 = 0$, $\lambda_4 = 0, \ldots, \lambda_l = 0$. Thus all the coefficients $\lambda_1, \lambda_2, \ldots, \lambda_l$ are zero.

This proves that the sequence $(R_1, R_2, \ldots, R_l)$ is L.I., and the result follows.

40. Bases and dimension

Throughout this section, the symbol V will denote a vector space over the field F.

A **basis** of V means a (finite) sequence of vectors in V which (1) is L.I. and (2) spans V.

If $V \neq \{\mathbf{0}\}$ and a basis of V is given, then (in view of 39.10) each vector in V can be expressed in *precisely one* way as a linear combination of the vectors in the basis, and hence all questions about vectors in V reduce in a *straightforward* manner to questions about linear combinations of the vectors in the basis. That is one easily understood reason for considering bases.

Perhaps the most familiar example of a basis is the sequence $(\mathbf{i}, \mathbf{j}, \mathbf{k})$ in E_3. This is a basis of E_3 because (1) it spans E_3 and (2) it is L.I. (facts that were remarked on in the course of §§38 and 39, respectively). The point made in the previous paragraph is illustrated by the fact (well seen in the work of chapter 1) that questions about vectors in E_3 reduce very straightforwardly to questions about linear combinations of $\mathbf{i}, \mathbf{j}, \mathbf{k}$.

It should be added that $(\mathbf{i}, \mathbf{j}, \mathbf{k})$ is by no means the only basis of E_3: for, in fact, every sequence of 3 non-coplanar vectors in E_3 is a basis of E_3. (A hint of this was given in remark (*b*) in §10.) Among this multiplicity of bases of E_3, there is one feature whose simplicity must not be allowed to hide its importance—the fact that *every* basis of E_3 is a sequence of the same length (namely 3). (One can show that no sequence of fewer than 3 vectors spans E_3, while no sequence of more than 3 vectors in E_3 can be L.I.)

One of the main objectives of this section is to show that something very similar is true of every f.d. vector space. In particular, as we shall prove, every f.d. vector space has a basis; and, for any given f.d. vector space V, all bases of V are sequences of the same length. (It is this common length of all the bases of V that we shall call the dimension of V.)

In an infinite-dimensional space V, there is (by the definition of "infinite-dimensional" in §38) no finite sequence of vectors that spans V, and so there can be no basis of V in the sense that we have defined the term. A theory of infinite bases of infinite-dimensional spaces can be developed, but lies outside the scope of this book.

Before we explore the general theory of bases, let us pause to consider some further specific examples of bases.

(1) Let n be any fixed positive integer. For each i in the range 1 to n, let $\mathbf{e}_i$ be the vector

$$\begin{array}{c}(0, 0, \ldots, 0, 1, 0, \ldots, 0)\\ \uparrow\\ i\text{th position}\end{array}$$

in F^n. Then the arbitrary vector $(\alpha_1, \alpha_2, \ldots, \alpha_n)$ in F^n can be expressed as $\sum_{i=1}^{n} \alpha_i \mathbf{e}_i$, and therefore the sequence $(\mathbf{e}_1, \mathbf{e}_2, \ldots, \mathbf{e}_n)$ spans F^n. It is also easy to prove that this sequence is L.I. Hence it is a basis of F^n. We call this basis $(\mathbf{e}_1, \mathbf{e}_2, \ldots, \mathbf{e}_n)$ the **standard basis** of F^n.

The same story, for $n \times 1$ columns rather than n-tuples, is that in $F_{n \times 1}$ the sequence $(E_1, E_2, \ldots, E_n)$ of columns of the identity matrix I_n is a basis of $F_{n \times 1}$—a basis which we call the standard basis of $F_{n \times 1}$.

(2) Let m, n be any given positive integers. In $F_{m \times n}$ there are mn matrix units E_{ik} as described in §17(3); and a basis of $F_{m \times n}$ is obtained by arranging these mn matrix units (in any order whatever) to form a sequence. (Cf. the remark in §17 that every matrix in $F_{m \times n}$ is uniquely expressible as a linear combination of the matrix units.)

(3) The empty sequence $\varnothing$ is a basis of $\{\mathbf{0}\}$, since, according to the conventions introduced in §§38, 39, $\text{sp}(\varnothing) = \{\mathbf{0}\}$ and $\varnothing$ is L.I. Note that the sequence $(\mathbf{0})$ is not a basis of $\{\mathbf{0}\}$, since this sequence is L.D. (cf. 39.6 or 39.8).

(4) Our work on echelon matrices, etc., enables us to see that:

40.1 The sequence of nonzero rows in an echelon matrix E is a basis of the row space of E and of every matrix row equivalent to E.

Proof. Consider first the exceptional case $E = O$. In this case of the sequence of nonzero rows of E is $\varnothing$; and so (cf. (3) above) the result holds in this case, since the only matrix row equivalent to E is O, whose row space is $\{O\}$.

Now consider the remaining case where $E \neq O$. It is clear that the sequence of nonzero rows in E spans the row space of E; and this sequence is L.I., by 39.11. Hence this sequence is a basis of the row space of E and, therefore, of the row space of every matrix row equivalent to E (cf. 38.5).

The stated result is now established.

Worked example. Find a basis of the subspace of $\mathbb{R}^4$ spanned by the sequence of vectors $((1, 0, 1, 2), (2, 3, 0, 1), (-1, 1, 1, -2), (1, 5, 3, -1))$.

Solution. We interpret the given vectors as row matrices. Accordingly, the subspace specified (let's call it W) is the row space of the matrix

$$A = \begin{bmatrix} 1 & 0 & 1 & 2 \\ 2 & 3 & 0 & 1 \\ -1 & 1 & 1 & -2 \\ 1 & 5 & 3 & -1 \end{bmatrix}.$$

We carry out on A the echelon reduction process described in chapter 3, so finding that A is row equivalent to the echelon matrix

$$E = \begin{bmatrix} 1 & 0 & 1 & 2 \\ 0 & 1 & 2 & 0 \\ 0 & 0 & 8 & 3 \\ 0 & 0 & 0 & 0 \end{bmatrix}.$$

By 40.1, the sequence of nonzero rows in E is a basis of W. So an answer to the question posed is the sequence

$$((1, 0, 1, 2), (0, 1, 2, 0), (0, 0, 8, 3)).$$

We come now to the statements and proofs of important general results about bases. Among these, 40.5 may be regarded as the highlight since it enables us to define the dimension of a f.d. vector space, but the other results too are illuminating and worthy of study.

40.2 Suppose that the finite sequence of vectors $L = (\mathbf{x}_1, \mathbf{x}_2, \ldots, \mathbf{x}_n)$, where $n \geqslant 1$, spans V. Then some subsequence of L (perhaps L itself) is a basis of V. (That is, either L is a basis of V, or else a basis of V can be obtained from L by deleting some of the $\mathbf{x}$s.)

Proof. The case $V = \{\mathbf{0}\}$ is trivial (since in that case we can obtain a basis of V (viz. $\varnothing$) by deleting all the $\mathbf{x}$s). And the case where L is L.I. (and so itself is a basis of V) is also trivial. Henceforth we leave these trivial cases aside and suppose that $V \neq \{\mathbf{0}\}$ and that L is L.D.

We make the observation (α) that every L.D. sequence of vectors spanning V must have length at least 2. (For, since $V \neq \{\mathbf{0}\}$, any sequence spanning V must contain at least one nonzero vector; and a sequence of just one nonzero vector is L.I. (cf. 39.8).)

The observation (α) applies to L, which, therefore, has length $\geqslant 2$. So, since L is L.D., there is (by 39.3) a vector in L expressible as a linear combination of the others. Let this vector be deleted from L to form a subsequence L'. By 38.4, L' also spans V.

If L' is L.I., then it is a basis of V as required. Otherwise, L' is L.D. and of length $\geqslant 2$ (by (α)); and in this case we can repeat the process of the deletion of an $\mathbf{x}$ to produce a subsequence L'' of L that has length $n-2$ and also spans V.

We simply repeat the process of deletion until we are left with a L.I. subsequence of L that spans V. This must happen after at most $n-1$ deletions since (by (α)) 2 is the minimum conceivable length for a L.D. sequence that spans V.

The truth of the stated result is thus apparent.

The appeals to 40.2 in subsequent proofs will make clear its theoretical importance. It is, however, not particularly helpful in numerical problems where a spanning sequence is given for a subspace and a basis of the subspace has to be found. The earlier worked example demonstrated a very efficient technique for dealing with such numerical problems, though it will be appreciated that this technique usually produces a basis which is *not* a subsequence of the given spanning sequence.

A simple but significant corollary of 40.2 is:

40.3 Every f.d. vector space has a (finite) basis.

Proof. Let V be an arbitrary f.d. vector space. By the definition of "finite-dimensional", there is a finite sequence L of vectors in V that spans V. If L is $\varnothing$ (and therefore $V = \{\mathbf{0}\}$), then L is a basis of V. Otherwise, by 40.2, some subsequence of L (perhaps L itself) is a basis of V. Thus, in any event, there exists a basis of V; and this proves the corollary.

It was mentioned earlier that E_3 has many different bases; and, in fact, it is true that (provided the field F contains infinitely many elements) every nonzero f.d. vector space over F has infinitely many different bases.

The next theorem is of crucial importance.

40.4 Suppose that V has a basis $(\mathbf{e}_1, \mathbf{e}_2, \ldots, \mathbf{e}_n)$ of length n. Then every sequence of more than n vectors in V is L.D.

Proof. Let $(\mathbf{y}_1, \mathbf{y}_2, \ldots, \mathbf{y}_m)$ be a sequence of m vectors in V, where $m > n$.

Since $(\mathbf{e}_1, \ldots, \mathbf{e}_n)$, being a basis of V, spans V, it follows that each $\mathbf{y}_j$ is expressible as a linear combination of the **e**s: say

$$\mathbf{y}_j = \sum_{i=1}^{n} \alpha_{ij}\mathbf{e}_i \qquad (j = 1, 2, \ldots, m).$$

Consider the homogeneous system of linear equations

$$\left\{\begin{array}{l} \alpha_{11}x_1 + \alpha_{12}x_2 + \ldots + \alpha_{1m}x_m = 0 \\ \quad\ldots \\ \alpha_{n1}x_1 + \alpha_{n2}x_2 + \ldots + \alpha_{nm}x_m = 0 \end{array}\right\}.$$

Here there are fewer equations than unknowns (n equations and m unknowns, and m is greater than n). So, by 28.4, there is a non-trivial solution—say $x_1 = \beta_1, x_2 = \beta_2, \ldots, x_m = \beta_m$. Thus we have

$$\sum_{j=1}^{m} \alpha_{ij}\beta_j = 0 \qquad (i = 1, 2, \ldots, n), \qquad (*)$$

while (the solution being non-trivial) not all the βs are zero.

We now note that

$$\sum_{j=1}^{m} \beta_j \mathbf{y}_j = \sum_{j=1}^{m} \beta_j \left(\sum_{i=1}^{n} \alpha_{ij} \mathbf{e}_i \right)$$

$$= \sum_{i=1}^{n} \left(\sum_{j=1}^{m} \alpha_{ij} \beta_j \right) \mathbf{e}_i \qquad \text{(on interchanging the order of summation)}$$

$$= \mathbf{0} \qquad \text{(by (*))}.$$

Since $\beta_1, \beta_2, \ldots, \beta_m$ are not all zero, this proves that the sequence $(\mathbf{y}_1, \ldots, \mathbf{y}_m)$ is L.D.; and the stated result follows.

Note. The alert student will perceive a defect in the above proof: it fails to cover the degenerate case where $n = 0$, the given basis is $\varnothing$, and $V = \{\mathbf{0}\}$, and where (significantly) the theorem is still true. As this degenerate case is a trivial one, the "defect" mentioned can be repaired by the insertion of one suitable sentence covering the degenerate case. However, there is a more general issue here than the tidying up of one particular proof: for the case where a space and/or subspace is zero is, though usually trivial, frequently a nuisance, requiring separate mention in a fully correct proof. To save space, the practice from now onwards in this textbook will simply be (as in the above proof of 40.4) to omit from proofs special arguments needed to cover the zero (sub)space case. It should be regarded as a standing exercise for the reader to consider whether such a special argument is necessary and, if so, to supply it.

With 40.4 proved, we are now in a position to show that:

40.5 In a f.d. vector space, all bases have the same length.

Proof. Suppose that L_1, L_2 are bases, with lengths m, n, respectively, of the f.d. vector space V. Since L_2 is a basis and the sequence L_1 is not L.D., it follows from 40.4 that $m \leqslant n$. But, by the same argument with the roles of L_1 and L_2 interchanged, $n \leqslant m$. Hence $m = n$. This proves the stated result.

Because of 40.3 and 40.5, we can define the **dimension** of a f.d. vector space V to be the length of each and every basis of V. We shall denote the dimension of V by dim V.

Note that dim $\{\mathbf{0}\} = 0$ ($\varnothing$ being the one and only basis of $\{\mathbf{0}\}$) and that the dimension of any nonzero f.d. vector space is a positive integer (since clearly any sequence of vectors spanning a nonzero vector space must have positive length).

Note also that a 1-dimensional vector space (i.e. a space whose dimension is 1) has a basis of the form $(\mathbf{e})$, where $\mathbf{e} \neq \mathbf{0}$, and thus consists of all the scalar multiples of the nonzero vector $\mathbf{e}$.

In view of examples of bases given earlier, we can tell the dimensions of certain vector spaces:

(*a*) $\dim(E_3) = 3$; (and incidentally, $\dim(E_2) = 2$, $(\mathbf{i}, \mathbf{j})$ being a basis of E_2);

(*b*) for each positive integer n, $\dim(F^n) = n$;

(*c*) for all positive integers m, n, $\dim(F_{m \times n}) = mn$.

For any matrix A, we give the names **row-rank** and **column-rank** to the dimensions of the row space and the column space, respectively, of A. In chapter 6 we shall be able to prove that the row-rank and column-rank of a matrix always coincide. Meanwhile notice that, as a consequence of 40.1,

40.6 If the matrix A is row equivalent to the echelon matrix E, then the row-rank of A is equal to the number of nonzero rows in E.

(This explains a remark in §24 that all echelon matrices row equivalent to a given matrix A have the same number of nonzero rows.)

We can also make further purposeful remarks on the earlier worked example in which we found a basis of a subspace W of $\mathbb{R}^4$ spanned by a given sequence of 4 vectors. It can now be appreciated that the work done there provides the answers to two other questions:

(1) What is $\dim W$? (The answer is 3, since the basis found for W was a sequence of 3 vectors.)

(2) Is the given sequence of 4 vectors L.I. or L.D.? (The answer is L.D. For, if the sequence were L.I., it would (since it spans W) be a basis of W—an impossibility since $\dim W = 3$.)

Thus the method deployed in the earlier worked example can be used to solve a variety of numerical problems.

Returning to generalities, we can quickly prove two simple theorems. First:

40.7 Let $(\mathbf{x}_1, \mathbf{x}_2, \ldots, \mathbf{x}_n)$ be any sequence of vectors in a vector space ($n \geqslant 1$). Then $\dim(\mathrm{sp}(\mathbf{x}_1, \mathbf{x}_2, \ldots, \mathbf{x}_n)) \leqslant n$.

Proof. Let $W = \mathrm{sp}(\mathbf{x}_1, \ldots, \mathbf{x}_n)$. Since the sequence $(\mathbf{x}_1, \ldots, \mathbf{x}_n)$ spans W, it follows from 40.2 that some subsequence (possibly the whole sequence) is a basis of W. The dimension of W is the length of this subsequence and so is $\leqslant n$. This proves the theorem.

Secondly, an informative theorem in two complementary halves:

40.8 Let V be a f.d. vector space. Then:

(i) any sequence of vectors that spans V must have length $\geqslant \dim V$;

(ii) any L.I. sequence of vectors in V must have length $\leqslant \dim V$.

Proof. (i) follows immediately from 40.7 (the case where $\mathrm{sp}(\mathbf{x}_1, \ldots, \mathbf{x}_n) = V$); and (ii) follows immediately from 40.4.

To round off this section, let us return to the theme of one of our first remarks on bases. Suppose that V is nonzero and f.d.: say $\dim V = n$ (n a positive integer). Let $L = (\mathbf{e}_1, \mathbf{e}_2, \ldots, \mathbf{e}_n)$ be some particular basis of V, and let $\mathbf{x}$ be an arbitrary vector in V. As pointed out early in the section, there is one and only one way of expressing $\mathbf{x}$ as a linear combination of the vectors in L. Suppose that this unique expression for $\mathbf{x}$ is

$$\mathbf{x} = \lambda_1 \mathbf{e}_1 + \lambda_2 \mathbf{e}_2 + \ldots + \lambda_n \mathbf{e}_n.$$

Thus we call the scalars $\lambda_1, \lambda_2, \ldots, \lambda_n$ the **components** of $\mathbf{x}$ with respect to L. Further we call the n-tuple $(\lambda_1, \lambda_2, \ldots, \lambda_n)$ (which is a member of F^n) the **component vector** of $\mathbf{x}$ with respect to L; and we call the corresponding column $\text{col}(\lambda_1, \lambda_2, \ldots, \lambda_n)$ the **component column** of $\mathbf{x}$ with respect to L.

It is clear that the idea of "components with respect to L" is a generalization of the idea of "components with respect to $\mathbf{i}, \mathbf{j}, \mathbf{k}$" introduced in §8. It is also clear that the result 8.5 generalizes in the obvious way: e.g., for $\mathbf{x}, \mathbf{y} \in V$, the component vector of $\mathbf{x} + \mathbf{y}$ with respect to L is the sum of the component vectors of $\mathbf{x}$ and of $\mathbf{y}$ with respect to L.

If L_1 and L_2 are two different bases of V, then, in general, a vector $\mathbf{x} \in V$ will have different component vectors with respect to L_1 and L_2. The question of how these two component vectors are related will be taken up in chapter 7.

In this connection one can see something of the importance of defining a basis to be a *sequence* of vectors (in a definite order), rather than just the set of vectors appearing in the sequence. It is necessary to regard a change in the order of the vectors in a basis as producing a different basis: for any such shuffling of the vectors in a basis of V will change the component vectors of most members of V. For example in E_3 the vector $\mathbf{i} + 2\mathbf{j} + 3\mathbf{k}$ has component vector $(1, 2, 3)$ with respect to the basis $(\mathbf{i}, \mathbf{j}, \mathbf{k})$, but has component vector $(2, 1, 3)$ with respect to the (different) basis $(\mathbf{j}, \mathbf{i}, \mathbf{k})$.

41. Further theorems about bases and dimension

Throughout this section, V will denote an *f.d.* vector space over F.

Certain questions naturally arise as follow-ups to the achievements of §40. In particular:

(1) Suppose that a sequence of vectors in V has the right length (i.e. $\dim V$) to be a basis of V, and suppose that we know either that the sequence spans V or that it is L.I. Must it be a basis of V?

(2) Given a sequence of vectors that spans V but is not a basis of V, we can (cf. 40.2) obtain a basis of V by deleting some of the vectors. If, however, we are given a sequence of vectors in V that is L.I. but is not a basis of V, can we necessarily obtain a basis of V by extending the given sequence?

(3) If W is a subspace of V, must W also be f.d.? (Not at all such a trivial question as it sounds!) And if so, need $\dim W$ be $\leqslant \dim V$? And, if W is not the whole of V, must $\dim W$ be $< \dim V$?

The theorems in this section will provide affirmative answers to all these questions. The following preliminary theorem opens the way by giving a new way of characterizing a basis.

41.1 Let $L = (\mathbf{e}_1, \mathbf{e}_2, \dots, \mathbf{e}_n)$ be a L.I. sequence of vectors in V. Then L is a basis of V if and only if every proper extension of L is L.D. (By a "proper extension" of L we mean a sequence of vectors in V produced from L by appending at least one further vector to it.)

Proof. (*a*) (The "$\Rightarrow$" half) Suppose that L is a basis of V.

Consider an arbitrary proper extension $L' = (\mathbf{e}_1, \dots, \mathbf{e}_n, \mathbf{f}_1, \dots, \mathbf{f}_m)$ of L, m being $\geqslant 1$. Because L is a basis of V and so spans V, $\mathbf{f}_1$ is expressible as a linear combination of $\mathbf{e}_1, \mathbf{e}_2, \dots, \mathbf{e}_n$. Hence (cf. 39.3) the sequence $(\mathbf{e}_1, \dots, \mathbf{e}_n, \mathbf{f}_1)$ is L.D.; and so therefore (cf. 39.7) is L'.

Thus every proper extension of L is L.D.

(*b*) (The "$\Leftarrow$" half) Suppose that every proper extension of L is L.D.

Let $\mathbf{x}$ be an arbitrary vector in V. Then $(\mathbf{e}_1, \mathbf{e}_2, \dots, \mathbf{e}_n, \mathbf{x})$ is a proper extension of L and so is L.D. Hence there are scalars $\alpha_1, \alpha_2, \dots, \alpha_n, \beta$, not all zero such that

$$\alpha_1\mathbf{e}_1 + \alpha_2\mathbf{e}_2 + \dots + \alpha_n\mathbf{e}_n + \beta\mathbf{x} = \mathbf{0}. \qquad (1)$$

If β were 0, the equation (1) would provide a contradiction by showing that $(\mathbf{e}_1, \dots, \mathbf{e}_n)$ is L.D. Hence $\beta \neq 0$, and so we can solve for $\mathbf{x}$ in (1) to obtain:

$$\mathbf{x} = \sum_{i=1}^{n} (-\alpha_i/\beta)\mathbf{e}_i.$$

Since $\mathbf{x}$ was arbitrary in V, this proves that L spans V. So, since L is L.I., L is a basis of V.

The stated result is now established.

The next theorem answers the first of the questions in the preamble.

41.2 Let L be a finite sequence of vectors in V. We may conclude that L is a basis of V if we know it possesses any two of the following properties:

(1) it spans V; (2) it is L.I.; (3) its length is dim V.

Proof. If L has properties (1) and (2), then it is a basis of V (by the definition of "basis"). It remains to prove that L is a basis of V (*a*) when L has properties (1) and (3), (*b*) when L has properties (2) and (3).

(*a*) Suppose that L has properties (1) and (3). Since L spans V, it follows from 40.2 that some subsequence L' of L (perhaps the whole of L) is a basis of V. But, in common with all bases of V, this subsequence must have length dim V, which is the length of L. Hence the subsequence L' must coincide with L, and thus L is a basis of V.

(*b*) Suppose that L has properties (2) and (3). By 41.1, if L were not a basis, there would exist a L.I. proper extension of L. But such a proper extension of L would be a L.I. sequence of vectors in V with length greater than dim V—an impossibility, by 40.8. Therefore L must be a basis of V.

This completes the proof of 41.2.

Next we answer question (2) from the preamble.

41.3 Let L be a L.I. sequence of vectors in V. Then either L is a basis of V or else some proper extension of L is a basis of V.

(It is helpful to simplify the conclusion to the statement "L can be extended to a basis of V", where it is to be understood that in the case where L itself is a basis, the extension will be the "trivial extension" in which no vectors are appended to L, i.e. L is left alone.)

Proof. Suppose that L is not a basis of V. (It will suffice to prove that in this case some proper extension of L is a basis of V.)

We use 41.1, which tells us that if any L.I. sequence of vectors in V is not a basis, that sequence can be extended to produce a (strictly) longer L.I. sequence of vectors in V. Accordingly, we can extend the sequence L to a longer L.I. sequence L' of vectors in V; and if L' is not a basis of V, we can extend it to a still longer L.I. sequence L'' of vectors in V. This process can be repeated until such time as a basis of V results—which must happen after a finite number of steps, since otherwise (as each step increases the length of the sequence by at least 1) we would eventually produce a L.I. sequence of vectors in V with length greater than dim V—a contradiction of 40.8. So there does exist an extension of L that is a basis of V.

The stated result follows.

As a very simple numerical illustration of 41.3, let's start with the L.I. sequence $((1,2,3,4),(0,0,5,6))$ of vectors in $\mathbb{R}^4$, and set ourselves the task of extending this sequence to a basis of $\mathbb{R}^4$.

As is usually helpful in such numerical problems, we interpret vectors in $\mathbb{R}^4$ as row matrices; and we solve this particular problem by writing down an echelon matrix whose rows include the given vectors, viz. $\begin{bmatrix} 1 & 2 & 3 & 4 \\ 0 & 1 & 0 & 0 \\ 0 & 0 & 5 & 6 \\ 0 & 0 & 0 & 1 \end{bmatrix}$. By 39.11, the rows of this matrix, written down in any order whatever (cf. 39.4) form a L.I. sequence of 4 vectors in $\mathbb{R}_{1\times 4}$ (which we here identify with $\mathbb{R}^4$); and since $\dim(\mathbb{R}^4) = 4$, it follows by 41.2 that this sequence is a basis of $\mathbb{R}^4$. Thus the sequence

$$((1,2,3,4),(0,1,0,0),(0,0,5,6),(0,0,0,1))$$

is a basis of $\mathbb{R}^4$ which is an extension of the given L.I. sequence.

Now comes a theorem answering all parts of question (3) of the preamble.

41.4 Let W be a subspace of the f.d. V. Then:

(i) W is f.d.;

(ii) $\dim W \leqslant \dim V$;

(iii) if $W \neq V$, then $\dim W < \dim V$;

(iv) any basis of W can be extended to produce a basis of V (the extension being trivial, i.e. no vectors appended, in the case $W = V$).

Proof. Any L.I. sequence of vectors in W can, of course, also be described as a L.I. sequence of vectors in V (since $W \subseteq V$). Any such sequence, therefore, must have length $\leqslant \dim V$ (by 40.8).

Let m ($\geqslant 0$) be the greatest possible length for a L.I. sequence of vectors in W, and let L be a L.I. sequence of vectors in W with this maximal length m. By the observation made in the previous paragraph, $m \leqslant \dim V$.

Every sequence of vectors in W that is a proper extension of L has length $> m$ and so, by the maximality of m, must be L.D. Therefore, by 41.1, L is a basis of W.

The existence of this basis of W shows that W is f.d. Moreover, since $\dim W$ = length of $L = m$ and since $m \leqslant \dim V$, we have $\dim W \leqslant \dim V$.

Parts (i) and (ii) are now proved.

To deal with part (iii), suppose that $W \neq V$. We know that $\dim W \leqslant \dim V$. With a view to obtaining a contradiction, suppose that $\dim W = \dim V$. Then the length of L equals $\dim V$; and so, since L is a L.I. sequence of vectors in V, it follows by 41.2 that L is a basis of V. Therefore, every vector in V is equal to some linear combination of the vectors in L, all of which belong to the subspace W. Hence (cf. 37.2) every vector in V belongs to W. Thus $V \subseteq W$ and so (since W is a subspace of V) $W = V$—a contradiction. Hence, in fact, $\dim W < \dim V$. This proves part (iii).

Finally, part (iv) is clear from 41.3: any basis of W is a L.I. sequence of vectors in V and so can be extended to a basis of V.

The last part of 41.4 is often useful, as will be seen, for example, in the proof of a major theorem in §42. It comes in when we want to arrange for a basis of a subspace to comprise the first few vectors in a basis of the whole space. We achieve this by *starting with a basis of the subspace* and invoking 41.4(iv) to extend it to a basis of the whole space.

A corollary of 41.4 is:

41.5 Suppose that S, T are f.d. subspaces in some vector space such that (1) $S \subseteq T$ and (2) $\dim S = \dim T$. Then $S = T$.

Proof. If $S \neq T$, then (by 41.4(iii) applied with $V = T$, $W = S$) $\dim S < \dim T$—a contradiction. Therefore $S = T$.

The result 41.5 provides a very useful method of proving the equality of two f.d. subspaces in a vector space: we need only prove that one subspace is contained in the other and that their dimensions are equal.

42. Sums of subspaces

Throughout this section V will denote a vector space (not necessarily f.d.) over F.

Suppose that S and T are subspaces of V.

Two similar questions that one might ask are:

(1) Is there a "largest" subspace (of V) contained in both S and T (i.e. a subspace contained in both S and T that contains all such subspaces)? And, if so, what is that "largest" subspace?

(2) Is there a "smallest" subspace (of V) containing both S and T (i.e. a subspace containing both S and T that is contained in all such subspaces)? And, if so, what is that "smallest" subspace?

The answer to (1) is clearly "Yes: $S \cap T$". For $S \cap T$ (a subspace by 37.6) is the largest *set* contained in both S and T. But (2) is not so easy to answer since $S \cup T$ (which is the smallest set containing both S and T) is in most cases not a subspace (a theme taken up in the exercises at the end of the chapter). The correct answer to (2) is "Yes: the subspace $S+T$ which we are about to define and consider", and the detailed justification of this claim will be given in the proposition 42.1 below.

The Venn diagram of figure XVII may be found helpful as we proceed into the discussion of "$S+T$".

The definition of $S+T$ (the **sum** of the subspaces S, T) is that $S+T$ is the subset

$$\{\mathbf{x}+\mathbf{y} : \mathbf{x} \in S, \mathbf{y} \in T\}$$

of V, i.e. the set of all vectors in V expressible in the form

$$(\text{a vector in } S)+(\text{a vector in } T).$$

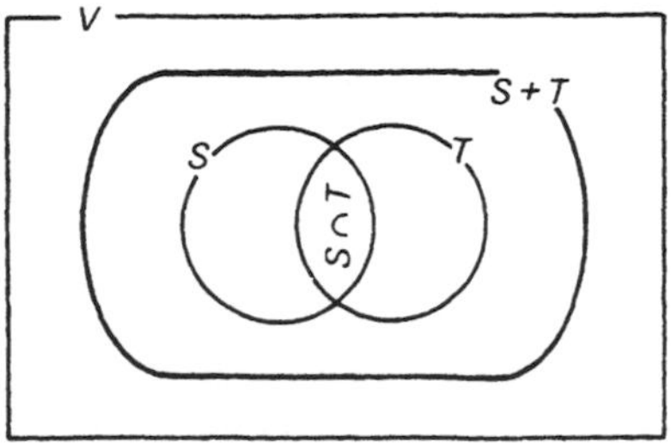

Figure XVII

As earlier promised, the following theorem justifies the claims made about $S+T$ in the preamble.

42.1 Let S, T be subspaces of V. Then:

(i) $S+T$ is also a subspace of V;

(ii) $S+T$ contains both S and T as subsets;

(iii) $S+T$ is (in the sense indicated in the preamble) the "smallest" subspace of V that contains both S and T.

Proof. (i) (1) Since S and T are subspaces, $\mathbf{0} \in S$ and $\mathbf{0} \in T$ (cf. 37.3). Hence, by the definition of $S+T$, $\mathbf{0}+\mathbf{0} \in S+T$ (i.e. $\mathbf{0} \in S+T$); and thus $S+T \neq \varnothing$.

(2) Let $\mathbf{x}, \mathbf{y} \in S+T$, and let $\lambda, \mu \in F$. Then

$$\mathbf{x} = \mathbf{u}_1+\mathbf{v}_1 \quad \text{and} \quad \mathbf{y} = \mathbf{u}_2+\mathbf{v}_2$$

for some $\mathbf{u}_1, \mathbf{u}_2 \in S$ and some $\mathbf{v}_1, \mathbf{v}_2 \in T$. Hence we have

$$\lambda\mathbf{x}+\mu\mathbf{y} = (\lambda\mathbf{u}_1+\mu\mathbf{u}_2)+(\lambda\mathbf{v}_1+\mu\mathbf{v}_2).$$

But, since S is a subspace, $\lambda\mathbf{u}_1+\mu\mathbf{u}_2 \in S$ (cf. 37.2); and, similarly, $\lambda\mathbf{v}_1+\mu\mathbf{v}_2 \in T$. Therefore $\lambda\mathbf{x}+\mu\mathbf{y} \in S+T$.

It now follows, by 37.1, that $S+T$ is a subspace of V.

(ii) If $\mathbf{x}$ is any vector in S, then, from the fact that $\mathbf{x} = \mathbf{x}+\mathbf{0}$ and $\mathbf{0} \in T$, it is apparent that $\mathbf{x} \in S+T$. Therefore $S \subseteq S+T$. Similarly, $T \subseteq S+T$. So $S+T$ contains both S and T.

(iii) Let W be any subspace of V containing S and T. By its closure under addition, W must contain every vector of the form $\mathbf{x}+\mathbf{y}$ with $\mathbf{x} \in S$ $(\subseteq W)$ and $\mathbf{y} \in T$ $(\subseteq W)$. Thus W contains every vector in $S+T$; i.e. $S+T \subseteq W$.

This proves that, as stated, $S+T$ is the "smallest" subspace of V that contains S and T.

In the light of 42.1(iii), it is obvious that:

42.2 (i) For any subspace S of V, $S+S = S$.

(ii) If S, T are subspaces of V such that $S \subseteq T$, then $S+T = T$.

It is not difficult to show that the operation of forming the sum of subspaces is associative and commutative. The essence of this statement is the fact that if $S_1, S_2, \ldots, S_n$ are subspaces of V, then

$$S_1+S_2+\ldots+S_n$$

is, irrespective of any bracketing that might be inserted and irrespective of the order of the summands, the set of all vectors in V expressible as

$$(\text{a vector in } S_1)+(\text{a vector in } S_2)+\ldots+(\text{a vector in } S_n).$$

We shall consider next the dimension of $S+T$ in the case where S, T are f.d. subspaces of V. If we introduce bases $(\mathbf{e}_1, \ldots, \mathbf{e}_m)$ and $(\mathbf{f}_1, \ldots, \mathbf{f}_n)$ of S and T, respectively, (m being dim S and n being dim T), then it is easy to show that the sequence $(\mathbf{e}_1, \ldots, \mathbf{e}_m, \mathbf{f}_1, \ldots, \mathbf{f}_n)$ spans $S+T$. Therefore $S+T$ is f.d. Moreover (by 40.8(i)) dim $(S+T)$ cannot exceed the length of this sequence, and so dim $(S+T) \leqslant m+n$; i.e.

$$\dim(S+T) \leqslant \dim S + \dim T.$$

A more sophisticated argument gives us the following more refined result.

42.3 Suppose that S, T are f.d. subspaces of V. Then

$$\dim(S+T) = \dim S + \dim T - \dim(S \cap T).$$

Proof. Let $m = \dim S, n = \dim T, p = \dim(S \cap T)$. (Of course, $S \cap T$ is also f.d., by 41.4(i)).

We start with a basis $(\mathbf{e}_1, \ldots, \mathbf{e}_p)$ of $S \cap T$. By 41.4(iv), this can be extended to a basis $(\mathbf{e}_1, \ldots, \mathbf{e}_p, \mathbf{f}_{p+1}, \ldots, \mathbf{f}_m)$ of S. Likewise, it can be extended to a basis $(\mathbf{e}_1, \ldots, \mathbf{e}_p, \mathbf{g}_{p+1}, \ldots, \mathbf{g}_n)$ of T. Let L be the sequence of vectors

$$(\mathbf{e}_1, \ldots, \mathbf{e}_p, \mathbf{f}_{p+1}, \ldots, \mathbf{f}_m, \mathbf{g}_{p+1}, \ldots, \mathbf{g}_n),$$

a sequence of length $m+n-p$. All the vectors in the sequence L are in $S+T$ (in view of 42.1(ii)). We shall prove that L is a basis of $S+T$ by showing that (1) L spans $S+T$ and (2) L is L.I.

(*Note*: There are various degenerate cases which we shall not digress to cover separately here—the cases where the number of **e**s or **f**s or **g**s is zero. It is left as an exercise for the student to convince himself that, in each such case, either the result is trivial or it cán be proved by an obvious adaptation of the present argument.)

(1) Let $\mathbf{x}$ be an arbitrary vector in $S+T$. Then

$$\begin{aligned} \mathbf{x} &= (\text{some vector in } S) + (\text{some vector in } T) \\ &= (\text{a linear combination of the } \mathbf{e}\text{s and } \mathbf{f}\text{s}) \\ &\quad + (\text{a linear combination of the } \mathbf{e}\text{s and } \mathbf{g}\text{s}) \\ &= \text{a linear combination of the vectors in } L. \end{aligned}$$

Thus L spans $S+T$.

(2) Suppose that

$$\sum_{i=1}^{p} \alpha_i \mathbf{e}_i + \sum_{j=p+1}^{m} \beta_j \mathbf{f}_j + \sum_{k=p+1}^{n} \gamma_k \mathbf{g}_k = \mathbf{0}, \qquad (*)$$

all the coefficients $\alpha_i, \beta_j, \gamma_k$ being scalars. Then we have

$$\sum_{i=1}^{p} \alpha_i \mathbf{e}_i + \sum_{j=p+1}^{m} \beta_j \mathbf{f}_j = -\sum_{k=p+1}^{n} \gamma_k \mathbf{g}_k = \mathbf{z}, \text{ say.}$$

This last line presents two expressions for the vector **z**, the first as a linear combination of the **e**s and **f**s, the second as a linear combination of the **g**s. Since the **e**s and **f**s all belong to the subspace S, the first expression shows (cf. 37.2) that $\mathbf{z} \in S$; and, since the **g**s all belong to the subspace T, the second expression shows that $\mathbf{z} \in T$. Therefore $\mathbf{z} \in S \cap T$, so that **z** must be expressible as a linear combination of the **e**s ($(\mathbf{e}_1, \dots, \mathbf{e}_p)$ being a basis of $S \cap T$). Hence, using the second of the above expressions for **z**, we have

$$(\mathbf{z}=) - \sum_{k=p+1}^{n} \gamma_k \mathbf{g}_k = \sum_{l=1}^{p} \delta_l \mathbf{e}_l$$

for some scalars $\delta_1, \dots, \delta_p$; and thus

$$\delta_1 \mathbf{e}_1 + \dots + \delta_p \mathbf{e}_p + \gamma_{p+1} \mathbf{g}_{p+1} + \dots + \gamma_n \mathbf{g}_n = \mathbf{0}.$$

Since $(\mathbf{e}_1, \dots, \mathbf{e}_p, \mathbf{g}_{p+1}, \dots, \mathbf{g}_n)$ is a basis of T and so is L.I., all the scalar coefficients in the last equation must be 0. In particular,

$$\gamma_{p+1} = \dots = \gamma_n = 0.$$

Feeding this information into the equation (*), we obtain

$$\alpha_1 \mathbf{e}_1 + \dots + \alpha_p \mathbf{e}_p + \beta_{p+1} \mathbf{f}_{p+1} + \dots + \beta_m \mathbf{f}_m = \mathbf{0}.$$

Hence, since $(\mathbf{e}_1, \dots, \mathbf{e}_p, \mathbf{f}_{p+1}, \dots, \mathbf{f}_m)$ is a basis of S and is therefore L.I., $\alpha_1 = \dots = \alpha_p = \beta_{p+1} = \dots = \beta_m = 0$.

It has now been proved that the equation (*) implies that all the coefficients $\alpha_i, \beta_j, \gamma_k$ are zero. Therefore L is L.I.

We can now conclude that L is a basis of $S+T$; and hence

$$\dim(S+T) = \text{length of } L = m+n-p;$$

i.e. $\dim(S+T) = \dim S + \dim T - \dim(S \cap T)$, as required.

Worked example. In F^{10}, S and T are two 6-dimensional subspaces. Prove that $\dim(S \cap T) \geqslant 2$.

Solution. $\dim(S \cap T) = \dim S + \dim T - \dim(S+T)$ (by 42.3)

$$= 6+6-\dim(S+T) = 12 - \dim(S+T).$$

But, since $S+T$ is a subspace of F^{10} (which has dimension 10), $\dim(S+T) \leqslant 10$. Hence $\dim(S \cap T) \geqslant 12-10 = 2$.

This example is, in a sense, a good advertisement for the idea of the sum of two subspaces: for, though the problem posed does not mention the sum of S and T, it is consideration of $S+T$ that is the key to solving the problem easily.

43. Direct sums of subspaces

As in §42, V will in this section denote a vector space over F.

Let $S_1, S_2, \ldots, S_n$ be subspaces of V. Then each vector in the sum $S_1+S_2+\ldots+S_n$ can be expressed in at least one way in the form

$$(\text{a vector in } S_1)+(\text{a vector in } S_2)+\ldots+(\text{a vector in } S_n).$$

In most cases one can expect there to be several ways of so expressing each vector in $S_1+S_2+\ldots+S_n$.

If (and only if) *each* vector in $S_1+S_2+\ldots+S_n$ can be expressed in *just one* way in the form

$$(\text{a vector in } S_1)+(\text{a vector in } S_2)+\ldots+(\text{a vector in } S_n),$$

we describe the sum of subspaces $S_1+S_2+\ldots+S_n$ as a **direct sum**; and we indicate this by writing the sum as

$$S_1 \oplus S_2 \oplus \ldots \oplus S_n.$$

For example, in E_3 each vector can be expressed *precisely one way* in the form

$$(\text{multiple of } \mathbf{i})+(\text{multiple of } \mathbf{j})+(\text{multiple of } \mathbf{k}).$$

Therefore (since "multiple of $\mathbf{i}$" is synonymous with "vector in sp($\mathbf{i}$)", etc.),

$$E_3 = \text{sp}(\mathbf{i}) \oplus \text{sp}(\mathbf{j}) \oplus \text{sp}(\mathbf{k}).$$

The easily seen generalization of this illustration is:

43.1 If V is f.d. and $(\mathbf{e}_1, \mathbf{e}_2, \ldots, \mathbf{e}_n)$ is a basis of V, then

$$V = \text{sp}(\mathbf{e}_1) \oplus \text{sp}(\mathbf{e}_2) \oplus \ldots \oplus \text{sp}(\mathbf{e}_n).$$

Remarks. (*a*) In more advanced work it is often helpful to have a given vector space expressed as the direct sum of two or more well understood subspaces.

(*b*) It should be noted that the statement

$$W = S_1 \oplus S_2 \oplus \ldots \oplus S_n$$

is essentially a two-fold statement. It asserts (1) that W is equal to the sum of the subspaces $S_1, S_2, \ldots, S_n$ *and* (2) that the sum $S_1+S_2+\ldots+S_n$ is a direct sum.

(*c*) When $S_1, S_2, \ldots, S_n$ denote subspaces of V, the combination of symbols "$S_1+S_2+\ldots+S_n$" always makes sense: it denotes a certain subspace of V as detailed in §42. But the combination of symbols "$S_1 \oplus S_2 \oplus \ldots \oplus S_n$" may or may not make sense: it makes sense only if the sum $S_1+S_2+\ldots+S_n$ actually is a direct sum, and whether that is true or not depends on what exactly

$S_1, S_2, \ldots, S_n$ are. Thus one has something to prove before one can write "$S_1 \oplus S_2 \oplus \ldots \oplus S_n$".

(*d*) The criterion for a sum of subspaces to be direct can, as we shall now see, be simplified to consideration of the zero vector.

43.2 Let $S_1, S_2, \ldots, S_n$ be subspaces of V. Suppose that the only way to express $\mathbf{0}$ in the form

$$\mathbf{u}_1 + \mathbf{u}_2 + \ldots + \mathbf{u}_n \qquad \text{with } \mathbf{u}_i \in S_i \text{ for each } i$$

is to take every $\mathbf{u}_i$ equal to $\mathbf{0}$. Then the sum $S_1 + S_2 + \ldots + S_n$ is a direct sum.

Proof. Let $\mathbf{x}$ be an arbitrary vector in $S_1 + \ldots + S_n$. Suppose that

$$\mathbf{x} = \mathbf{u}_1 + \mathbf{u}_2 + \ldots + \mathbf{u}_n = \mathbf{v}_1 + \mathbf{v}_2 + \ldots + \mathbf{v}_n,$$

where, for each i, $\mathbf{u}_i \in S_i$ and $\mathbf{v}_i \in S_i$. We shall show that these two expressions for $\mathbf{x}$ must be identical. Subtracting the second expression from the first, we obtain

$$\mathbf{0} = (\mathbf{u}_1 - \mathbf{v}_1) + (\mathbf{u}_2 - \mathbf{v}_2) + \ldots + (\mathbf{u}_n - \mathbf{v}_n),$$

which (in view of 37.2) expresses $\mathbf{0}$ in the form

$$(\text{a vector in } S_1) + (\text{a vector in } S_2) + \ldots + (\text{a vector in } S_n).$$

Hence, by the hypothesis that there is only one way to express $\mathbf{0}$ in this form, $\mathbf{u}_i - \mathbf{v}_i = \mathbf{0}$ for each i; i.e. $\mathbf{u}_i = \mathbf{v}_i$ for each i. So our two expressions for $\mathbf{x}$ are indeed identical.

This proves the unique expressibility, in the form in question, of each $\mathbf{x} \in S_1 + S_2 + \ldots + S_n$. Hence this sum of subspaces is a direct sum, as claimed.

In the special case of the sum of two subspaces, there is a very simple criterion, given in the next result, for the sum to be a direct sum. Beware of imagining that naïve generalizations of this result are true for sums of more than two subspaces: they are not!

43.3 Let S, T be two subspaces of V. The sum $S + T$ is a direct sum if and only if $S \cap T = \{\mathbf{0}\}$.

Proof. (*a*) (The "$\Rightarrow$" half) Suppose that $S + T$ is a direct sum.

Let $\mathbf{x}$ be an arbitrary vector in $S \cap T$. Then $\mathbf{x} \in S$, and (since T is a subspace) it is also clear that $-\mathbf{x} \in T$. Hence the equation $\mathbf{0} = \mathbf{x} + (-\mathbf{x})$ expresses $\mathbf{0}$ as the sum of a vector in S and a vector in T. So, of course, does the equation $\mathbf{0} = \mathbf{0} + \mathbf{0}$. But, the sum $S + T$ being direct, there can only be one way of expressing $\mathbf{0}$ as the sum of a vector in S and a vector in T. Therefore $\mathbf{x}$ (and $-\mathbf{x}$) must be $\mathbf{0}$.

This proves that no vector other than $\mathbf{0}$ can belong to $S \cap T$. So, since $S \cap T$ (being a subspace) certainly contains $\mathbf{0}$ (cf. 37.6 and 37.3), it follows that $S \cap T = \{\mathbf{0}\}$.

(*b*) (The "$\Leftarrow$" half) Suppose that $S \cap T = \{\mathbf{0}\}$.

Suppose further that $\mathbf{0} = \mathbf{u}+\mathbf{v}$, with $\mathbf{u} \in S$ and $\mathbf{v} \in T$. Then $\mathbf{u} = -\mathbf{v}$; and hence (since $\mathbf{v} \in T$ and T is a subspace) $\mathbf{u} \in T$. It follows that $\mathbf{u} \in S \cap T$, and hence ($S \cap T$ being $\{\mathbf{0}\}$) $\mathbf{u} = \mathbf{0}$. Hence also $\mathbf{v}$ $(= -\mathbf{u}) = \mathbf{0}$.

This proves that the only way to express $\mathbf{0}$ as the sum of a vector in S and a vector in T is to take both these vectors equal to $\mathbf{0}$. By 43.2, it follows that the sum $S+T$ is a direct sum.

Parts (*a*) and (*b*) together establish the stated result.

Note. Part (*a*) of the above proof illustrates a standard procedure for proving that a subspace (W, say) is zero: since certainly $\mathbf{0} \in W$, the conclusion $W = \{\mathbf{0}\}$ follows when one shows that an arbitrary vector in W must be equal to $\mathbf{0}$.

Consider now the case where the sum $S+T$ of the subspaces S, T is a direct sum, so that $S \cap T = \{\mathbf{0}\}$, by 43.3, and where both S and T are f.d. Let $(\mathbf{f}_1, \ldots, \mathbf{f}_m)$ and $(\mathbf{g}_1, \ldots, \mathbf{g}_n)$ be bases of S and T, respectively, and let $L = (\mathbf{f}_1, \ldots, \mathbf{f}_m, \mathbf{g}_1, \ldots, \mathbf{g}_n)$. This leads us to precisely the situation of the proof of 42.3 in the degenerate case $p = 0$. A suitably modified version of the proof (simpler than the typical case) is valid for the case $p = 0$; and so the above sequence L is a basis of $S \oplus T$, and $\dim(S \oplus T)$ is $m+n$. We thus prove the following theorem:

43.4 If the sum of the f.d. subspaces S, T is a direct sum and if $(\mathbf{f}_1, \ldots, \mathbf{f}_m)$ and $(\mathbf{g}_1, \ldots, \mathbf{g}_n)$ are bases of S and T, respectively, then:

(i) $(\mathbf{f}_1, \ldots, \mathbf{f}_m, \mathbf{g}_1, \ldots, \mathbf{g}_n)$ is a basis of $S \oplus T$;

(ii) $\dim(S \oplus T) = \dim S + \dim T$.

This theorem, unlike 43.3, does generalize in the obvious manner to direct sums of more than two f.d. subspaces.

Worked example. Let S, T be subspaces of the f.d. vector space V. Show that $V = S \oplus T$ if any two of the following conditions hold: (1) $V = S+T$, (2) $S \cap T = \{\mathbf{0}\}$, (3) $\dim V = \dim S + \dim T$.

Solution. (*a*) If (1) and (2) hold, then, by 43.3, the sum $S+T$ is a direct sum, and so it follows from (1) that $V = S \oplus T$.

(*b*) Suppose that (1) and (3) hold. Then

$$\begin{aligned} \dim S + \dim T &= \dim V = \dim(S+T) \qquad \text{(by (3) and (1))} \\ &= \dim S + \dim T - \dim(S \cap T) \qquad \text{(by 42.3).} \end{aligned}$$

Hence $\dim(S \cap T) = 0$, and so $S \cap T = \{\mathbf{0}\}$. Thus (2) also holds, and hence (by part (*a*)) $V = S \oplus T$.

(*c*) Suppose that (2) and (3) hold. By (2) and 43.3, $S+T$ can be written as $S \oplus T$ (i.e. it is a direct sum). So we have

$$\begin{aligned} \dim(S \oplus T) &= \dim S + \dim T \qquad \text{(by 43.4)} \\ &= \dim V \qquad \text{(by condition (3))}. \end{aligned}$$

Since $S \oplus T$ is a subspace of V, it follows by 41.5 that $V = S \oplus T$.

All parts of the set task are now accomplished.

Part (*c*).may be considered especially instructive because it provides a first illustration of the use of 41.5 to establish the equality of two subspaces.

EXERCISES ON CHAPTER FIVE

(In view of the length of chapter 5, the student is advised that in this set of exercises:
(i) numbers 1 to 13 are based on the material up to the end of §39;
(ii) only those from number 29 onwards require knowledge of §§42 and 43.)

1. Prove that, in $\mathbb{R}_{n\times n}$, each of the following subsets is a subspace: (*a*) the set of symmetric matrices; (*b*) the set of skew-symmetric matrices; (*c*) the set of diagonal matrices.

2. Which of the following are subspaces of $\mathbb{R}^3$? Justify your answers.
(i) $\{(x_1, x_2, x_3): x_1+2x_2+3x_3 = 0\}$.
(ii) $\{(x_1, x_2, x_3): x_1+2x_2+3x_3 = 6\}$.
(iii) $\{(x_1, x_2, x_3): x_1+2x_2+3x_3 \leqslant 0\}$.

3. Let A be a given matrix in $\mathbb{R}_{n\times n}$, and let

$$C(A) = \{X \in \mathbb{R}_{n\times n}: AX = XA\}.$$

Prove that $C(A)$ is a subspace of $\mathbb{R}_{n\times n}$.

4.* Let S, T be subspaces of a vector space such that neither $S \subseteq T$ nor $T \subseteq S$ is true. Prove that $S \cup T$ is not a subspace. (*Hint*: with a view to obtaining a contradiction, suppose that $S \cup T$ is a subspace; introduce vectors $\mathbf{x}, \mathbf{y}$ such that $\mathbf{x} \in S$ while $\mathbf{x} \notin T$ and $\mathbf{y} \in T$ while $\mathbf{y} \notin S$; consider $\mathbf{x}+\mathbf{y}$.)

5. Let $\mathbf{x}, \mathbf{y}, \mathbf{z}$ be vectors in a vector space. Suppose that $\alpha\mathbf{x}+\beta\mathbf{y}+\gamma\mathbf{z} = \mathbf{0}$, where α, β, γ are scalars with β and γ both nonzero. Prove that $\text{sp}(\mathbf{x}, \mathbf{y}) = \text{sp}(\mathbf{x}, \mathbf{z})$.

6. In $\mathbb{R}^{\mathbb{R}}$ (the vector space of all real-valued functions with domain $\mathbb{R}$) show that $\sin^4 x \in \text{sp}(1, \cos 2x, \cos 4x)$.

7. Consider a system of linear equations $AX = K$, where the notational details are as usual. Show that the system is consistent if and only if K belongs to the column space of A.

8.* Let W be the vector space of all formal polynomials in an indeterminate t with coefficients in the field F. Prove that W is infinite-dimensional. (*Hint*: suppose otherwise, so that some finite sequence of polynomials $(f_1, \ldots, f_n)$ spans W; consider the maximum of the degrees of $f_1, \ldots, f_n$ and the degrees of linear combinations of $f_1, \ldots, f_n$.)

9. Let $\mathbf{w}, \mathbf{x}, \mathbf{y}, \mathbf{z}$ denote vectors in a vector space over $\mathbb{R}$. Prove that:
(i) the sequence $(\mathbf{w}+\mathbf{x}, \mathbf{x}+\mathbf{y}, \mathbf{y}+\mathbf{z}, \mathbf{z}+\mathbf{w})$ is L.D.;
(ii) if the sequence $(\mathbf{w}, \mathbf{x}, \mathbf{y}, \mathbf{z})$ is L.I., then so is the sequence $(\mathbf{w}+\mathbf{x}+\mathbf{y}, \mathbf{x}+\mathbf{y}+\mathbf{z}, \mathbf{w}+2\mathbf{z})$.

10. Let V be a vector space over $\mathbb{R}$. Suppose that $(\mathbf{x}, \mathbf{y}, \mathbf{z})$ is a L.I. sequence of vectors in V. Let $\mathbf{u} = \mathbf{x}-\mathbf{y}, \mathbf{v} = \mathbf{y}-\mathbf{z}, \mathbf{w} = \mathbf{z}+\alpha\mathbf{x}$, where α is a scalar. Prove that the sequence $(\mathbf{u}, \mathbf{v}, \mathbf{w})$ is L.D. if and only if $\alpha = -1$.

11. Write down three sequences of three vectors in E_3 according to the following specification. Each sequence must be L.D.; in the first sequence all of the three subsequences of length 2 are to be L.I.; in the second sequence there are to be precisely two L.I. subsequences of length 2; in the third sequence there is to be precisely one L.I. subsequence of length 2.

12. In a vector space, the sequence of vectors $(\mathbf{x}_1, \mathbf{x}_2, \ldots, \mathbf{x}_n)$ is given to be L.I.; and $\mathbf{y} = \sum_{i=1}^{n} \alpha_i \mathbf{x}_i$, where $\alpha_1, \alpha_2, \ldots, \alpha_n$ are scalars with $\alpha_1 \neq 0$. Prove that the sequence $(\mathbf{y}, \mathbf{x}_2, \mathbf{x}_3, \ldots, \mathbf{x}_n)$ is also L.I.

13. Suppose that the sequence of columns $(X_1, X_2, \ldots, X_m)$ is L.I. in $F_{n\times 1}$ and that A is a nonsingular matrix in $F_{n\times n}$. Prove that the sequence of columns $(AX_1, AX_2, \ldots, AX_m)$ is also L.I. in $F_{n\times 1}$.

14. Find a basis of, and the dimension of, the subspace

$$\{(x_1, x_2, x_3, x_4) : x_1 + x_2 + x_3 + x_4 = 0\}$$

of F^4.

15. Working over $\mathbb{R}$, find a basis of the solution set [solution space, we may call it, in the light of 37.5] of the homogeneous system of equations

$$\begin{Bmatrix} x_1 + x_2 - x_3 + x_4 = 0 \\ 2x_1 - x_2 \qquad + 2x_4 = 0 \end{Bmatrix}.$$

16. In $\mathbb{R}^4$, let $\mathbf{w} = (1, -1, 0, 3)$, $\mathbf{x} = (2, 1, 1, -1)$, $\mathbf{y} = (4, -1, 1, 3)$, $\mathbf{z} = (1, -4, -1, 8)$, and let $S = \operatorname{sp}(\mathbf{w}, \mathbf{x}, \mathbf{y}, \mathbf{z})$. By performing a suitable echelon reduction, find a basis of S. What is dim S? Is the sequence $(\mathbf{w}, \mathbf{x}, \mathbf{y}, \mathbf{z})$ L.I. or L.D.? Extend the basis of S that you found to a basis of $\mathbb{R}^4$. Find also a basis of $S \cap T$, where T is the subspace

$$\{(x_1, x_2, x_3, x_4) : x_1 + x_2 + x_3 + x_4 = 0\}$$

of $\mathbb{R}^4$.

17. In $\mathbb{R}_{2\times 2}$, let $A = \begin{bmatrix} 1 & 0 \\ 1 & 2 \end{bmatrix}$, and let $C(A)$ be as defined in question 3. Find a basis of $C(A)$.

(*) Prove that, for *every* $A \in \mathbb{R}_{n\times n}$ where $n \geqslant 2$, $\dim(C(A)) \geqslant 2$.

18. Determine whether or not the sequence of vectors

$$((1, -1, 0, 2), (2, 1, -1, 1), (0, 1, 3, -1), (3, 2, 5, 1))$$

spans $\mathbb{R}^4$.

19. In $\mathbb{R}_{2\times 2}$, find bases of the subspaces of symmetric matrices, skew-symmetric matrices, and diagonal matrices (cf. question 1). State the dimension of each of these subspaces. Generalize to $\mathbb{R}_{n\times n}$.

20.* Suppose that A is a matrix in $F_{n\times n}$ such that $A^p = O$ but $A^{p-1} \neq O$, where p is an integer greater than 1. Show that there exists a column $X \in F_{n\times 1}$ such that $A^{p-1}X \neq O$ and that, for any such column X, the sequence $(X, AX, A^2X, \ldots, A^{p-1}X)$ is L.I. in $F_{n\times 1}$. Deduce that $p \leqslant n$.

21. Find, for each possible value of the scalar a, the row-rank of the matrix

$$\begin{bmatrix} 1 & 1 & 2 & 0 \\ 2 & a+1 & 3 & a-1 \\ -3 & a-2 & a-5 & a+1 \\ a+2 & 2 & a+4 & -2a \end{bmatrix}.$$

(See 40.6.) Note that one must avoid division by any scalar that might be zero.

22. Suppose that V is a vector space in which there is a spanning sequence of length n, but no shorter spanning sequence. Prove that dim $V = n$.

23. In E_3, find the component column of the vector $2\mathbf{i} + 3\mathbf{j} - \mathbf{k}$ with respect to the basis $(\mathbf{i} + \mathbf{j} + \mathbf{k}, \mathbf{j} + \mathbf{k}, \mathbf{k})$.

24. Using 40.6 and 26.2, prove that an $n \times n$ matrix is nonsingular if and only if its row-rank is n.

25. In a vector space $V, \mathbf{x}_1, \mathbf{x}_2, \ldots, \mathbf{x}_n$ are n vectors; and $\mathbf{y}_1, \mathbf{y}_2, \ldots, \mathbf{y}_m$ all belong to $\mathrm{sp}(\mathbf{x}_1, \mathbf{x}_2, \ldots, \mathbf{x}_n)$. Prove that if $m > n$, then (whether or not the sequence $(\mathbf{x}_1, \ldots, \mathbf{x}_n)$ is L.I.) the sequence $(\mathbf{y}_1, \mathbf{y}_2, \ldots, \mathbf{y}_m)$ is L.D.

26. Use 41.4 to prove that F^{∞} is infinite-dimensional.

27. Let S be a subspace of a vector space V, and let $\mathbf{x}$ be a vector in V that does not belong to S. Prove that $S \cap \mathrm{sp}(\mathbf{x}) = \{\mathbf{0}\}$.

28. In a vector space $V, \mathbf{x}_1, \mathbf{x}_2, \mathbf{x}_3$ are vectors such that the sequence $(\mathbf{x}_1, \mathbf{x}_2)$ is L.I. and $\mathbf{x}_1 + 2\mathbf{x}_2 - \mathbf{x}_3 = \mathbf{0}$. Let $S = \mathrm{sp}(\mathbf{x}_1, \mathbf{x}_2, \mathbf{x}_3)$. Prove that $\dim S = 2$.

Given further that $\mathbf{y}_1 = \mathbf{x}_1 + \mathbf{x}_3$, $\mathbf{y}_2 = 2\mathbf{x}_1 + \mathbf{x}_2 - \mathbf{x}_3$ and $T = \mathrm{sp}(\mathbf{y}_1, \mathbf{y}_2)$, use 41.5 to prove that $S = T$.

If $\mathbf{z}_1 = \mathbf{x}_1 + 2\mathbf{x}_2 - 2\mathbf{x}_3$, $\mathbf{z}_2 = 2\mathbf{x}_1 + 4\mathbf{x}_2 - \mathbf{x}_3$ and $U = \mathrm{sp}(\mathbf{z}_1, \mathbf{z}_2)$, does U equal S?

29.* Let S, T, U be subspaces of a vector space V. Prove that

$$(S \cap T) + (S \cap U) \subseteq S \cap (T + U).$$

Give a specific example of subspaces S, T, U of E_2 such that

$$(S \cap T) + (S \cap U) \neq S \cap (T + U).$$

30. Given that S, T are subspaces of a f.d. vector space and that $\dim(S \cap T) = \dim(S + T)$, prove that $S = T$.

31. Suppose that the subspaces S, T, U of a f.d. vector space are such that (1) $S \cap T = S \cap U$, (2) $S + T = S + U$, and (3) $T \subseteq U$. Prove that $T = U$. Show by means of a specific example of subspaces in E_2 that hypothesis (3) cannot be omitted.

32. It is given that S, T are subspaces of $\mathbb{R}^8$ with dimensions 4, 6, respectively, and that S is not contained in T. What can be said about $\dim(S \cap T)$?

33. Let S, T be unequal subspaces, each of dimension $n-1$, in an n-dimensional vector space V. Prove that $S + T = V$ and that $\dim(S \cap T) = n-2$.

34. Let $\mathbf{x}_1, \mathbf{x}_2, \ldots, \mathbf{x}_n, \mathbf{x}_{n+1}$ be vectors in a vector space; let $S = \mathrm{sp}(\mathbf{x}_1, \mathbf{x}_2, \ldots, \mathbf{x}_n)$ and $T = \mathrm{sp}(\mathbf{x}_1, \mathbf{x}_2, \ldots, \mathbf{x}_n, \mathbf{x}_{n+1})$. Prove that $\dim T = \dim S + \varepsilon$, where ε is 0 if $\mathbf{x}_{n+1} \in S$ and 1 if $\mathbf{x}_{n+1} \notin S$.

35. In $\mathbb{R}_{n \times n}$, let S, T be the subspaces consisting of the symmetric matrices and the skew-symmetric matrices, respectively. Prove that $\mathbb{R}_{n \times n} = S \oplus T$.

36. Let $(\mathbf{e}_1, \mathbf{e}_2, \ldots, \mathbf{e}_n)$ be a basis of an n-dimensional vector space V. Prove that for each s in the range 1 to $n-1$, inclusive,

$$V = \mathrm{sp}(\mathbf{e}_1, \ldots, \mathbf{e}_s) \oplus \mathrm{sp}(\mathbf{e}_{s+1}, \ldots, \mathbf{e}_n).$$

37.* Let S be a subspace of a f.d. vector space V. Using the result of question 36, establish the existence of a *direct complement* of S in V, i.e. a subspace T of V such that $V = S \oplus T$.

Show that if T_1, T_2 are two such direct complements of S in V, then

$$\dim V - 2 \dim S \leqslant \dim(T_1 \cap T_2) \leqslant \dim V - \dim S.$$

CHAPTER SIX

LINEAR MAPPINGS

44. Introduction

The convention that F denotes an arbitrary field of scalars continues in this chapter.

Throughout the chapter, as the title suggests, we are very much concerned with mappings. As notations, technical terms and general results to do with mappings arise in the course of the chapter, reference may be made by the student (when necessary) to the Appendix, where a brief but self-contained summary of all these things is given. Typically, we shall be considering a mapping $a: V \to W$, where V and W are vector spaces over the same field of scalars. In this situation, for each $\mathbf{x} \in V$, $a(\mathbf{x})$ makes sense and stands for a vector in W (the image of $\mathbf{x}$ under a). From the start, we shall simplify notation slightly by preferring to use $a\mathbf{x}$ instead of $a(\mathbf{x})$ in this context of mappings between vector spaces. Our interest is not in arbitrary mappings from one vector space to another, but rather in mappings which have nice properties in relation to the operations (addition and multiplication by scalars) defined on the vector spaces. This brings us to the fundamental definition of the chapter.

Let V, W be vector spaces over the same field F. A mapping $a: V \to W$ is described as a **linear mapping** if and only if it has the following two properties:

(1) $a(\mathbf{x}+\mathbf{y}) = a\mathbf{x}+a\mathbf{y}$ for all $\mathbf{x}, \mathbf{y} \in V$;

(2) $a(\lambda\mathbf{x}) = \lambda(a\mathbf{x})$ for all $\mathbf{x} \in V$ and all $\lambda \in F$.

As linear mappings occur very widely in mathematics and its applications, they certainly provide a natural and important subject for study. Our knowledge of vector space theory from chapter 5 (bases, subspaces and their dimensions, etc.) equips us well for the study of this subject-matter, and the foremost facts about linear mappings will soon be covered in the course of this chapter.

An important theme, both in this chapter and chapter 7, is the close connection between matrices on the one hand, and, on the other, linear mappings from one nonzero f.d. vector space to another. Early in §45 it will be seen that every matrix A gives rise, in a simple way, to such a linear mapping (the linear mapping induced by A, we shall call it). Through the strategy of applying general results about linear mappings to the linear mappings induced by matrices, we shall, later in this chapter, greatly enlarge our understanding of the world of matrices. One detail of the story is that the key concept of the *rank* of a linear mapping (introduced in §49) is closely related to the concept of the row-rank of a matrix, which, in turn, (cf. 40.6) ties up with some most productive discussions of matrices and systems of linear equations in chapter 3. From the point of view of seeking new light on basic questions about matrices and systems of linear equations, the contents of §§50, 51, 52 will be found particularly interesting.

45. Some examples of linear mappings

(*a*) Let A be an arbitrary matrix in $F_{m\times n}$. The mapping $m_A: F_{n\times 1} \rightarrow F_{m\times 1}$ defined by

$$m_A X = AX \qquad (X \in F_{n\times 1})$$

is a linear mapping. (Here $m_A X$ means $m_A(X)$: cf. remark about notation in §44.) The mapping m_A is linear because (by its definition and the elementary properties of matrices):

(1) for all $X, Y \in F_{n\times 1}$, $m_A(X+Y) = A(X+Y) = AX + AY = m_A X + m_A Y$; and (2) for all $X \in F_{n\times 1}$ and $\lambda \in F$, $m_A(\lambda X) = A(\lambda X) = \lambda(AX) = \lambda(m_A X)$.

This mapping is called the **linear mapping induced by** A, and we shall use m_A as the standard notation for it. Its action, one sees, is simply the premultiplication (of columns) by A. An indication was given above (in §44) of the importance of such mappings.

There are other ways of describing the mapping m_A. Suppose that $A = [\alpha_{ik}]_{m\times n}$. If the column $X (\in F_{n\times 1})$ is $\text{col}(x_1, \ldots, x_n)$, then $m_A X$ is $\text{col}(y_1, \ldots, y_m)$, where

$$y_i = \sum_{k=1}^{n} \alpha_{ik} x_k \qquad (i = 1, 2, \ldots, m).$$

So m_A could be defined by

$$\text{col}(x_1, \ldots, x_n) \mapsto \text{col}(y_1, \ldots, y_m) \qquad (x_1, \ldots, x_n \in F),$$

where $y_i = \sum_{k=1}^{n} \alpha_{ik} x_k$ $(i = 1, 2, \ldots, m)$.

The same story, told in terms of n-tuples and m-tuples instead of columns, is the mapping from F^n to F^m given by

$$(x_1, \ldots, x_n) \mapsto (y_1, \ldots, y_m) \qquad \text{where } y_i = \sum_{k=1}^{n} \alpha_{ik} x_k \qquad (1 \leqslant i \leqslant m).$$

With a little further knowledge (especially the result 46.2 in the next section) the student will have no difficulty in proving that *every* linear mapping from F^n to F^m is of this form, i.e.

$$\left\{\begin{array}{c}(x_1,\ldots,x_n)\mapsto(y_1,\ldots,y_m)\\ \text{where each } y_i \text{ is a linear combination of } x_1,\ldots,x_n.\end{array}\right\} \qquad (*)$$

Moreover, it will emerge in chapter 7 that, when expressed in terms of component vectors, every linear mapping from an n-dimensional vector space to an m-dimensional vector space is of the form (*). So we have in (*) something of very great generality—in particular a pattern that may be stored in the mind as the general pattern for a linear mapping between nonzero f.d. vector spaces.

(*b*) Let V, W be any vector spaces whatever over F. The mapping $O: V \to W$ defined by

$$O\mathbf{x} = \mathbf{0} \qquad \text{for all } \mathbf{x} \in V$$

is easily seen to be a linear mapping. This mapping is called the **zero mapping** from V to W.

(*c*) Linear mappings are to be found in interesting places in calculus. Let V denote the subspace of $\mathbb{R}^{\mathbb{R}}$ consisting of all the real-valued functions that are differentiable at all points of $\mathbb{R}$. Then "differentiation", i.e. the mapping from V to $\mathbb{R}^{\mathbb{R}}$ given by

$$f \mapsto f' \, (= \text{the derivative of } f) \qquad (f \in V)$$

is a linear mapping. In informal terms, this mapping could be denoted by d/dx, and its action indicated by $f(x) \mapsto df(x)/dx$. It is a linear mapping because of two basic properties of differentiation, viz.

$$\frac{d}{dx}(f(x)+g(x)) = \frac{d}{dx}(f(x)) + \frac{d}{dx}(g(x)) \quad \text{and}$$

$$\frac{d}{dx}(\lambda f(x)) = \lambda\frac{d}{dx}(f(x)) \quad (\lambda \in \mathbb{R}).$$

In the same way, other more complicated "differential operators" can be recognized as linear mappings. For example, the differential equation

$$\alpha\frac{d^2y}{dx^2} + \beta\frac{dy}{dx} + \gamma y = k(x) \qquad (1)$$

(where α, β, γ are real constants and $k(x)$ is a given real function of x) can be re-expressed as

$$a(y(x)) = k(x),$$

where a is the "differential operator" $\alpha(d^2/dx^2) + \beta(d/dx) + \gamma$, which is a linear mapping (in the same way as d/dx).

It is instructive to notice that both the differential equation (1) and the typical system of linear equations $AX = K$ can be put in the form

$$\text{linear mapping acting on "unknown"} = \text{given object}. \qquad (2)$$

(The linear mapping is m_A in the case of the system $AX = K$.) In §51 we shall see that there is, as might be expected, something illuminating to be said about all equations of the form (2).

(*d*) For any vector space V, a linear mapping from V to itself may be called a **linear transformation** of V. A simple but important example is the identity mapping i_V of V to itself (cf. 14 in Appendix). The linearity of this mapping is clear; and in this context it may be called the **identity** (linear) **transformation** of V. Notice that:

45.1 The linear mapping m_I induced by the identity $n \times n$ matrix is the identity transformation of $F_{n \times 1}$.

(*e*) As explained in §10 (remark (*c*)), geometrical transformations of 2- or 3-dimensional space can be represented as algebraic transformations $\mathbf{x} \mapsto \ldots$ of E_2 or E_3. The algebraic version results from the identification of points with their position vectors, and so it expresses what the transformation is in terms of its effect on the position vectors of points. Often the algebraic transformations arising in this way are linear transformations (of E_2 or E_3). Specific instances include rotations about the origin in 2 dimensions and rotations about axes through the origin in 3 dimensions. It is an interesting diversion to think through why these geometrical transformations give rise to *linear* transformations of E_2 and E_3: the parallelogram law of vector addition may be found useful. We shall use rotations in illustrations in future chapters, and, without further ado, shall regard them as linear transformations (of E_2 or E_3, as the case may be).

46. Some elementary facts about linear mappings

Throughout this section V and W will denote vector spaces over the field F.

46.1 Let $a: V \to W$ be a linear mapping. Then:

(i) $a\mathbf{0} = \mathbf{0}$ (i.e. a maps the zero of V to the zero of W);

(ii) $a(-\mathbf{x}) = -a\mathbf{x}$ for all $\mathbf{x} \in V$.

These follow immediately from property (2) in the definition of a linear mapping on taking $\mathbf{x}$ to be an arbitrary vector in V and λ to be 0 and -1, respectively.

The result 46.1 presents just two aspects of the fact that every linear mapping $a: V \to W$ has nice properties additional to those explicitly stated in the definition of a linear mapping. For example, if $\mathbf{x}_1, \mathbf{x}_2, \ldots$ denote vectors in V

and $\lambda_1, \lambda_2, \ldots$ denote scalars, then, by repeated use of one part or other of the definition, we obtain:

$$a(\mathbf{x}_1+\mathbf{x}_2+\mathbf{x}_3) = a((\mathbf{x}_1+\mathbf{x}_2)+\mathbf{x}_3) = a(\mathbf{x}_1+\mathbf{x}_2)+a\mathbf{x}_3 = a\mathbf{x}_1+a\mathbf{x}_2+a\mathbf{x}_3;$$

and hence

$$\begin{aligned} a(\lambda_1\mathbf{x}_1+\lambda_2\mathbf{x}_2+\lambda_3\mathbf{x}_3) &= a(\lambda_1\mathbf{x}_1)+a(\lambda_2\mathbf{x}_2)+a(\lambda_3\mathbf{x}_3) \\ &= \lambda_1(a\mathbf{x}_1)+\lambda_2(a\mathbf{x}_2)+\lambda_3(a\mathbf{x}_3). \end{aligned}$$

The general result along these lines is

$$a\left(\sum_{i=1}^{n} \lambda_i\mathbf{x}_i\right) = \sum_{i=1}^{n} \lambda_i(a\mathbf{x}_i).$$

Notice also that: $a(\mathbf{x}_1-\mathbf{x}_2) = a\mathbf{x}_1 - a\mathbf{x}_2$. (Cf. $a(\mathbf{x}_1+(-1)\mathbf{x}_2)$.) In future we shall take all this for granted as part and parcel of what is entailed in describing the mapping a as linear.

Worked example. A linear mapping $a: \mathbb{R}^3 \to \mathbb{R}^2$ is such that

$$a(1,0,0) = (1,2), a(0,1,0) = (3,4), a(0,0,1) = (5,6).$$

Find a general formula for $a(x_1,x_2,x_3)$((x_1,x_2,x_3) arbitrary in $\mathbb{R}^3$).

Solution. Let $(\mathbf{e}_1,\mathbf{e}_2,\mathbf{e}_3)$ be the standard basis of $\mathbb{R}^3$ (as described in §40—e.g. $\mathbf{e}_1 = (1,0,0)$). Then

$$\begin{aligned} a(x_1,x_2,x_3) &= a(x_1\mathbf{e}_1+x_2\mathbf{e}_2+x_3\mathbf{e}_3) \\ &= x_1(a\mathbf{e}_1)+x_2(a\mathbf{e}_2)+x_3(a\mathbf{e}_3) \qquad \text{(by the linearity of } a\text{)} \\ &= x_1(1,2)+x_2(3,4)+x_3(5,6) \\ &= (x_1+3x_2+5x_3, 2x_1+4x_2+6x_3). \end{aligned}$$

The above worked example illustrates an important general truth—that (in the case of a f.d. domain) a linear mapping can be specified satisfactorily and completely by prescribing its effect on the vectors in a basis of the domain. The following result gives a more formal expression of this general truth.

46.2 Suppose that V is f.d. with basis $(\mathbf{e}_1,\ldots,\mathbf{e}_n)$, n being dim V, and let $\mathbf{y}_1,\mathbf{y}_2,\ldots,\mathbf{y}_n$ be n arbitrary vectors (not necessarily all different) in W. Then there exists a unique linear mapping $a: V \to W$ such that $a\mathbf{e}_i = \mathbf{y}_i$ for $i = 1,2,\ldots,n$.

Proof. The mapping $a: V \to W$ defined by

$$\underbrace{a(\lambda_1\mathbf{e}_1+\lambda_2\mathbf{e}_2+\ldots+\lambda_n\mathbf{e}_n)}_{\text{general member of } V} = \lambda_1\mathbf{y}_1+\lambda_2\mathbf{y}_2+\ldots+\lambda_n\mathbf{y}_n$$

is easily seen to be linear and to have the property that $a\mathbf{e}_i = \mathbf{y}_i$ for every relevant i. So there does exist a linear mapping a as described.

Moreover, there is but one possibility for a linear mapping with the stated property: for, if a is to be a linear mapping such that $a\mathbf{e}_i = \mathbf{y}_i$ for all relevant i, then we must have

$$a(\lambda_1\mathbf{e}_1 + \ldots + \lambda_n\mathbf{e}_n) = \sum_{i=1}^{n} \lambda_i(a\mathbf{e}_i) = \sum_{i=1}^{n} \lambda_i\mathbf{y}_i \qquad (\text{for all } \lambda_1,\ldots,\lambda_n \in F);$$

and this shows that a is uniquely determined.

The result is now established.

The process of inferring the whole story of a linear mapping a from its effect on the vectors in a basis of the domain (by using $a(\sum \lambda_i\mathbf{e}_i) = \sum \lambda_i(a\mathbf{e}_i)$) is often described by the phrase "extending linearly".

A result in the same vein as 46.2 is:

46.3 Suppose that $V = S \oplus T$, S and T being subspaces of V, and that $b: S \to W$ and $c: T \to W$ are given linear mappings. Then there exists a unique linear mapping $a: V \to W$ such that $a|_S = b$ and $a|_T = c$.

(See 6 in Appendix for the meaning of the restriction $a|_S$.)

The "unique linear mapping a" in 46.3 is given by $a(\mathbf{u}+\mathbf{v}) = b\mathbf{u}+c\mathbf{v}$ ($\mathbf{u} \in S, \mathbf{v} \in T$). It is left as an instructive exercise for the student to think through the details of the proof.

47. New linear mappings from old

In this section our primary interest is in how further linear mappings can be constructed from one or more given linear mappings. Throughout the section U, V, W will denote vector spaces over the same field F. With one exception we shall omit proofs of assertions that mappings are linear. The omitted proofs are straightforward verifications that properties (1) and (2) in the definition of "linear mapping" hold for the mappings in question.

General ideas about mappings that come into this section are the restriction $a|_S$ of a mapping a to a subset S of its domain, the composition of two mappings, and the inverse of a bijection. (See 6, 8, 13, 16 in Appendix.) In this context, we denote the composition $a \circ b$ of two linear mappings a, b simply by ab: thus, for $\mathbf{x}$ in the domain of b, $(ab)\mathbf{x}$ means $(a \circ b)\mathbf{x}$, i.e. $a(b\mathbf{x})$. Further, for any linear transformation c, we use the notation c^2 for $c \circ c$, c^3 for $c \circ c \circ c$, etc.

Among the following three results, the first is trivial, and proof of the second also is omitted.

47.1 The restriction of a linear mapping to a subspace of its domain is also linear.

47.2 A composition of two linear mappings is also linear: i.e. if $a: V \to W$ and $b: U \to V$ are linear mappings, then so is ab.

47.3 The inverse of a bijective linear mapping is also linear.

Proof. Let $a: V \to W$ be a bijective linear mapping. Its inverse a^{-1} is a mapping from W to V.

(1) Let $\mathbf{x}, \mathbf{y} \in W$. Let $a^{-1}\mathbf{x} = \mathbf{u}, a^{-1}\mathbf{y} = \mathbf{v}$, so that $\mathbf{x} = a\mathbf{u}, \mathbf{y} = a\mathbf{v}$. Hence

$$\mathbf{x}+\mathbf{y} = a\mathbf{u}+a\mathbf{v} = a(\mathbf{u}+\mathbf{v}) \qquad \text{(since } a \text{ is linear)},$$

and so: $a^{-1}(\mathbf{x}+\mathbf{y}) = \mathbf{u}+\mathbf{v} = a^{-1}\mathbf{x}+a^{-1}\mathbf{y}$.

(2) Let $\mathbf{x} \in W$ and $\lambda \in F$. Let $a^{-1}\mathbf{x} = \mathbf{u}$, so that $\mathbf{x} = a\mathbf{u}$. Hence

$$\lambda\mathbf{x} = \lambda(a\mathbf{u}) = a(\lambda\mathbf{u}) \qquad \text{(since } a \text{ is linear)},$$

and so: $a^{-1}(\lambda\mathbf{x}) = \lambda\mathbf{u} = \lambda(a^{-1}\mathbf{x})$.

From (1) and (2) it follows that a^{-1} is linear.

Now suppose that a and b are linear mappings from V to W and that λ is a scalar. Then mappings $a+b$ and λa, both from V to W, may be defined by

$$(a+b)\mathbf{x} = a\mathbf{x}+b\mathbf{x}(\mathbf{x} \in V) \quad \text{and} \quad (\lambda a)\mathbf{x} = \lambda(a\mathbf{x})(\mathbf{x} \in V).$$

It is easy to verify that:

47.4 If a, b are linear mappings from V to W and λ is a scalar, then $a+b$ and λa (as just defined) are also linear mappings.

The next proposition is about ab, $a+b$, λa and a^{-1} when a, b are linear mappings induced by suitable matrices. It makes visible a close parallelism between operations on linear mappings and the corresponding operations on matrices. To a great extent this parallelism is an expression of the motivatory thoughts underlying the definitions of operations on matrices. For example, remark (*b*) in §15 suggested that the definition of matrix multiplication is designed so that there will be a connection between matrix multiplication and mapping composition: part (*a*) of the following proposition is really just a precise and tidy expression of that connection.

47.5 Let A, B denote matrices over F and λ a scalar in F.

(*a*) If the product AB exists, then $m_{AB} = m_A m_B$.

(*b*) If A and B are of the same type, then $m_{A+B} = m_A + m_B$.

(*c*) $m_{\lambda A} = \lambda m_A$.

(*d*) If A is nonsingular, then m_A is bijective and $(m_A)^{-1} = m_{A^{-1}}$.

Proof. (*a*) Let $A \in F_{l \times m}, B \in F_{m \times n}$, so that $AB \in F_{l \times n}$. Both $m_A m_B$ and m_{AB} are mappings from $F_{n \times 1}$ to $F_{l \times 1}$. Further, for all $X \in F_{n \times 1}$,

$$\begin{aligned}(m_A m_B)X &= m_A(m_B X) \qquad \text{(since } m_A m_B \text{ means } m_A \circ m_B\text{)}\\ &= m_A(BX) = A(BX) = (AB)X = m_{AB}X\end{aligned}$$

(by the meanings of m_B, etc., and the associativity of matrix multiplication). Therefore $m_{AB} = m_A m_B$.

This proves part (*a*). Parts (*b*) and (*c*) are easily proved in similar fashion.

(*d*) Suppose that A is a nonsingular matrix in $F_{n\times n}$, so that A^{-1} exists (and is also a matrix in $F_{n\times n}$). Both m_A and $m_{A^{-1}}$ are linear transformations of V, where $V = F_{n\times 1}$. And:

$$\begin{aligned} m_A m_{A^{-1}} &= m_{AA^{-1}} && \text{(by part } (a)) \\ &= m_I && (I \text{ being } I_n) \\ &= i_V && \text{(by 45.1)}, \end{aligned}$$

while, similarly, $m_{A^{-1}}m_A = i_V$. Hence it follows (cf. 18 in Appendix) that m_A is bijective and that $(m_A)^{-1}$ is $m_{A^{-1}}$.

This proves part (*d*).

48. Image space and kernel of a linear mapping

This section draws attention to two important subspaces associated with an arbitrary linear mapping—its image space (which is a subspace of its codomain) and its kernel (which is a subspace of its domain).

Throughout the section, V and W will denote vector spaces over the same field F.

The first result involves the idea of the image under a mapping of a subset of its domain and the particular case of the image set of the mapping (see 5 in the Appendix). In the present context, dealing with mappings from one vector space to another, we use the notation aS (rather than $a(S)$) for the image under mapping a of the subset S of its domain.

48.1 Let $a: V \to W$ be a linear mapping. Then, for every subspace S of V, aS is a subspace of W. In particular (the case $S = V$) im a is a subspace of W.

Proof. Let S be an arbitrary subspace of V.

(1) Because $S \neq \varnothing$, it is clear that $aS \neq \varnothing$.

(2) Let $\mathbf{x}, \mathbf{y} \in aS$ and let $\lambda, \mu \in F$. Because $\mathbf{x}, \mathbf{y} \in aS$, $\mathbf{x} = a\mathbf{u}$ and $\mathbf{y} = a\mathbf{v}$ for some $\mathbf{u}, \mathbf{v} \in S$. Hence

$$\lambda\mathbf{x} + \mu\mathbf{y} = \lambda(a\mathbf{u}) + \mu(a\mathbf{v}) = a(\lambda\mathbf{u} + \mu\mathbf{v}) \qquad \text{(since } a \text{ is linear)}.$$

But, since S is a subspace to which $\mathbf{u}, \mathbf{v}$ belong, $\lambda\mathbf{u} + \mu\mathbf{v} \in S$ (cf. 37.2). Therefore, $\lambda\mathbf{x} + \mu\mathbf{y} \in aS$.

It now follows, by 37.1, that aS is a subspace of W; and thus the stated result is established.

In the notation of 48.1, the subspace im a (which may equally well be denoted by aV) is called the **image space** of a.

As we now point out, im a in the case $a = m_A$ is a space that has already engaged our attention:

48.2 Let A be an arbitrary matrix. Then $\text{im}(m_A)$ is the column space of A. (For: if $A \in F_{m \times n}$, $\text{im}(m_A)$ is the set of all objects of the form

$$m_A(\text{col}(x_1, \ldots, x_n)), \quad \text{i.e.} \quad A \times \text{col}(x_1, \ldots, x_n),$$

with $x_1, \ldots, x_n \in F$; and this (by 21.4) is the set of all linear combinations of the columns of A).

Later in the section there will be a numerical example in which the image space of a specific linear mapping will be considered. Our immediate concern is with two results supplementary to the general case of 48.1.

48.3 Let $a: V \to W$ be a linear mapping, and let S, T be subspaces of V (so that aS and aT are subspaces of W). Then

$$a(S+T) = aS + aT.$$

(For: it is easily seen that both sides are the set of all vectors in W expressible in the form $a(\mathbf{y}+\mathbf{z})$ (which is the same thing as $a\mathbf{y}+a\mathbf{z}$) with $\mathbf{y} \in S$ and $\mathbf{z} \in T$.)

The second result gives information about the dimension of "aS" in the case where S is f.d.

48.4 Let $a: V \to W$ be a linear mapping, and let S be a f.d. subspace of V.

(*a*) If $(\mathbf{e}_1, \ldots, \mathbf{e}_n)$ is a basis of S (n being dim S), then the sequence $(a\mathbf{e}_1, \ldots, a\mathbf{e}_n)$ spans aS.

(*b*) aS is f.d., and $\dim(aS) \leqslant \dim S$.

Proof. Suppose that $(\mathbf{e}_1, \ldots, \mathbf{e}_n)$ is a basis of S, so that $n = \dim S$ and the vectors $a\mathbf{e}_1, \ldots, a\mathbf{e}_n$ all belong to aS.

Let $\mathbf{y}$ be an arbitrary vector in aS. Then

$$\begin{aligned}
\mathbf{y} &= a\mathbf{u} && \text{for some } \mathbf{u} \in S \\
&= a\left(\sum_{i=1}^{n} \lambda_i \mathbf{e}_i\right) && \text{for some } \lambda_1, \ldots, \lambda_n \in F \\
&= \sum_{i=1}^{n} \lambda_i (a\mathbf{e}_i) && (a \text{ being linear}) \\
&= \text{a linear combination of } a\mathbf{e}_1, a\mathbf{e}_2, \ldots, a\mathbf{e}_n.
\end{aligned}$$

Thus the sequence $(a\mathbf{e}_1, \ldots, a\mathbf{e}_n)$ spans aS. The existence of such a spanning sequence proves that aS is f.d. Moreover, (cf. 40.8(i))

$$\dim(aS) \leqslant \text{length of this spanning sequence} = n = \dim S.$$

All parts of the result are now proved.

Next we come to the second important subspace associated with a linear mapping $a: V \to W$. For any such linear mapping a, we define the **kernel** of a

(denoted by ker a) to be the subset

$$\{\mathbf{x} \in V : a\mathbf{x} = \mathbf{0}\}$$

of the domain V, i.e. the set of all vectors mapped by a to zero (or, as we sometimes put it, "annihilated by a").

As hinted in the previous paragraph:

48.5 The kernel of a linear mapping is a subspace of its domain.

Proof. Let $a: V \to W$ be a linear mapping, so that ker a is a subset of V.
(1) Since $a\mathbf{0} = \mathbf{0}$ (cf. 46.1), $\mathbf{0} \in \ker a$. So $\ker a \neq \emptyset$.
(2) Let $\mathbf{x}, \mathbf{y} \in \ker a$ and let $\lambda, \mu \in F$. Then

$$\begin{aligned} a(\lambda\mathbf{x} + \mu\mathbf{y}) &= \lambda(a\mathbf{x}) + \mu(a\mathbf{y}) \qquad \text{(since } a \text{ is linear)} \\ &= \lambda\mathbf{0} + \mu\mathbf{0} \qquad \text{(since } \mathbf{x}, \mathbf{y} \in \ker a\text{)} \\ &= \mathbf{0}, \end{aligned}$$

and so $\lambda\mathbf{x} + \mu\mathbf{y} \in \ker a$.
By 37.1 it follows that ker a is a subspace of V.

A glance back at 37.5 will reveal that it is a particular case (the case $a = m_A$) of 48.5. Notice for future reference that:

48.6 For $A \in F_{m \times n}$, the solution set of the system of homogeneous equations $AX = O$ is ker m_A.

Henceforth let us speak of the solution *space* of a homogeneous system of linear equations, in recognition of the fact that the solution is a vector space.

Worked example (to illustrate the concepts of image space and kernel). Let a be the linear transformation of $\mathbb{R}^3$ given by

$$a(x_1, x_2, x_3) = (x_1 + x_2 + 2x_3, 2x_1 + x_2 + x_3, 3x_1 - x_2 - 6x_3).$$

Find bases of (i) im a, (ii) ker a.

Solution. (i) im a is the set of all vectors of the form

$$\begin{gathered} (x_1 + x_2 + 2x_3, 2x_1 + x_2 + x_3, 3x_1 - x_2 - 6x_3), \\ \text{i.e. } x_1(1, 2, 3) + x_2(1, 1, -1) + x_3(2, 1, -6), \end{gathered}$$

with $x_1, x_2, x_3 \in \mathbb{R}$. Thus im $a = \operatorname{sp}((1, 2, 3), (1, 1, -1), (2, 1, -6))$.
A basis of this space may be found through an echelon reduction, as described in §40. We have:

$$\begin{bmatrix} 1 & 2 & 3 \\ 1 & 1 & -1 \\ 2 & 1 & -6 \end{bmatrix} \sim \begin{bmatrix} 1 & 2 & 3 \\ 0 & -1 & -4 \\ 0 & -3 & -12 \end{bmatrix} \quad \begin{matrix} (R_2 \to R_2 - R_1 \\ \text{then} \\ R_3 \to R_3 - 2R_1) \end{matrix}$$

$$\sim \begin{bmatrix} 1 & 2 & 3 \\ 0 & 1 & 4 \\ 0 & 0 & 0 \end{bmatrix} \begin{matrix} (R_2 \to (-1)R_2 \\ \text{then} \\ R_3 \to R_3 + 3R_2). \end{matrix}$$

Hence im a is 2-dimensional, and a basis of it is

$$((1,2,3),(0,1,4)).$$

(ii) For $(x_1, x_2, x_3) \in \mathbb{R}^3$,

$$(x_1, x_2, x_3) \in \ker a \Leftrightarrow a(x_1, x_2, x_3) = (0,0,0) \qquad \text{(the zero of } \mathbb{R}^3\text{)}$$

$$\Leftrightarrow \begin{Bmatrix} x_1 + x_2 + 2x_3 = 0 \\ 2x_1 + x_2 + \;\; x_3 = 0 \\ 3x_1 - x_2 - 6x_3 = 0 \end{Bmatrix}.$$

We proceed (cf. §28) by transforming this system of equations to reduced echelon form. Hence we obtain:

$$\begin{aligned} (x_1, x_2, x_3) \in \ker a &\Leftrightarrow \begin{Bmatrix} x_1 - \;\; x_3 = 0 \\ x_2 + 3x_3 = 0 \end{Bmatrix} \\ &\Leftrightarrow x_3 = \alpha, x_2 = -3\alpha, x_1 = \alpha \qquad \text{for some } \alpha \in \mathbb{R} \\ &\Leftrightarrow (x_1, x_2, x_3) = \alpha(1, -3, 1) \qquad \text{for some } \alpha \in \mathbb{R}. \end{aligned}$$

Thus ker a is a 1-dimensional space, with basis $((1, -3, 1))$.

This section concludes with a useful result concerning the kernel. (For the meaning of the word "injective", see 12 in the Appendix.)

48.7 Let $a: V \to W$ be a linear mapping. Then a is injective if and only if $\ker a = \{\mathbf{0}\}$.

Proof. (1) (The "$\Rightarrow$" half) Suppose that a is injective. If $\mathbf{x} \in \ker a$, then $a\mathbf{x} = \mathbf{0}$, i.e. $a\mathbf{x} = a\mathbf{0}$; and hence, a being injective, $\mathbf{x} = \mathbf{0}$. It follows that the subspace ker a is $\{\mathbf{0}\}$. (Cf. remark after the proof of 43.3 on how to prove a subspace zero.)

(2) (The "$\Leftarrow$" half) This time suppose that $\ker a = \{\mathbf{0}\}$. Then, for $\mathbf{x}, \mathbf{y} \in V$,

$$\begin{aligned} a\mathbf{x} = a\mathbf{y} &\Rightarrow a\mathbf{x} - a\mathbf{y} = \mathbf{0} \Rightarrow a(\mathbf{x} - \mathbf{y}) = \mathbf{0} \qquad (a \text{ being linear}) \\ &\Rightarrow \mathbf{x} - \mathbf{y} \in \ker a \qquad (\text{by definition of } \ker a) \\ &\Rightarrow \mathbf{x} - \mathbf{y} = \mathbf{0} \qquad (\text{since } \ker a = \{\mathbf{0}\}) \\ &\Rightarrow \mathbf{x} = \mathbf{y}. \end{aligned}$$

Thus a is injective.

The stated result is now proved.

49. Rank and nullity

Once again in this section V and W will denote vector spaces over the field F. Our concern here is with a linear mapping $a: V \to W$ in the case where the domain V is f.d. In this case im a $(=aV)$ and ker a are both f.d. (by 48.4 and 41.4(i), respectively), and so the following definitions can be made.

(1) The **rank** of a (denoted by $r(a)$) is dim (im a), i.e. dim (aV).

(2) The **nullity** of a (denoted by $s(a)$) is dim (ker a).

From 48.2 and 48.6 it follows that:

49.1 For any matrix A, (1) $r(m_A)$ is the column-rank of A (i.e. the dimension of its column space), and (2) $s(m_A)$ is the dimension of the solution space of the homogeneous system of equations $AX = O$.

It will be seen on looking back that in the worked example in §48 the rank and nullity of the linear transformation discussed were 2 and 1, respectively. The fact that these add up to the dimension of the domain of the transformation is one illustration of the following cardinally important theorem.

49.2 Let $a: V \to W$ be a linear mapping, the domain V being f.d. Then $r(a)+s(a) = \dim V$.

Proof. Let s stand for $s(a)$ and n for dim V. Since s is the dimension of a subspace of V (viz. ker a), $0 \leqslant s \leqslant n$.

Strictly speaking, the present proof will ignore the extreme cases $s = 0$ and $s = n$, and the task of covering these cases is left for the student to think about. (The latter is a trivial case; the former demands just a tidy up of the present proof to take account of the degeneracy.)

We introduce a basis $(\mathbf{e}_1, \ldots, \mathbf{e}_s)$ of ker a, and (cf. 41.4(iv)) we extend this basis of ker a to a basis $(\mathbf{e}_1, \ldots, \mathbf{e}_s, \mathbf{e}_{s+1}, \ldots, \mathbf{e}_n)$ of V. By 48.4(a) the sequence $(a\mathbf{e}_1, \ldots, a\mathbf{e}_n)$ spans aV. In this sequence the first s vectors $a\mathbf{e}_1, \ldots, a\mathbf{e}_s$ are all zero (since $\mathbf{e}_1, \ldots, \mathbf{e}_s \in \ker a$). So, obviously, we shall still have a sequence spanning aV if we delete these vectors to leave the sequence

$$L = (a\mathbf{e}_{s+1}, \ldots, a\mathbf{e}_n)$$

of length $n-s$.

Suppose, with a view to proving that L is L.I., that

$$\lambda_{s+1}(a\mathbf{e}_{s+1}) + \ldots + \lambda_n(a\mathbf{e}_n) = \mathbf{0} \qquad (\lambda_{s+1}, \ldots, \lambda_n \in F).$$

Then, a being linear,

$$a(\lambda_{s+1}\mathbf{e}_{s+1} + \ldots + \lambda_n\mathbf{e}_n) = \mathbf{0},$$
$$\text{i.e.} \quad \lambda_{s+1}\mathbf{e}_{s+1} + \ldots + \lambda_n\mathbf{e}_n \in \ker a.$$

So, since $(\mathbf{e}_1, \ldots, \mathbf{e}_s)$ is a basis of ker a,

$$\lambda_{s+1}\mathbf{e}_{s+1} + \ldots + \lambda_n\mathbf{e}_n = \mu_1\mathbf{e}_1 + \ldots + \mu_s\mathbf{e}_s$$

for some $\mu_1, \ldots, \mu_s \in F$. The last equation may be rewritten

$$-\mu_1\mathbf{e}_1 - \ldots - \mu_s\mathbf{e}_s + \lambda_{s+1}\mathbf{e}_{s+1} + \ldots + \lambda_n\mathbf{e}_n = \mathbf{0}.$$

Here all the scalar coefficients must be 0 since the sequence $(\mathbf{e}_1, \ldots, \mathbf{e}_n)$ is L.I., being a basis of V. In particular, $\lambda_{s+1}, \ldots, \lambda_n$ are all zero.

It follows that L is L.I. So, since (as pointed out when L was introduced) L spans aV, L is a basis of aV; and hence

$$\dim(aV) = \text{length of } L = n - s;$$
$$\text{i.e.} \quad r(a) = \dim V - s(a),$$

which proves the stated result.

The theorem 49.2 has many applications, one of which will soon appear as we start to think, in terms of rank and nullity, about the possibility of a linear mapping being surjective or injective or both. (See 11 in the Appendix for the meaning of "surjective".)

From 48.7 it follows straight away that:

49.3 If $a: V \to W$ is a linear mapping and V is f.d., then a is injective if and only if $s(a) = 0$.

A corresponding result about surjectivity is:

49.4 Suppose that $a: V \to W$ is a linear mapping, V and W being f.d. Then a is surjective if and only if $r(a) = \dim W$.

Proof. If $\dim(aV) = \dim W$, then (aV being a subspace of W) $aV = W$, by 41.5. The converse is trivially true, and thus

$$aV = W \Leftrightarrow \dim(aV) = \dim W,$$
$$\text{i.e. } a \text{ is surjective} \Leftrightarrow r(a) = \dim W,$$

as asserted.

Using 49.3 and 49.4 together, along with 49.2, we deduce the following theorem.

49.5 Let a be a linear transformation of the f.d. vector space V. If we know *either* that a is injective *or* that a is surjective, then we may conclude that a is bijective.

Proof. It will suffice to show that each of the statements "a is injective" and

"a is surjective" implies the other. And this is true because

$$\begin{aligned} a \text{ is injective} &\Leftrightarrow s(a) = 0 && \text{(by 49.3)} \\ &\Leftrightarrow r(a) = \dim V && \text{(by 49.2)} \\ &\Leftrightarrow a \text{ is surjective} && \text{(by 49.4 in the case } V = W\text{).} \end{aligned}$$

It should be noted that 49.5 is about a linear transformation of a f.d. vector space and that (as demonstrated in an exercise at the end of the chapter) it does not extend to infinite-dimensional spaces.

By analogy with the theory of square matrices, we use the word "nonsingular" to describe a linear transformation with an inverse, i.e. a bijective linear transformation.

It can be seen from the above results that:

49.6 A linear transformation a of a f.d. vector space V is nonsingular if and only if $r(a) = \dim V$.

$$\begin{aligned} \text{(For: } a \text{ is nonsingular} &\Leftrightarrow a \text{ is surjective} && \text{("}\Leftarrow\text{" by 49.5)} \\ &\Leftrightarrow r(a) = \dim V && \text{(by 49.4).)} \end{aligned}$$

Worked example. The linear transformation a of a f.d. vector space V is such that $r(a^2) = r(a)$. Prove that $V = aV \oplus \ker a$. (This is quite a formidable problem, included here to illustrate some relatively subtle lines of thought.)

Solution. We first prove, by using 41.5, that $\ker a^2 = \ker a$.

Because $r(a^2) = r(a)$, we have: $\dim V - r(a^2) = \dim V - r(a)$; i.e. (by 49.2) $s(a^2) = s(a)$; i.e. $\dim(\ker a^2) = \dim(\ker a)$.

But $\ker a \subseteq \ker a^2$: for if $\mathbf{x} \in \ker a$, then $a\mathbf{x} = \mathbf{0}$ and so $a^2\mathbf{x} = a(a\mathbf{x}) = a\mathbf{0} = \mathbf{0}$, showing that $\mathbf{x}$ also belongs to $\ker a^2$.

Therefore, by 41.5, $\ker a^2 = \ker a$.

The next step is to prove that $aV \cap \ker a = \{\mathbf{0}\}$.

For this, let $\mathbf{x}$ be an arbitrary vector in $aV \cap \ker a$. Since $\mathbf{x} \in aV$, $\mathbf{x} = a\mathbf{z}$ for some $\mathbf{z} \in V$. Further, since $\mathbf{x} \in \ker a$,

$$\mathbf{0} = a\mathbf{x} = a(a\mathbf{z}) = a^2\mathbf{z},$$

and so $\mathbf{z} \in \ker a^2$. Hence, since $\ker a^2 = \ker a$, $\mathbf{z} \in \ker a$; i.e. $a\mathbf{z} = \mathbf{0}$; i.e. $\mathbf{x} = \mathbf{0}$. This establishes that

$$aV \cap \ker a = \{\mathbf{0}\}. \tag{2}$$

But, by 49.2, it is also true that

$$\dim(aV) + \dim(\ker a) = \dim V. \tag{3}$$

Hence it follows (cf. part (c) of the solution of the worked example at the end of §43) that $V = aV \oplus \ker a$.

Postscript. One of the vital ingredients in the above solution as the fact (true for any linear transformation a) that $\ker a \subseteq \ker a^2$. By the same simple kind of argument one can show, more generally, that (for any linear transformation a) the kernels of the powers of a form an "ascending chain"; i.e.

$$\ker a \subseteq \ker a^2 \subseteq \ker a^3 \subseteq \ker a^4 \subseteq \ldots.$$

It is also easy to prove (again for an arbitrary linear transformation a) that the image spaces of the powers of a form a "descending chain"; i.e.

$$aV \supseteq a^2V \supseteq a^3V \supseteq a^4V \supseteq \ldots.$$

50. Row- and column-rank of a matrix

When the row- and column-rank of a matrix were defined in §40, it was stated that in due course we would show that these two numbers always coincide. The following theorem fulfils that promise.

50.1 Let A be an arbitrary matrix (in $F_{m\times n}$), and let r, c be the row- and column-rank, respectively, of A. Then $r = c$.

Proof. Let $R_1, R_2, \ldots, R_m$ be the rows of A. Since the sequence $(R_1, R_2, \ldots, R_m)$ spans the row space of A, there is (cf. 40.2) some subsequence that is a basis of the row space. Any such subsequence must have length r (the dimension of the row space). Let one such subsequence be $(R_{i_1}, R_{i_2}, \ldots, R_{i_r})$, $i_1, i_2, \ldots, i_r$ being r integers in the range 1 to m (inclusive). Note before proceeding that every row of A is a linear combination of $R_{i_1}, \ldots, R_{i_r}$.

Let B be the $r \times n$ matrix with rows $R_{i_1}, R_{i_2}, \ldots, R_{i_r}$.

For arbitrary $X \in F_{n\times 1}$,

$$AX = \operatorname{col}(R_1X, R_2X, \ldots, R_mX) \quad \text{and} \quad BX = \operatorname{col}(R_{i_1}X, R_{i_2}X, \ldots, R_{i_r}X).$$

Suppose that $BX = O$ $(X \in F_{n\times 1})$. Then $R_{i_1}X, \ldots, R_{i_r}X$ are all zero, and so every scalar expressible in the form

$$(\alpha_1 R_{i_1} + \alpha_2 R_{i_2} + \ldots + \alpha_{i_r} R_r)X \qquad (\text{each } \alpha_j \in F)$$

is 0. But this applies to every entry of AX, since every row of A is a linear combination of $R_{i_1}, \ldots, R_{i_r}$. Hence $AX = O$.

This has proved that: $BX = O \Rightarrow AX = O$ $(X \in F_{n\times 1})$. The converse is obviously true (the entries of BX being some of the entries of AX). Thus

$$AX = O \Leftrightarrow BX = O \qquad (X \in F_{n\times 1}).$$

Hence: $\{X \in F_{n\times 1} : AX = O\} = \{X \in F_{n\times 1} : BX = O\}$; i.e. $\ker m_A = \ker m_B$. On taking dimensions, we deduce that $s(m_B) = s(m_A)$. So, since both m_B and m_A

have the n-dimensional space $F_{n \times 1}$ as domain, it follows by 49.2 that

$$n - r(m_B) = n - r(m_A); \qquad \text{i.e. } r(m_B) = r(m_A).$$

But, by 49.1(1), $r(m_A) = c$. Hence $c = r(m_B)$.

We next observe that, because $\operatorname{im}(m_B)$ is a subspace of the codomain of m_B (viz. $F_{r \times 1}$), $\dim[\operatorname{im}(m_B)] \leqslant \dim(F_{r \times 1})$; i.e. $r(m_B) \leqslant r$. Hence (cf. conclusion of previous paragraph) $c \leqslant r$. That is

$$\text{column-rank of } A \leqslant \text{row-rank of } A. \tag{1}$$

This inequality has been proved for an arbitrary matrix A, and so it follows that, equally,

$$\text{column-rank of } A^T \leqslant \text{row-rank of } A^T. \tag{2}$$

But, if we think of columns in A^T and rows in A as n-tuples (disregarding whether they are written horizontally or vertically), then we can say that the columns of A^T are the same n-tuples as the rows of A. And therefore

$$\dim(\text{column space of } A^T) = \dim(\text{row space of } A),$$
$$\text{i.e.} \quad \text{column-rank of } A^T = r. \tag{3}$$

Similarly, the row-rank of A^{-T} is c; and so, from (2), it follows that

$$r \leqslant c.$$

By this and (1), we obtain $r = c$, which is the stated result.

Having proved 50.1, we can henceforth forget the distinction between the row-rank and the column-rank of a matrix A. So we refer to their common value simply as the **rank** of A; and for this number we shall in future use the notation $r(A)$.

The rest of the section is devoted to a variety of results that are all now easy to obtain.

In the proof of 50.1, we saw (at (3)) that, for an arbitrary matrix A,

$$\text{column-rank of } A^T = \text{row-rank of } A.$$

From this it now follows that:

50.2 For every matrix A, $r(A^T) = r(A)$.

Next, 40.6 can be re-stated as:

50.3 For every matrix A, $r(A)$ equals the number of nonzero rows in any echelon matrix row-equivalent to A.

In the same vein, it is a corollary of 38.5 that:

50.4 Matrices that are row equivalent to each other have the same rank.

Then from 49.1(1) we have:

50.5 For every matrix A, $r(m_A) = r(A)$.

As exercise 24 on chapter 5 pointed out, 26.2 and 40.6 combine to tell us that an $n \times n$ matrix is nonsingular if and only if its row-rank is n. Re-stating this, we have:

50.6 If $A \in F_{n \times n}$, then A is nonsingular if and only if $r(A) = n$.

Using this, we can improve part of 47.5(d) to:

50.7 If A is a square matrix, A is nonsingular if and only if m_A is nonsingular.

Proof. Let $A \in F_{n \times n}$, so that m_A is a linear transformation of the n-dimensional vector space $F_{n \times 1}$. Then we have:

$$\begin{aligned} A \text{ is nonsingular} &\Leftrightarrow r(A) = n \quad &&\text{(by 50.6)} \\ &\Leftrightarrow r(m_A) = n \quad &&\text{(by 50.5)} \\ &\Leftrightarrow m_A \text{ is nonsingular} \quad &&\text{(by 49.6);} \end{aligned}$$

and the result is established.

A further corollary of 50.6 is:

50.8 For $A \in F_{n \times n}$, the following three statements are equivalent to one another:

(1) A is nonsingular;
(2) the rows of A form a L.I. sequence;
(3) the columns of A form a L.I. sequence.

Proof. Consider first a sequence L of n vectors in any vector space. If L is L.I., then it is a basis of the space it spans and so that space has dimension n. If, on the other hand, L is L.D., then it follows by 40.2 that the space spanned by the sequence L will have dimension less than n. Therefore, the space spanned by the sequence L has dimension n if and only if L is L.I.

Taking the n vectors in L to be the rows of A, we deduce that $r(A) = n$ if and only if the rows of A form a L.I. sequence. The equivalence of (1) and (2) now follows by 50.6; and the equivalence of (1) and (3) is proved by applying the same argument to columns instead of rows. Hence, as claimed, all three statements are equivalent.

The final result is one whose truth has been hinted at in exercises on chapters 2 and 3. It can now be seen in a new perspective and proved very simply.

50.9 Let $A \in F_{n \times n}$. Then there exists a nonzero column X such that $AX = O$ if and only if A is singular.

Proof. There exists nonzero $X \in F_{n \times 1}$ such that $AX = O$ (i.e. $m_A X = O$)

$\Leftrightarrow \ker m_A$ contains a nonzero vector
$\Leftrightarrow s(m_A) > 0$
$\Leftrightarrow r(m_A) < n$ (by 49.2, $F_{n \times 1}$ being the domain of m_A)
$\Leftrightarrow r(A) < n$ (by 50.5)
$\Leftrightarrow A$ is singular (by 50.6).

51. Systems of linear equations revisited

In this section we return to consideration of an arbitrary system of linear equations $AX = K$, where the notational details are as usual (cf. §27).

Further to 48.6, the following proposition concisely describes the main features of the homogeneous case (i.e. the case $K = O$).

51.1 For the homogeneous system $AX = O$,
(i) the solution space (i.e. $\ker m_A$) has dimension $n - r(A)$;
(ii) there is a non-trivial solution if and only if $r(A) < n$.

Proof. (i) $\dim(\ker m_A) = s(m_A)$

$= n - r(m_A)$ (by 49.2, $F_{n \times 1}$ being the domain of m_A)
$= n - r(A)$ (by 50.5).

(ii) follows at once because there will be a non-trivial solution (i.e. a nonzero column in the solution space) if and only if this dimension is positive.

The earlier result 28.4 can now be seen in a broader perspective. The rank of A, being the dimension of the space spanned by the m rows of A, cannot exceed m (by 40.7). So if the number of equations is less than the number of unknowns (i.e. $m < n$), then $r(A)$ must be less than n, and thus 51.1 guarantees the existence of a non-trivial solution (as asserted by 28.4 and proved there by a much less sophisticated method).

Returning from these remarks about homogeneous systems to discussion of a general system $AX = K$, we show in the next proposition how the criterion 28.2 for consistency/inconsistency can be re-expressed in terms of ranks.

51.2 The system $AX = K$ is consistent if and only if $r([A \;\; K]) = r(A)$.

Proof. Let the augmented matrix $[A \;\; K]$ be transformed by e.r.o.s to a reduced echelon matrix $[B \;\; L]$ ($B \in F_{m \times n}, L \in F_{m \times 1}$), so that (cf. remark following 28.1) B is a reduced echelon matrix row equivalent to A. By 50.3, $r([A \;\; K])$ and $r(A)$ are the numbers of nonzero rows in $[B \;\; L]$ and B, respectively.

Clearly $[B \;\; L]$ has at least as many nonzero rows as B. Moreover, it is easily seen that their numbers of nonzero rows are the same except in the case where the last nonzero row of $[B \quad L]$ is

$$[0 \quad \dots \quad 0 \quad 0 \mid 1],$$

in which case $[B \quad L]$ has one more nonzero row than B.

Therefore, $r([A \quad K]) = r(A)$ if and only if the last nonzero row of $[B \quad L]$ is *not* $[0 \quad \dots \quad 0 \quad 0 \mid 1]$.

The stated result now follows by 28.2.

The truth of 51.2 can also be proved nicely through consideration of column spaces (cf. exercise 7 on chapter 5).

There are interesting things to say about the shape of the general solution of $AX = K$ in the consistent case and its relationship to ker m_A, the solution space of $AX = O$ (which system we term the "corresponding homogeneous system").

51.3 Suppose that the system $AX = K$ is consistent and that X_0 is one particular solution (i.e. a specific column with the property that $AX_0 = K$). Then the solution set of the system is

$$\{X_0 + Z : Z \in \ker m_A\}.$$

(*Note*: in the non-homogeneous case the solution set is *not* a subspace of $F_{n\times 1}$ since it does not contain O.)

Proof of 51.3. For $X \in F_{n\times 1}$,

$$\begin{aligned} X \text{ is a solution} &\Leftrightarrow AX = K \Leftrightarrow AX = AX_0 \qquad (X_0 \text{ being a solution}) \\ &\Leftrightarrow A(X - X_0) = O \Leftrightarrow X - X_0 \in \ker m_A \\ &\Leftrightarrow X - X_0 = Z \text{ for some } Z \in \ker m_A \\ &\Leftrightarrow X = X_0 + Z \text{ for some } Z \in \ker m_A. \end{aligned}$$

Hence, as claimed, the solution set is the set of all columns of the form $X_0 + Z$ with $Z \in \ker m_A$.

Since ker m_A is the solution space of the corresponding homogeneous system, 51.3 can be re-expressed as follows:

$$\begin{pmatrix} \text{typical solution} \\ \text{of } AX = K \end{pmatrix} = \begin{pmatrix} \text{particular} \\ \text{solution } X_0 \end{pmatrix} + \begin{pmatrix} \text{typical solution} \\ \text{of C.H.S.} \end{pmatrix},$$

("C.H.S." standing for "corresponding homogeneous system").

A specific example should bring this to life. For the simple system

$$\left\{ \begin{aligned} x_1 - 2x_2 \qquad + x_4 &= 1 \\ x_3 - x_4 &= 2 \end{aligned} \right\},$$

the general solution is

$$x_4 = \alpha,\ x_3 = 2 + \alpha,\ x_2 = \beta,\ x_1 = 1 - \alpha + 2\beta \qquad (\alpha, \beta \in F).$$

This can be expressed as

$$X = \text{col}(1-\alpha+2\beta, \beta, 2+\alpha, \alpha) \qquad (\alpha, \beta \in F),$$

$$\text{i.e.} \quad X = \underbrace{\text{col}(1,0,2,0)}_{\substack{\text{This is a}\\ \text{particular}\\ \text{solution.}}} + \underbrace{\alpha\,\text{col}(-1,0,1,1) + \beta\,\text{col}(2,1,0,0)}_{\substack{\text{This is the typical solution of}\\ \text{the C.H.S. [its solution space}\\ \text{being the space spanned by}\\ (\text{col}(-1,0,1,1),\ \text{col}(2,1,0,0))].}} \qquad (\alpha, \beta \in F).$$

And thus the pattern of 51.3 is made apparent.

Remarks. (*a*) With the aid of this illustration, the student should be able to see that 51.3 and the fact that $\dim(\ker m_A) = n - r(A)$ together shed new light on the earlier result 28.3 about the number of parameters in a general solution in the consistent case.

(*b*) Further to a remark in §45(*c*), let us now consider a generalization of 51.3 that gives information about all equations of the form

linear mapping acting on "unknown" = given vector.

The generalization (proved by the obvious adaptation of the proof of 51.3) is as follows.

51.4 Let $a: V \to W$ be a linear mapping and $\mathbf{k}$ a specific vector in W; and consider the equation $a\mathbf{x} = \mathbf{k}$. Suppose that this equation has at least one solution and that $\mathbf{x}_0$ is one particular solution (i.e. one particular vector in V satisfying $a\mathbf{x}_0 = \mathbf{k}$). Then the set of all solutions is $\{\mathbf{x}_0 + \mathbf{z} : \mathbf{z} \in \ker a\}$.

This can be expressed as:

$$\begin{pmatrix}\text{typical solution}\\ \text{of } a\mathbf{x} = \mathbf{k}\end{pmatrix} = \begin{pmatrix}\text{particular}\\ \text{solution } \mathbf{x}_0\end{pmatrix} + \begin{pmatrix}\text{typical solution of}\\ \text{the C.H.E. } a\mathbf{x} = \mathbf{0}\end{pmatrix}$$

("C.H.E." standing for "corresponding homogeneous equation").

It is probable that the student will have encountered the application of this to certain differential equations. Let us illustrate with a specific case—the equation $d^2y/dx^2 - 5(dy/dx) + 6y = e^x$, which can be written as $a(y(x)) = e^x$, where $y(x)$ is the "unknown" and a is the linear mapping (or linear differential operator) $d^2/dx^2 - 5(d/dx) + 6$. In this case, one particular solution (for $y(x)$) is $\frac{1}{2}e^x$, while the typical solution of the C.H.E. (i.e. of $d^2y/dx^2 - 5(dy/dx) + 6y = 0$) is $\alpha e^{2x} + \beta e^{3x}$. (This typical solution of the C.H.E. is often called the "complementary function" in this context.) In conformity with 51.4, the general solution of the given differential equation is

$$y(x) = \tfrac{1}{2}e^x + \alpha e^{2x} + \beta e^{3x}.$$

After this digression, let us return to the arbitrary system of linear equations

$AX = K$ and consider three cases as follows:

(i) $r([A \quad K]) \neq r(A)$;

(ii) $r([A \quad K]) = r(A) = n$;

(iii) $r([A \quad K]) = r(A) < n$.

These cases cover all possibilities ($r(A)$ being necessarily $\leqslant n$, since $n - r(A)$ is dim (ker m_A) [cf. 51.1(i)]).

In case (i), the system has no solutions (by 51.2). In cases (ii) and (iii) the system is consistent (again by 51.2). In case (ii), ker m_A is $\{O\}$ (by 51.1(i)), and so by 51.3 the system has precisely one solution. In case (iii), dim (ker m_A) > 0 (by 51.1(i)), and so by 51.3 the system has more than one solution. Thus the cases (i), (ii), (iii) correspond exactly to the system having no solutions, precisely one solution, and more than one solution, respectively. This information is conveniently summarized in the following table.

51.5

Number of solutions of $AX = K$	Ranks of $[A \quad K]$ and A
None	$r([A \quad K]) \neq r(A)$
Precisely one	$r([A \quad K]) = r(A) = n$
More than one	$r([A \quad K]) = r(A) < n$

(*Note*: in referring to the table, remember that n is our notation for the number of unknowns.)

Worked example. For which value(s) of c does the following system of equations (over $\mathbb{R}$) have more than one solution?

$$\begin{aligned} x_1 + x_2 + x_3 &= 0 \\ x_1 + 2x_2 + cx_3 &= 1 \\ x_1 + cx_2 + (2-4c)x_3 &= 3(c+1). \end{aligned}$$

Solution. We shall employ the usual $[A \quad K]$ notation for the augmented matrix of the system. By a straightforward echelon reduction, we find that

$$[A \quad K] \sim \left[\begin{array}{ccc|c} 1 & 1 & 1 & 0 \\ 0 & 1 & c-1 & 1 \\ 0 & 0 & -c(c+2) & 2(c+2) \end{array}\right].$$

(*Note.*(1) Since our interest is in ranks, transformation to echelon form is sufficient (cf. 50.3); transformation to reduced echelon form would be an unnecessary refinement. (2) In the echelon matrix that we have produced, the first 3 columns form an echelon matrix row equivalent to A.)

From the above echelon matrix, it is apparent (by 50.3) that $r(A)$ is 2 if $c = 0$ or -2, while $r(A)$ is 3 for all other values of c. Further, when $c = 0$, $r([A \quad K]) = 3$; and when $c = -2$, $r([A \quad K]) = 2$. Hence $c = -2$ is the one and only case where $r(A)$ and $r([A \quad K])$ are equal and have a value less than 3.

It follows (cf. 51.5) that the system has more than one solution if and only if $c = -2$.

The final result in this section gives a simple criterion for $AX = K$ to have precisely one solution in the case where A is square.

51.6 Suppose that A is square, of type $n \times n$. Then the system $AX = K$ has precisely one solution if and only if A is nonsingular.

Proof. If the system $AX = K$ has precisely one solution, then $r(A) = n$ (by 51.5) and so A is nonsingular (by 50.6). Conversely, if A is nonsingular, then (cf. 20.6) the system has $X = A^{-1}K$ as its one and only solution. These observations prove the stated result.

(*Note.* If A is square and singular, the system $AX = K$ may either have no solutions or have more than one solution.)

52. Rank inequalities

In this section, U, V, W will denote f.d. vector spaces over F. The section contains a number of interesting theorems among which inequalities about ranks (of linear mappings or matrices) are predominant.

We begin with a very simple result.

52.1 For a linear mapping $a: V \to W$, $r(a) \leqslant \dim V$ and $r(a) \leqslant \dim W$.

Proof. (i) $\dim(aV) \leqslant \dim V$ (by 48.4(*b*)); and thus $r(a) \leqslant \dim V$.

(ii) $\dim(aV) \leqslant \dim W$, by 41.4(ii), aV being a subspace of W; and thus $r(a) \leqslant \dim W$.

The next result is less trivial and of considerable importance.

52.2 Let $a: V \to W$ and $b: U \to V$ be linear mappings. Then

$$\text{(i) } r(ab) \leqslant r(a), \text{ and (ii) } r(ab) \leqslant r(b).$$

Proof. (i) Since $bU \subseteq V$, it is clear that $a(bU) \subseteq aV$; i.e. (cf. 9 in Appendix) $(ab)U$ is a subspace of aV; i.e. $\text{im}(ab)$ is a subspace of $\text{im } a$. On taking dimensions (cf. 41.4(ii)) we deduce $r(ab) \leqslant r(a)$.

(ii) By 48.4(*b*), $\dim(a(bU)) \leqslant \dim(bU)$; i.e. $\dim((ab)U) \leqslant \dim(bU)$; i.e. $\dim(\text{im}(ab)) \leqslant \dim(\text{im } b)$; i.e. $r(ab) \leqslant r(b)$.

The basic message of 52.2 is:

$r(\text{product}) \leqslant r(\text{any factor})$ ("product" meaning "composition" here).

By suitable applications of the result 52.2 itself, we can obtain many variations on this theme: e.g.

$$r(abc) \leqslant r(c); \; r(abc) \leqslant r(bc); \; r(abcd) \leqslant r(bc).$$

By clever use of 52.2 we are able to prove the following interesting theorem.

52.3 Let p, a, q denote linear mappings with f.d. domains and codomains such that the compositions pa and aq exist (and hence paq exists too). Suppose that p and q are bijective. Then

$$r(pa) = r(aq) = r(paq) = r(a).$$

Proof. By 52.2, $r(pa) \leqslant r(a)$.

But, since p^{-1} exists (p being bijective) we can make pa play the role of factor and a the role of product by noting that $a = p^{-1}(pa)$; and hence, by 52.2, $r(a) \leqslant r(pa)$.

It now follows that:

$$r(a) = r(pa). \tag{1}$$

By an obvious similar argument,

$$r(a) = r(aq). \tag{2}$$

Hence further:

$$\begin{aligned} r(paq) &= r(pa) \qquad \text{(by (2) with } a \text{ replaced by } pa) \\ &= r(a) \qquad \text{(by (1))}. \end{aligned}$$

The full proposition is now proved.

The basic message of 52.3 is that multiplication by (i.e. composition with) a bijective linear mapping (in particular a nonsingular linear transformation) does not change the rank.

A simple consequence is:

52.4 If $a: V \to W$ is a linear mapping and λ a nonzero scalar, then $r(\lambda a) = r(a)$.

(This follows from 52.3 because, if a and λ are as described, $\lambda a = (\lambda i_W)a$ and λi_W is nonsingular, having inverse $(1/\lambda)i_W$.)

As a preliminary to a further result, notice that, by 49.6 and 52.1:

52.5 A singular linear transformation of an n-dimensional vector space has rank less than n.

The further result is:

52.6 Suppose that a, b are linear transformations of V, at least one of which is singular. Then ab is also singular.

Proof. Let $\dim V = n$. By 52.5, at least one of a, b has rank less than n, and hence, by 52.2, $r(ab) < n$. So, by 49.6, ab is singular.

(*Note.* In 52.6, as throughout the section, V denotes a f.d. space. The result 52.6 does not extend to infinite-dimensional spaces.)

In contrast to 52.2 (which gave an inequality of the form $r(ab) \leqslant \ldots$), the next theorem gives an inequality of the form $r(ab) \geqslant \ldots$.

52.7 Let $a: V \to W$ and $b: U \to V$ be linear mappings. Then

$$r(ab) \geqslant r(a)+r(b)-\dim V.$$

Proof. Consider $a|_{bU}$, whose image space is $a(bU)\,(=(ab)U)$ and whose rank is, therefore, $r(ab)$. Hence

$$\begin{aligned} r(ab) &= \dim(\text{domain of } a|_{bU}) - s(a|_{bU}) \qquad \text{(by 49.2)} \\ &= \dim(bU) - s(a|_{bU}) = r(b) - \dim(\ker(a|_{bU})). \end{aligned}$$

But $\ker(a|_{bU}) = \{\mathbf{x} \in bU : a\mathbf{x} = \mathbf{0}\} = bU \cap \ker a$, which is a subspace of $\ker a$. Hence

$$\dim(\ker(a|_{bU})) \leqslant \dim(\ker a) = s(a).$$

From the above it now follows that

$$\begin{aligned} r(ab) &\geqslant r(b) - s(a) \\ &= r(b) - (\dim V - r(a)) \qquad \text{(by 49.2)} \\ &= r(a) + r(b) - \dim V, \end{aligned}$$

and thus the stated result is established.

As a last rank inequality for linear mappings, here now is an inequality for the sum of two linear mappings.

52.8 Let a, b be linear mappings from V to W. Then

$$r(a+b) \leqslant r(a)+r(b).$$

Proof. We begin by proving that $(a+b)V \subseteq aV+bV$.

Let $\mathbf{y} \in (a+b)V$. Then, for some $\mathbf{x} \in V$,

$$\mathbf{y} = (a+b)\mathbf{x} = a\mathbf{x} + b\mathbf{x} \qquad \text{(cf. meaning of } a+b\text{)};$$

and thus $\mathbf{y} \in aV+bV$.

This proves that $(a+b)V \subseteq aV+bV$. Hence

$$\dim((a+b)V) \leqslant \dim(aV+bV); \text{ i.e. } r(a+b) \leqslant \dim(aV+bV).$$

But

$$\begin{aligned} \dim(aV+bV) &\leqslant \dim(aV)+\dim(bV) \qquad \text{(by 42.3)} \\ &= r(a)+r(b). \end{aligned}$$

Hence, as stated, $r(a+b) \leqslant r(a)+r(b)$.

By applying all these results in the case where $a, b, \ldots$ are linear mappings induced by matrices and by using certain obviously relevant results (50.5, 47.5, 50.7), we easily obtain matrix versions of the above results. These matrix versions are as follows.

52.1M If $A \in F_{m \times n}$, then $r(A) \leqslant m$ and $r(A) \leqslant n$.

52.2M If the matrix product AB exists, then $r(AB) \leqslant r(A)$ and $r(AB) \leqslant r(B)$.

52.3M If $P \in F_{m \times m}$, $A \in F_{m \times n}$, $Q \in F_{n \times n}$, and if P and Q are nonsingular, then $r(PA) = r(AQ) = r(PAQ) = r(A)$.

52.4M If A is a matrix and λ a nonzero scalar, then $r(\lambda A) = r(A)$.

52.5M If A is a singular $n \times n$ matrix, then $r(A) < n$.

52.6M If $A, B \in F_{n \times n}$ and at least one of A, B is singular, then AB is singular.

52.7M If $A \in F_{l \times m}$ and $B \in F_{m \times n}$, then $r(AB) \geqslant r(A)+r(B)-m$.

52.8M If $A, B \in F_{m \times n}$, then $r(A+B) \leqslant r(A)+r(B)$.

As an illustration of how these may be deduced straightforwardly from the previous results 52.1, …, 52.8, here is a proof of 52.8M.

Let $A, B \in F_{m \times n}$. Then

$$\begin{aligned} r(A+B) &= r(m_{A+B}) \qquad \text{(by 50.5)} \\ &= r(m_A + m_B) \qquad \text{(by 47.5}(b)) \\ &\leqslant r(m_A)+r(m_B) \qquad \text{(by 52.8)} \\ &= r(A)+r(B) \qquad \text{(by 50.5).} \end{aligned}$$

Note that 52.6M is the same as 26.7. It is interesting that this result on matrices corresponds to a result on linear transformations that is valid only in the finite-dimensional case.

Worked example (on the use of rank inequalities). An $n \times n$ matrix A is such that $A^2 = I$. Show that

$$r(I+A)+r(I-A) = n.$$

Solution. $(I+A)(I-A) = I - A^2 = O$ (since $A^2 = I$). Hence:

$$\begin{aligned} 0 &= r[(I+A)(I-A)] \\ &\geqslant r(I+A)+r(I-A)-n \qquad \text{(by 52.7M),} \end{aligned}$$

and so: $r(I+A)+r(I-A) \leqslant n$. (α)

On the other hand, $(I+A)+(I-A) = 2I$, which is nonsingular and so has rank n. Hence:

$$n = r[(I+A)+(I-A)]$$
$$\leqslant r(I+A)+r(I-A) \qquad \text{(by 52.8M)};$$

and thus: $r(I+A)+r(I-A) \geqslant n$. (β)

From (α) and (β) it follows that $r(I+A)+r(I-A) = n$, as required.

53. Vector spaces of linear mappings

Let V, W be any two vector spaces over F. We denote by $\hom(V \to W)$ the set of all linear mappings from V to W. (Here "hom" stands for "homomorphism". This word arises in broader discussions of mappings between algebraic systems, but, in the narrow context of vector space theory, may be regarded as a synonym for "linear mapping").

We have already, in §47, given a meaning to $a+b$ when $a, b \in \hom(V \to W)$ and to λa when $a \in \hom(V \to W)$ and $\lambda \in F$; and in every case $a+b$ and λa also belong to $\hom(V \to W)$. Thus addition and multiplication have been defined on the set $\hom(V \to W)$, and this set is closed under both these operations. One can go further and prove that:

53.1 Under the operations of addition and multiplication by scalars that have been introduced, $\hom(V \to W)$ is a vector space (over F).

To prove this in full, we would have to verify that all the vector space axioms (A1),...,(A4), (M1),...,(M4) hold. ((The closure axioms (A0) and (M0) have already been commented on.) The eight remaining axioms take some time to work through, but in each case the verification is straightforward provided one appreciates that the equality of two mappings from V to W is established by showing that the mappings have the same effect on every vector in V. The axiom (A3) is shown to hold by checking that the zero mapping from V to W acts as a zero object in $\hom(V \to W)$.

As a non-trivial illustration of the axiom-verification process, let us now prove that (M1) holds for $\hom(V \to W)$. Our goal is to prove that

$$\lambda(a+b) = \lambda a + \lambda b \text{ for all } a, b \in \hom(V \to W) \text{ and for all } \lambda \in F.$$

So we proceed by introducing arbitrary $a, b \in \hom(V \to W)$ and $\lambda \in F$. Both $\lambda(a+b)$ and $\lambda a + \lambda b$, then, are mappings from V to W. Further, for all $\mathbf{x} \in V$,

$$\begin{aligned}
(\lambda(a+b))\mathbf{x} &= \lambda((a+b)\mathbf{x}) && \text{(by meaning of } \lambda c \text{ for } c \in \hom(V \to W))\\
&= \lambda(a\mathbf{x}+b\mathbf{x}) && \text{(by meaning of } a+b)\\
&= \lambda(a\mathbf{x})+\lambda(b\mathbf{x}) && \text{(by the property (M1) of } V)\\
&= (\lambda a)\mathbf{x}+(\lambda b)\mathbf{x} && \text{(by meaning of } \lambda a \text{ and } \lambda b)\\
&= (\lambda a+\lambda b)\mathbf{x} && \text{(by meaning of sum of two linear mappings).}
\end{aligned}$$

This proves that $\lambda(a+b) = \lambda a + \lambda b$; and since a, b, λ were arbitrary, our goal is achieved.

A special case of hom $(V \to W)$ is hom $(V \to F)$, the vector space of all linear mappings from V (a vector space over F) into F $(=F^1)$. This vector space is called the **dual space** of V, and its elements are often called **linear functionals** (with domain V). In the case $V = F^n$, it can be shown that each linear functional has the form

$$(x_1, x_2, \ldots, x_n) \mapsto \alpha_1 x_1 + \alpha_2 x_2 + \ldots + \alpha_n x_n$$

$(\alpha_1, \alpha_2, \ldots, \alpha_n$ fixed elements of $F)$.

Another interesting special case of hom $(V \to W)$ is hom $(V \to V)$, whose member objects are the linear transformations of the arbitrary vector space V. The word "endomorphism" can be used for "linear transformation", and accordingly we shall use the notation end V for hom $(V \to V)$. In view of 53.1, end V is certainly a vector space. Notice further that it is a vector space whose member objects can be multiplied together: if $a, b \in$ end V, then ab (in accordance with our established use of notation) means $a \circ b$, which is also a member of end V.

For students familiar with the meaning of the word "ring" in algebra, the situation can be summarized by saying that the system end V is both a vector space and a ring and that the operations of multiplying members of end V and of multiplying them by scalars behave nicely together in the sense that

$$(\lambda a)b = a(\lambda b) = \lambda(ab) \qquad \text{for all } a, b \in \text{end } V \text{ and all } \lambda \in F.$$

For students unfamiliar with the term "ring", it is perhaps best to say that the properties of the system end V with regard to the three operations (addition, multiplication by scalars, and multiplication) are exactly parallel to the properties of $F_{n \times n}$ with regard to the three operations defined on it (addition, multiplication by scalars, and multiplication).

Whichever account of the properties of the system end V is preferred, one should certainly pause to recognize that the above accounts make a number of claims about multiplication (i.e. composition) of linear transformations of V, not all of which can be dismissed as obvious. For example, there are the distributive laws:

$$a(b+c) = ab + ac \qquad \text{and} \qquad (b+c)a = ba + ca$$

for all $a, b, c \in$ end V. It is instructive to work through a proof of one of these. Let's take the first.

Let $a, b, c \in$ end V. Then both $a(b+c)$ and $ab + ac$ are mappings from V to V.

Further, for all $x \in V$,

$$
\begin{aligned}
(a(b+c))\mathbf{x} &= a((b+c)\mathbf{x}) && \text{(since } a(b+c) \text{ means } a \circ (b+c)) \\
&= a(b\mathbf{x}+c\mathbf{x}) && \text{(by meaning of } b+c) \\
&= a(b\mathbf{x})+a(c\mathbf{x}) && \text{(since } a \text{ is linear)} \\
&= (ab)\mathbf{x}+(ac)\mathbf{x} && (ab \text{ meaning } a \circ b, \text{ etc.)} \\
&= (ab+ac)\mathbf{x} && \text{(by meaning of sum of two linear mappings).}
\end{aligned}
$$

Hence it follows that $a(b+c) = ab+ac$, as claimed.

Systems like end V and $F_{n\times n}$, where there are three operations (viz. addition of member objects, multiplication of member objects by scalars, and multiplication of member objects) and where these operations have properties conforming to the pattern familiar in the case of $F_{n\times n}$, are called **associative algebras**.

EXERCISES ON CHAPTER SIX

1. Let A be a fixed matrix in $F_{n\times n}$. Show that the mapping $b: F_{n\times n} \to F_{n\times n}$ given by $b(X) = AX - XA$ ($X \in F_{n\times n}$) is a linear transformation of $F_{n\times n}$.

2. (i) A linear mapping $a: \mathbb{R}^3 \to \mathbb{R}^4$ is such that $a(1,0,0) = (2,-1,0,4)$, $a(0,1,0) = (1,3,-4,7)$, $a(0,0,1) = (0,0,5,2)$. Obtain a general formula for $a(x_1, x_2, x_3)$.

(ii) A linear transformation b of $\mathbb{R}^3$ is such that $b(1,1,1) = (1,-1,1)$, $b(1,1,0) = (-2,1,-1)$, $b(1,0,0) = (3,1,0)$. Obtain a general formula for $b(x_1, x_2, x_3)$.

(iii) Is there a linear transformation c of $\mathbb{R}^2$ such that $c(1,0) = (2,3)$, $c(1,1) = (2,4)$ and $c(1,-1) = (2,2)$?

(iv) Is there a linear transformation d of $\mathbb{R}^3$ such that $d(1,1,0) = (1,2,3)$, $d(1,-1,1) = (-1,0,2)$ and $d(1,-3,2) = (-3,-2,0)$?

3. Let $A, B \in F_{m\times n}$. Show that if $m_A = m_B$, then $A = B$.

4. Having noted that $\begin{bmatrix} 0 & 0 \\ 1 & 0 \end{bmatrix}^2$ is O, write down a nonzero linear transformation a of $\mathbb{R}^2$ such that $a^2 = O$.

5. Find bases of the images and kernels of the linear transformations a, b of $\mathbb{R}^3$ defined by

$$a(x_1, x_2, x_3) = (x_1 + 2x_2 + x_3, x_1 + 2x_2 + x_3, 2x_1 + 4x_2 + 2x_3),$$
$$b(x_1, x_2, x_3) = (x_1 + 2x_2 + 3x_3, x_1 - x_2 + x_3, x_1 + 5x_2 + 5x_3).$$

6. Suppose that a, b are linear transformations of a f.d. vector space V such that $ab = i_V$. Prove that ba is also equal to i_V. (*Hint*: prove a surjective and apply 49.5.) (*Note*: this supplies the analogue, for linear transformations of f.d. vector spaces, of 26.5.)

7. Using the (clearly linear) transformations c, d of F^∞ defined by

$$c(x_1, x_2, x_3, x_4, x_5, \ldots) = (x_2, x_3, x_4, x_5, x_6, \ldots),$$
$$d(x_1, x_2, x_3, x_4, x_5, \ldots) = (0, x_1, x_2, x_3, x_4, \ldots),$$

show that the following statements about arbitrary linear transformations a, b of a vector space V, though true if V is f.d., do not extend to the case where V is infinite-dimensional:

(i) if a is injective, then a is bijective;
(ii) if a is surjective, then a is bijective;
(iii) if a is singular, ab is also singular;
(iv) if $ab = i_V$, then $ba = i_V$.

8. Let a be a linear transformation of the vector space V. Without assuming V to be f.d., show that if a is idempotent (i.e. $a^2 = a$) then $V = aV \oplus \ker a$. (*Hint*: $\mathbf{x} = a\mathbf{x} + (\mathbf{x} - a\mathbf{x})$.)

9.* Let a be a linear transformation of the f.d. vector space V such that $a^3 = a^2$ and $r(a) = r(a^2)$. Show that a is idempotent.

10.* Let b, c denote linear transformations of the f.d. vector space V. Show that if $r(bc) = r(c)$, then $\ker(bc) = \ker c$ and $cV \cap \ker b = \{\mathbf{0}\}$. Deduce that

$$r(b) = r(c) = r(bc) \Rightarrow V = cV \oplus \ker b.$$

Prove also the converse of this statement.

11.* Suppose that a is a linear transformation of the n-dimensional vector space V such that $a^n = O$ but $a^{n-1} \neq O$. By considering strict containments in the descending chain $V \supseteq aV \supseteq a^2V \supseteq a^3V \supseteq \ldots$, show that $r(a^j) = n-j$ $(j = 1, 2, \ldots, n)$.

12. Use 50.8 to prove that if A, B are square matrices and B is singular, then also singular are all matrices of the forms

$$\begin{bmatrix} A & O \\ X & B \end{bmatrix} \quad \text{and} \quad \begin{bmatrix} A & Y \\ O & B \end{bmatrix}.$$

(Cf. exercise 8 on chapter 4.)

13. Establish the truth or falsehood of each of the following assertions about an arbitrary system of linear equations $AX = K$.

(i) If the system has more equations than unknowns, there cannot be more than one solution. (ii) If $r(A)$ equals the number of equations, then the system is consistent. (iii) If the system is consistent and there are fewer equations than unknowns, then there is more than one solution.

14. Show that the system of equations

$$\begin{aligned} x_1 - 3x_2 + x_3 + cx_4 &= b \\ x_1 - 2x_2 + (c-1)x_3 - x_4 &= 2 \\ 2x_1 - 5x_2 + (2-c)x_3 + (c-1)x_4 &= 3b+4 \end{aligned}$$

is consistent for all values of b if $c \neq 1$. Find the value of b for which the system is consistent if $c = 1$, and find the general solution when b has this value and $c = 1$.

15.* Show that the system of equations

$$\begin{aligned} x_1 + x_2 + x_3 &= 1 \\ x_1 + cx_2 + cx_3 &= b \\ x_1 + c^2x_2 + 2cx_3 &= bc \end{aligned}$$

is consistent whenever $b = c$. If $b \neq c$, can the system have more than one solution?

16. Let a, b denote linear mappings from V to W (V, W being vector spaces over F) and let S denote a subspace of V. Show that it need not be true that $(a+b)S = aS + bS$.

17. (i) Let a be a linear transformation of the f.d. vector space V. Prove that $r(a^2) \geqslant 2r(a) - \dim V$ and $r(a^3) \geqslant 3r(a) - 2(\dim V)$. Write down the corresponding general result of the form $r(a^k) \geqslant \ldots$, and note this result for future reference.

(ii) Prove that if $A \in F_{10 \times 10}$ and $r(A) = 8$, then $A^4 \neq O$.

18. (i) Let a, b be linear mappings from V to W (V, W being f.d. vector spaces over F). Prove that $r(a+b) \geqslant |r(a) - r(b)|$.

(ii) Suppose that the matrices A, B, C, D in $F_{8 \times 8}$ have ranks 7, 6, 5, 3, respectively. Prove that $r(AB + CD) \geqslant 2$.

(iii)* Suppose that $P, Q \in F_{n \times n}$, that P is nonsingular, and that $(P+Q)^3 = O$. Prove that $r(Q) \geqslant \frac{1}{3}n$.

19. Let A be an $m \times n$ matrix, where $m \neq n$. Show that there cannot exist both an $n \times m$ matrix P such that $PA = I_n$ and an $n \times m$ matrix R such that $AR = I_m$.

20.* Suppose that the linear transformations a, b of a f.d. vector space V are such that $aba = a$. Prove that $r(a) = r(ab)$. Suppose further that $r(a) = r(b)$. Show that $a|_{bV}$ is injective, and deduce that $bab = b$.

21. Suppose that $A \in F_{m \times n}$ and $B \in F_{n \times m}$, where $m \geqslant n$, and that BA is nonsingular. Prove that $r(A) = r(B) = r(AB) = n$.

22. Let a be a linear transformation of a f.d. vector space V. Prove that, for every subspace W of V,

$$\dim(\ker a \cap W) = \dim W - \dim(aW).$$

By applying this with $W = aV$ and with $W = a^2V$, prove that $r(a^2) \leqslant \frac{1}{2}(r(a)+r(a^3))$.
Deduce that if $\dim V = 5$ and $a^3 = O$ while $a^2 \neq O$, then $r(a^2) = 1$ and $r(a) = 2$ or 3.

23.* The linear transformations a, b of the f.d. vector space V are such that $a^2 = b^2$ and $a+b$ is nonsingular. Prove that

$$r(a) = r(b) \geqslant \tfrac{1}{2}(\dim V).$$

24.* Let b, c be linear transformations of an n-dimensional vector space V such that bc is idempotent. Prove that $(cb)^3 = (cb)^2$. By considering $r((bc)^3)$, prove also that $r((cb)^2) = r(bc)$.

Now suppose that $r(bc) = n-1$. Show that $r(cb)$ is also $n-1$; and deduce, using question 9 above, that cb is idempotent too.

25.* Let V be a f.d. vector space over $\mathbb{R}$.

(i) Show that if a and b are linear transformations of V such that $r(a+b) = r(a)+r(b)$, then $aV \cap bV = \{\mathbf{0}\}$.

(ii) Now suppose that a is a linear transformation of V with the property that $r(i_V+a)+r(i_V-a) = n$, where $n = \dim V$. By applying part (i), show that $(i_V+a)V \cap (i_V-a)V = \{\mathbf{0}\}$; and, by considering $(i_V-a^2)\mathbf{x}$ for arbitrary $\mathbf{x} \in V$, deduce that $a^2 = i_V$. (Note that the corresponding matrix result to this is the converse of that proved in the worked example of §52.)

26.* Let a be a linear transformation of the f.d. vector space V. Establish the existence of a positive integer j such that

$$a^jV = a^{j+1}V = a^{j+2}V = a^{j+3}V = \ldots,$$

and prove that, for the same integer j,

(i) $\ker(a^j) = \ker(a^{j+1}) = \ker(a^{j+2}) = \ker(a^{j+3}) = \ldots,$

(ii) $V = a^jV \oplus \ker(a^j)$.

CHAPTER SEVEN

MATRICES FROM LINEAR MAPPINGS

54. Introduction

In addition to the now very familiar convention that F denotes an arbitrary field of scalars, a further convention will apply in this chapter: throughout the chapter the symbols U, V, W, wherever they appear, will denote arbitrary *nonzero finite-dimensional* vector spaces over F. A smaller point to be explained is that the abbreviation w.r.t. will be used for "with respect to".

In chapter 6 we saw that a matrix $A \in F_{m \times n}$ gives rise to a linear mapping $m_A : F_{n \times 1} \to F_{m \times 1}$ with the fundamental property that

$$\underbrace{\text{image of column } X \text{ under } m_A}_{\text{a column}} = A \times X \tag{1}$$

for every $X \in F_{n \times 1}$. This chapter is concerned with developing the reverse idea—obtaining from a given linear mapping $a : V \to W$ a matrix M_a which in some sense corresponds to, or represents, a.

Our early thoughts on this "reverse idea" are inevitably tentative and exploratory. The first step is to introduce bases $L_V = (\mathbf{e}_1, \ldots, \mathbf{e}_n)$ of V and $L_W = (\mathbf{f}_1, \ldots, \mathbf{f}_m)$ of W, n being dim V and m being dim W. (The matrix that we shall eventually take to be M_a depends on our choice of bases at this point—a matter pursued in detail later in the chapter (§§57, 58, 59).) Since the obvious columns corresponding to vectors in V and W are their component columns w.r.t. L_V and L_W, respectively, analogy with (1) suggests that a fundamental property required of M_a should be

$$\begin{pmatrix} \text{component column of } a\mathbf{x} \\ \text{w.r.t. } L_W \end{pmatrix} = M_a \times \begin{pmatrix} \text{component column of } \mathbf{x} \\ \text{w.r.t. } L_V \end{pmatrix} \tag{2}$$

for every $\mathbf{x} \in V$. Analysis of what (2) entails soon reveals that it allows just one possibility for M_a—viz. the $m \times n$ matrix (A, let's call it) whose 1st, 2nd, 3rd, … columns are the component columns of $a\mathbf{e}_1, a\mathbf{e}_2, a\mathbf{e}_3, \ldots$, respectively, w.r.t. L_W. This discovery signals the end of the tentative exploratory phase; and soon, in §55, we shall begin the proper logical development of our idea by *defining* M_a

(the matrix of a w.r.t. the bases L_V and L_W) to be the aforementioned $m \times n$ matrix A. Working from this definition, we then prove properties of "M_a", and the equation (2) appears among these at an early stage.

As the story unfolds, we see that the correspondence $a \leftrightarrow M_a$ is a one-to-one correspondence between the mappings in $\hom(V \to W)$ and the matrices in $F_{m \times n}$ and that virtually anything that can be said about mappings $a, b, \ldots$ in $\hom(V \to W)$ is mirrored by exactly corresponding properties of the corresponding matrices $M_a, M_b, \ldots$, and vice versa. In due course we are led to feel that (as some features of chapter 6 foreshadowed) working with linear mappings $a, b, \ldots$ is effectively the same as working with corresponding matrices $M_a, M_b, \ldots$ or even that the mappings and the matrices are much the same thing appearing in two different guises. A more precise formal expression of this "feeling" can be given through the concept of isomorphism, to which we come in §60.

55. The main definition and its immediate consequences

Let $a: V \to W$ be a linear mapping. Let $n = \dim V$, $m = \dim W$; let $L_V = (\mathbf{e}_1, \ldots, \mathbf{e}_n)$ be a basis of V, and let $L_W = (\mathbf{f}_1, \ldots, \mathbf{f}_m)$ be a basis of W. Then the **matrix of** a w.r.t. L_V and L_W (also called the **matrix representing** a w.r.t. L_V and L_W) is defined to be the $m \times n$ matrix whose kth column is the component column of $a\mathbf{e}_k$ w.r.t. L_W ($k = 1, 2, \ldots, n$). We shall denote this matrix by $M(a; L_V, L_W)$ (indicating its dependence on the choice of bases L_V, L_W, as well as on a); but the shorter notation M_a may be used for the matrix when there is no doubt as to what bases of V and W are in use.

The following proposition records the most basic properties of $M(a; L_V, L_W)$.

55.1 Let $a: V \to W$, $L_V = (\mathbf{e}_1, \ldots, \mathbf{e}_n)$, $L_W = (\mathbf{f}_1, \ldots, \mathbf{f}_m)$ be as in the above definition. Write M_a for $M(a; L_V, L_W)$, and let α_{ik} be the (i, k)th entry of this matrix. Then:

(i) M_a is an $m \times n$ matrix, i.e. a $(\dim W) \times (\dim V)$ matrix;

(ii) for each $\mathbf{e}_k$ in L_V (i.e. for $k = 1, \ldots, n$), $a\mathbf{e}_k = \sum_{i=1}^{m} \alpha_{ik}\mathbf{f}_i$.

(iii) for every $\mathbf{x} \in V$,

$$\begin{pmatrix} \text{component column} \\ \text{of } a\mathbf{x} \text{ w.r.t. } L_W \end{pmatrix} = M_a \times \begin{pmatrix} \text{component column} \\ \text{of } \mathbf{x} \text{ w.r.t. } L_V \end{pmatrix};$$

(iv) property (iii) characterizes M_a—i.e. if an $m \times n$ matrix N is such that

$$\begin{pmatrix} \text{component column} \\ \text{of } a\mathbf{x} \text{ w.r.t. } L_W \end{pmatrix} = N \times \begin{pmatrix} \text{component column} \\ \text{of } \mathbf{x} \text{ w.r.t. } L_V \end{pmatrix}$$

for *all* $\mathbf{x} \in V$, then N must coincide with M_a.

Proof. (i) merely records one feature of the definition.

(ii) For each $\mathbf{e}_k$, the component column of $a\mathbf{e}_k$ w.r.t. L_W is (by the definition of M_a) the kth column of M_a, which is $\text{col}(\alpha_{1k}, \alpha_{2k}, \ldots, \alpha_{mk})$; and therefore

$$a\mathbf{e}_k = \alpha_{1k}\mathbf{f}_1 + \alpha_{2k}\mathbf{f}_2 + \ldots + \alpha_{mk}\mathbf{f}_m = \sum_{i=1}^{m} \alpha_{ik}\mathbf{f}_i.$$

(iii) Let $\mathbf{x}$ be an arbitrary vector in V; let the component column of $\mathbf{x}$ w.r.t. L_V be $\text{col}(x_1, \ldots, x_n)$; and let the component column of $a\mathbf{x}$ w.r.t. L_W be $(y_1, \ldots, y_m)$. Then

$$\begin{aligned}
\sum_{i=1}^{m} y_i\mathbf{f}_i = a\mathbf{x} &= a\left(\sum_{k=1}^{n} x_k\mathbf{e}_k\right) \\
&= \sum_{k=1}^{n} x_k(a\mathbf{e}_k) \qquad \text{(by linearity of } a\text{)} \\
&= \sum_{k=1}^{n} x_k\left(\sum_{i=1}^{m} \alpha_{ik}\mathbf{f}_i\right) \qquad \text{(by part (ii))} \\
&= \sum_{i=1}^{m}\left(\sum_{k=1}^{n} \alpha_{ik}x_k\right)\mathbf{f}_i \qquad \text{(on interchanging order of summation).}
\end{aligned}$$

By 39.10 it follows that for each i in the range 1 to m,

$$y_i = \sum_{k=1}^{n} \alpha_{ik}x_k;$$

and thus $\text{col}(y_1, \ldots, y_m) = M_a \times \text{col}(x_1, \ldots, x_n)$.

Since $\mathbf{x}$ was arbitrary in V, this proves part (iii).

(iv) Suppose that the $m \times n$ matrix N has the property described in the statement. Then, since every column in $F_{n\times 1}$ arises as the component column of some vector in V, it must be the case (in view of part (iii)) that $NX = M_aX$ for *every* column $X \in F_{n\times 1}$. Hence, by 21.3, $N = M_a$. This proves part (iv).

Remarks. (*a*) Part (iii) of 55.1 will be recognized as the desired property used in §54 to motivate the definition of M_a.

(*b*) If we write out 55.1(ii) in the form of a sequence of equations for $a\mathbf{e}_1, a\mathbf{e}_2, \ldots, a\mathbf{e}_n$ in terms of $\mathbf{f}_1, \ldots, \mathbf{f}_m$, i.e. if we write out

$$\left\{\begin{array}{l} a\mathbf{e}_1 = \alpha_{11}\mathbf{f}_1 + \alpha_{21}\mathbf{f}_2 + \ldots + \alpha_{m1}\mathbf{f}_m \\ a\mathbf{e}_2 = \alpha_{12}\mathbf{f}_1 + \alpha_{22}\mathbf{f}_2 + \ldots + \alpha_{m2}\mathbf{f}_m \\ \quad\ldots \end{array}\right\},$$

then the matrix of coefficients that confronts us is M_a^T (and not M_a itself). The situation may be remembered schematically as

$$\text{col}(a\mathbf{e}_1, \ldots, a\mathbf{e}_n) = M_a^T \times \text{col}(\mathbf{f}_1, \ldots, \mathbf{f}_m).$$

(*c*) In considering a linear transformation of V, we would normally have in mind just one basis—a basis L of V, the only vector space involved. This basis L would fill both the role of L_V and the role of L_W in the previous discussion. So if a is a linear transformation of V, we simply speak of "the matrix of a w.r.t. L" (denoted by $M(a; L)$ or by M_a), meaning what we have formerly denoted by $M(a; L, L)$.

Example 1. Consider the linear mapping a from $\mathbb{R}^3$ to $\mathbb{R}^2$ given by

$$a(x_1, x_2, x_3) = (x_1 + 2x_2 + 3x_3, 4x_1 + 5x_2 + 6x_3).$$

Let us obtain the matrix M_a of a w.r.t. the standard bases of $\mathbb{R}^3$ and $\mathbb{R}^2$. We shall denote these standard bases by $(\mathbf{e}_1, \mathbf{e}_2, \mathbf{e}_3)$ and $(\mathbf{f}_1, \mathbf{f}_2)$, respectively.

Method A. We here use the definition which tells us that the entries in the kth column of M_a are the components of $a\mathbf{e}_k$ w.r.t. $(\mathbf{f}_1, \mathbf{f}_2)$.

Since $a\mathbf{e}_1 = a(1,0,0) = (1,4) = 1\mathbf{f}_1 + 4\mathbf{f}_2$, it follows that the 1st column of M_a is col $(1,4)$. Similarly, $a\mathbf{e}_2$ and $a\mathbf{e}_3$ work out to be $2\mathbf{f}_1 + 5\mathbf{f}_2$ and $3\mathbf{f}_1 + 6\mathbf{f}_2$, and so the 2nd and 3rd columns of M_a are col $(2,5)$ and col $(3,6)$, respectively.

Hence $M_a = \begin{bmatrix} 1 & 2 & 3 \\ 4 & 5 & 6 \end{bmatrix}$.

Method B. In view of 55.1, we want the (one and only) 2×3 matrix M_a such that, in every case,

$$\begin{pmatrix} \text{component column of} \\ a(x_1, x_2, x_3) \text{ w.r.t. } (\mathbf{f}_1, \mathbf{f}_2) \end{pmatrix} = M_a \times \begin{pmatrix} \text{component column of} \\ (x_1, x_2, x_3) \text{ w.r.t. } (\mathbf{e}_1, \mathbf{e}_2, \mathbf{e}_3) \end{pmatrix},$$

$$\text{i.e. } \operatorname{col}(x_1 + 2x_2 + 3x_3, 4x_1 + 5x_2 + 6x_3) = M_a \times \operatorname{col}(x_1, x_2, x_3)$$

Hence it is apparent that $M_a = \begin{bmatrix} 1 & 2 & 3 \\ 4 & 5 & 6 \end{bmatrix}$.

Example 2. Let a be rotation of 2-dimensional space (the x, y-coordinate plane) through angle θ in the positive direction about the origin. (By "the positive direction" we mean anti-clockwise in the conventional way of drawing the x, y-plane.) It was explained in §45(*e*) that a may be regarded as a linear transformation of E_2. Let us obtain the matrix M_a of a w.r.t. the basis $(\mathbf{i}, \mathbf{j})$ of E_2.

Figure XVIII is helpful: the circle drawn is the circle with centre the origin and radius 1. We see that, under a, the point with position vector $\mathbf{i}$ goes to the point with coordinates $(\cos\theta, \sin\theta)$ and thus with position vector $(\cos\theta)\mathbf{i} + (\sin\theta)\mathbf{j}$. Thus

$$a\mathbf{i} = (\cos\theta)\mathbf{i} + (\sin\theta)\mathbf{j}.$$

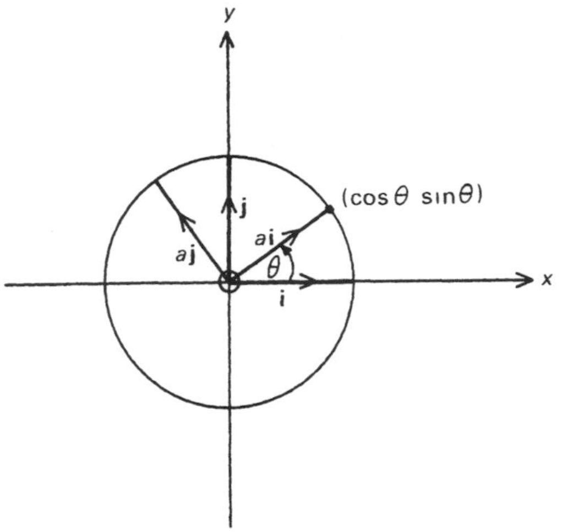

Figure XVIII

It is clear that $a\mathbf{j}$ is the same thing with θ replaced by $\theta+\frac{1}{2}\pi$. So

$$a\mathbf{j} = \cos(\theta+\tfrac{1}{2}\pi)\mathbf{i}+\sin(\theta+\tfrac{1}{2}\pi)\mathbf{j};$$
$$\text{i.e. } a\mathbf{j} = (-\sin\theta)\mathbf{i}+(\cos\theta)\mathbf{j}.$$

The equations for $a\mathbf{i}$ and $a\mathbf{j}$ give us the 1st and 2nd columns of M_a (cf. definition of M_a); and thus

$$M_a = \begin{bmatrix} \cos\theta & -\sin\theta \\ \sin\theta & \cos\theta \end{bmatrix}.$$

Postscript. In view of 55.1(iii), if the point (x, y) goes under a to the point (x', y'), then

$$\text{col}(x', y') = M_a \times \text{col}(x, y);$$
$$\text{i.e. } \begin{Bmatrix} x' = (\cos\theta)x-(\sin\theta)y \\ y' = (\sin\theta)x+(\cos\theta)y \end{Bmatrix}.$$

So by a nice simple method we have discovered what happens (in terms of coordinates) to any point under a rotation about the origin.

Consider next the identity transformation of vector space V.

55.2 For any basis L of V, $M(i_V; L) = I_n$, where $n = \dim V$.

Proof. This is clear by 55.1(iv) since, for every $\mathbf{x} \in V$,

$$\text{component column of } i_V\mathbf{x} \text{ w.r.t. } L = \text{component column of } \mathbf{x} \text{ w.r.t. } L$$
$$= I_n \times (\text{component column of } \mathbf{x} \text{ w.r.t. } L).$$

There now follows a very significant proposition.

55.3 Let dim $V = n$, dim $W = m$; let L_V and L_W be given bases of V and W; and let A be an arbitrary matrix in $F_{m \times n}$. Then there is precisely one linear mapping $a: V \to W$ such that $M(a; L_V, L_W) = A$.

Proof. Let $L_V = (\mathbf{e}_1, \ldots, \mathbf{e}_n)$; let $L_W = (\mathbf{f}_1, \ldots, \mathbf{f}_m)$; and let $A = [\alpha_{ik}]_{m \times n}$. By 46.2, there is a linear mapping $a: V \to W$ such that

$$a\mathbf{e}_k = \sum_{i=1}^{m} \alpha_{ik}\mathbf{f}_i \qquad (k = 1, 2, \ldots, n). \tag{*}$$

For this mapping a, it is apparent (from (*) and the definition of $M(a; L_V, L_W)$) that in every case the kth column of $M(a; L_V, L_W)$ is $\text{col}(\alpha_{1k}, \ldots, \alpha_{mk})$ and that therefore $M(a; L_V, L_W) = A$.

Suppose now that the linear mapping $b: V \to W$ is such that $M(b; L_V, L_W) = A$. Then (cf. 55.1(ii)), for each $\mathbf{e}_k$,

$$b\mathbf{e}_k = \sum_{i=1}^{m} \alpha_{ik}\mathbf{f}_i = a\mathbf{e}_k.$$

But (cf. the word "unique" in 46.2) there is only one linear mapping from V to W with any particular effect on $\mathbf{e}_1, \ldots, \mathbf{e}_n$. Therefore $b = a$. Thus a is the only linear mapping from V to W having matrix (w.r.t. L_V and L_W) equal to A.

The truth of the stated result is now apparent.

The proposition 55.3 is important because it shows that (when the bases L_V and L_W in use have been fixed) the correspondence

$$a \leftrightarrow M_a \qquad [M_a \text{ standing for } M(a; L_V, L_W)]$$

is a *one-to-one* correspondence between the sets $\text{hom}(V \to W)$ and $F_{m \times n}$ (where $m = \dim W$ and $n = \dim V$): our unambiguous definition of M_a lets us say that each linear mapping $a \in \text{hom}(V \to W)$ gives rise to precisely one corresponding matrix M_a, and 55.3 lets us add that each matrix $A \in F_{m \times n}$ arises from precisely one mapping in $\text{hom}(V \to W)$.

In chapter 6 we used the idea of the linear mapping $m_A: F_{n \times 1} \to F_{m \times 1}$ induced by a matrix $A \in F_{m \times n}$. The definition of this section gives a meaning to the matrix of m_A w.r.t. any bases of $F_{n \times 1}$ and $F_{m \times 1}$ that we might decide to use. While this latter matrix could be different from A if we use non-standard bases, the final result in this section reveals a straightforward situation when the standard bases of $F_{n \times 1}$ and $F_{m \times 1}$ are used.

55.4 Let A be an arbitrary matrix in $F_{m \times n}$. Then the matrix of the linear mapping m_A w.r.t. the standard bases of $F_{n \times 1}$ and $F_{m \times 1}$ is A itself.

Proof. Suppose it stipulated that the standard bases of $F_{n \times 1}$ and $F_{m \times 1}$ are in use. Observe, then, that, for every $X \in F_{n \times 1}$, the component column of X is simply X itself! And the same goes for vectors in $F_{m \times 1}$. So in this discussion the distinction between a vector and its component column can be forgotten. Therefore, since

$$m_A X = A \times X$$

for every $X \in F_{n \times 1}$, it follows (by 55.1(iv)) that the matrix of m_A w.r.t. the bases in use is A. This proves the result.

56. Matrices of sums, etc. of linear mappings

The following propositions demonstrate further nice properties of the mapping $\leftrightarrow$ matrix correspondence described in §55—that sums correspond to sums, products to products, etc. In these propositions L_U, L_V, L_W denote arbitrary bases of the vector spaces U, V, W.

56.1 Let a, b be linear mappings from V to W. Then

$$M(a+b; L_V, L_W) = M(a; L_V, L_W) + M(b; L_V, L_W)$$

(or, in short, $M_{a+b} = M_a + M_b$).

56.2 Let a be a linear mapping from V to W and λ a scalar. Then

$$M(\lambda a; L_V, L_W) = \lambda M(a; L_V, L_W)$$

(or, in short, $M_{\lambda a} = \lambda M_a$).

56.3 Let $a: V \to W$ and $b: U \to V$ be linear mappings. Then

$$M(ab; L_U, L_W) = M(a; L_V, L_W) M(b; L_U, L_V)$$

(or, in short, $M_{ab} = M_a M_b$).

56.4 Let a be a linear transformation of V, and let M_a stand for $M(a; L_V)$. Then M_a is a nonsingular matrix if and only if a is a nonsingular linear transformation.

In the first three propositions, the short versions are the simple stories that emerge when one can take as understood the basis being used in each space concerned.

In an important sense, none of these results is surprising. For example, we would scarcely be exaggerating if we reworded remark (b) in §15 (on the connection between matrix multiplication and mapping composition) to say that the definition of matrix multiplication is designed to make 56.3 true.

All of 56.1, 56.2, 56.3 are easy to prove by using 55.1(iv). As an illustration, a proof of 56.3 follows. After that, 56.4 is proved.

Proof of 56.3. We shall write M_a for $M(a; L_V, L_W)$, M_b for $M(b; L_U, L_V)$, and M_{ab} for $M(ab; L_U, L_W)$.

Let $\mathbf{x}$ be an arbitrary vector in U. Let the component columns of $\mathbf{x}$, $b\mathbf{x}$ and $ab\mathbf{x}$ (w.r.t. L_U, L_V, L_W, respectively) be X, Y and Z, respectively. In view of 55.1(iv), since $\mathbf{x}$ is arbitrary in U, the result will follow if we can prove that $Z = (M_a M_b)X$. But that is easily done: for, by 55.1(iii), $Y = M_b X$ and $Z = M_a Y$, and hence $Z = (M_a M_b)X$.

Proof of 56.4. (1) (The "$\Rightarrow$" half) Suppose that M_a is nonsingular, so that M_a^{-1} exists. Like M_a, it too is an $n \times n$ matrix, where $n = \dim V$. By 55.3, there is a linear transformation b of V such that $M_b = M_a^{-1}$. Hence we have

$$\begin{aligned} M_{ab} &= M_a M_b \qquad \text{(by 56.3)} \\ &= M_a M_a^{-1} = I_n = M_{i_V} \qquad \text{(by 55.2)}; \end{aligned}$$

and because (cf. 55.3) there is only one linear transformation of V whose matrix is I_n, it follows that $ab = i_V$. At this point one could appeal to exercise 6 on chapter 6 or simply remark that, obviously, a similar argument will prove that $ba = i_V$. By either means, it is apparent that a is nonsingular (with inverse b).

(2) (The "$\Leftarrow$" half) Suppose that a is nonsingular, so that a^{-1} exists (and is a linear transformation of V). Then we have

$$\begin{aligned} M_a M_{a^{-1}} &= M_{aa^{-1}} \qquad \text{(by 56.3)} \\ &= M_{i_V} = I_n \qquad \text{(by 55.2)}; \end{aligned}$$

and hence (cf. 26.5) it follows that M_a is nonsingular (with inverse $M_{a^{-1}}$).

Paragraphs (1) and (2) together prove the stated result.

A supplementary detail that emerged in the proof is worth recording.

56.5 In the notation of 56.4, if a and M_a are nonsingular, then $M_{a^{-1}} = M_a^{-1}$.

At this point the student should pause to appreciate that results proved in §§55 and 56 lead to a powerfully useful general principle—viz. that from any relationship between linear mappings one can deduce the exactly parallel relationship between the corresponding matrices, and conversely.

The general idea should be clear from a variegated illustration. Suppose that a, b, c denote linear transformations of V, a being nonsingular, and that A, B, C denote the corresponding matrices M_a, M_b, M_c, the matrices of a, b, c w.r.t. some chosen basis of V. Then, given that $a^{-1}b + 2i_V = c^2$, one can immediately deduce that $A^{-1}B + 2I = C^2$; and conversely, given that $A^{-1}B + 2I = C^2$, one can immediately deduce that $a^{-1}b + 2i_V = c^2$. To see why this is so, one should first note that, by results in this section and 55.2,

$$M_{a^{-1}b+2i_V} = M_{a^{-1}b} + M_{2i_V} = M_{a^{-1}}M_b + 2M_{i_V} = M_a^{-1}M_b + 2I = A^{-1}B + 2I$$

and $\qquad M_{c^2} = M_{c \times c} = M_c \times M_c = C^2.$

Therefore we have

$$\begin{aligned} a^{-1}b + 2i_V = c^2 &\Leftrightarrow M_{a^{-1}b+2i_V} = M_{c^2} \qquad \text{("}\Leftarrow\text{" by 55.3)} \\ &\Leftrightarrow A^{-1}B + 2I = C^2. \end{aligned}$$

Any other example can be justified in detail in the same way. In future, we shall regard the general principle as understood and switch without fuss from a statement about linear mappings to the corresponding statement about matrices, or vice versa.

57. Change of basis

Suppose that dim $V = n$ and that $L = (\mathbf{e}_1, \mathbf{e}_2, \dots, \mathbf{e}_n)$ is one basis of V.

Instead of starting straight away with a second basis of V, let us consider an arbitrary second sequence $\hat{L} = (\hat{\mathbf{e}}_1, \hat{\mathbf{e}}_2, \dots, \hat{\mathbf{e}}_n)$ of n vectors in V and look into the question of when exactly $\hat{L}$ is a basis.

In these circumstances, a linear transformation of V and an $n \times n$ matrix come fairly naturally to the attention. By 46.2, there is a linear transformation p of V such that

$$p\mathbf{e}_1 = \hat{\mathbf{e}}_1, p\mathbf{e}_2 = \hat{\mathbf{e}}_2, \dots, p\mathbf{e}_n = \hat{\mathbf{e}}_n. \qquad (*)$$

At the same time, let us introduce the matrix P ($\in F_{n \times n}$) whose 1st, 2nd, 3rd,... columns are, respectively, the component columns of $\hat{\mathbf{e}}_1, \hat{\mathbf{e}}_2, \hat{\mathbf{e}}_3, \dots$ w.r.t. L, i.e. the component columns of $p\mathbf{e}_1, p\mathbf{e}_2, p\mathbf{e}_3, \dots$ w.r.t. L. From the definition of "$M(p; L)$", it is apparent that

$$P = M(p; L).$$

Since the length of the sequence $\hat{L}$ is dim V, it follows from 41.2 that $\hat{L}$ is a basis of V if and only if $\hat{L}$ spans V. But (cf. (*)) $\hat{L} = (p\mathbf{e}_1, p\mathbf{e}_2, \dots, p\mathbf{e}_n)$, and hence, by 48.4($a$), the subspace of V spanned by $\hat{L}$ is pV. Therefore $\hat{L}$ is a basis of V if and only if $pV = V$. Pursuing this line of thought, we have:

$\hat{L}$ is a basis of $V \Leftrightarrow p$ is surjective
$\Leftrightarrow p$ is nonsingular ("$\Rightarrow$" by 49.5)
$\Leftrightarrow P$ is nonsingular (by 56.4, P being $M(p; L)$).

The mapping p can now be forgotten, leaving the following conclusion to be recorded.

57.1 Let $L = (\mathbf{e}_1, \mathbf{e}_2, \dots, \mathbf{e}_n)$ be a basis of V (n being dim V); let $\hat{L} = (\hat{\mathbf{e}}_1, \hat{\mathbf{e}}_2, \dots, \hat{\mathbf{e}}_n)$ be a second sequence of n vectors in V; and let P be the $n \times n$ matrix whose kth column is the component column of $\hat{\mathbf{e}}_k$ w.r.t. L ($k = 1, 2, \dots, n$). Then $\hat{L}$ is a basis of V if and only if P is nonsingular.

In the case where $\hat{L}$ is a second basis and P is nonsingular, we call P the **matrix of the change of basis** from L to $\hat{L}$, and we denote it by $M(L \rightarrow \hat{L})$. It plays a second role in connecting the component columns w.r.t. L and $\hat{L}$ of vectors in V.

57.2 Let V have bases L and $\hat{L}$, and let $P = M(L \to \hat{L})$. Let an arbitrary vector $\mathbf{x} \in V$ have component columns X and $\hat{X}$ w.r.t. L and $\hat{L}$, respectively. Then $X = P\hat{X}$.

Proof. Let $L = (\mathbf{e}_1, \ldots, \mathbf{e}_n)$ and $\hat{L} = (\hat{\mathbf{e}}_1, \ldots, \hat{\mathbf{e}}_n)$; and let $P = [\pi_{ik}]_{n \times n}$, $X = \text{col}(x_1, \ldots, x_n)$, $\hat{X} = \text{col}(\hat{x}_1, \ldots, \hat{x}_n)$. Then

$$\sum_{i=1}^{n} x_i \mathbf{e}_i = \mathbf{x} = \sum_{k=1}^{n} \hat{x}_k \hat{\mathbf{e}}_k$$

$$= \sum_{k=1}^{n} \hat{x}_k \left(\sum_{i=1}^{n} \pi_{ik} \mathbf{e}_i \right) \quad \text{(since the } k\text{th column of } P \text{ is the component column of } \hat{\mathbf{e}}_k \text{ w.r.t. } L)$$

$$= \sum_{i=1}^{n} \left(\sum_{k=1}^{n} \pi_{ik} \hat{x}_k \right) \mathbf{e}_i \quad \text{(on interchanging the order of summation).}$$

Hence (cf. 39.10) $x_i = \sum_{k=1}^{n} \pi_{ik} \hat{x}_k$ for each relevant i; and therefore $X = P\hat{X}$, as claimed.

There now follows a sequence of remarks in which we continue to use the notation of 57.2.

(*a*) It is not difficult to show that the property given in 57.2 characterizes $M(L \to \hat{L})$: i.e. if the $n \times n$ matrix Q is such that component column of $\mathbf{x}$ w.r.t. $L = Q$ (component vector of $\mathbf{x}$ w.r.t. $\hat{L}$) for every $\mathbf{x} \in V$, then Q must equal $M(L \to \hat{L})$.

(*b*) If we write out the equations giving $\hat{\mathbf{e}}_1, \hat{\mathbf{e}}_2, \ldots, \hat{\mathbf{e}}_n$ in terms of $\mathbf{e}_1, \mathbf{e}_2, \ldots, \mathbf{e}_n$, then the array of coefficients that confronts us is P^T (not P itself). This is because it is the kth *column* of P that contains the components of $\hat{\mathbf{e}}_k$ w.r.t. L. The situation can be remembered schematically as

$$\text{col}(\hat{\mathbf{e}}_1, \hat{\mathbf{e}}_2, \ldots, \hat{\mathbf{e}}_n) = P^T \times \text{col}(\mathbf{e}_1, \mathbf{e}_2, \ldots, \mathbf{e}_n).$$

(*c*) If we think of L as the "old" or "original" basis and $\hat{L}$ as the "new" basis, then the matrix $P\,(= M(L \to \hat{L}))$ contains coefficients appearing when the new basis vectors are expressed in terms of the old (cf. note (*b*) for the detail). But, in 57.2, where we relate the component columns of a typical vector $\mathbf{x}$, P comes into the expression for the old component vector in terms of the new. (It is $X = P\hat{X}$; and the corresponding "new in terms of old" equation is $\hat{X} = P^{-1}X$.)

(*d*) At any point, any nonsingular matrix in $F_{n \times n}$ can be used to define a change of basis in the n-dimensional V. More precisely:

57.3 Let L be an arbitrary basis of the n-dimensional vector space V, and let P be an arbitrary nonsingular matrix in $F_{n \times n}$. Then there is a basis $\hat{L}$ of V such that $M(L \to \hat{L}) = P$.

Outline of proof. Let $L = (\mathbf{e}_1, \ldots, \mathbf{e}_n)$, and let $P = [\pi_{ik}]_{n \times n}$. For $k = 1, 2, \ldots, n$, let $\hat{\mathbf{e}}_k$ be defined as $\sum_{i=1}^{n} \pi_{ik}\mathbf{e}_i$; and let $\hat{L} = (\hat{\mathbf{e}}_1, \ldots, \hat{\mathbf{e}}_n)$. Then it is easy to see that (1) $\hat{L}$ is a basis of V (by 57.1 and the fact that P is nonsingular) and (2) $M(L \to \hat{L}) = P$.

58. Matrices of a linear mapping w.r.t. different bases

We noted in §55 that, for a linear mapping $a: V \to W$, the matrix $M(a; L_V, L_W)$ will, in general, depend not only on a but also on the choice of bases L_V, L_W. In this section, we explore the relationship between the matrices that represent a given linear mapping w.r.t. different bases of its domain and codomain. The first proposition gives the fundamental result.

58.1 Let $a: V \to W$ be a linear mapping; let L_V and $\hat{L}_V$ be two bases of V, and let L_W and $\hat{L}_W$ be two bases of W; let $A = M(a; L_V, L_W)$ and let $\hat{A} = M(a; \hat{L}_V, \hat{L}_W)$. Then

$$\hat{A} = R^{-1}AP,$$

where $P = M(L_V \to \hat{L}_V)$ and $R = M(L_W \to \hat{L}_W)$.

Proof. Let $\mathbf{x}$ be an arbitrary vector in V; let X and $\hat{X}$ be the component columns of $\mathbf{x}$ w.r.t. L_V and $\hat{L}_V$; and let Y and $\hat{Y}$ be the component columns of $a\mathbf{x}$ w.r.t. L_W and $\hat{L}_W$. In view of 55.1(iv), since $\mathbf{x}$ is arbitrary in V, the result will follow if we can prove that $\hat{Y} = (R^{-1}AP)\hat{X}$. This is easily achieved as follows:

$$\begin{aligned} \hat{Y} &= R^{-1}Y && \text{(by 57.2 (cf. end of remark (c))} \\ &= R^{-1}AX && \text{(by 55.1(iii), since } A = M(a; L_V, L_W)\text{)} \\ &= R^{-1}AP\hat{X} && \text{(by 57.2).} \end{aligned}$$

An important special case is that of a linear transformation a of V, where we are interested in the relationship between the matrices $M(a; L)$ and $M(a; \hat{L})$ of a w.r.t. two bases L and $\hat{L}$ of V. This is the subject of the next result, which is an immediate corollary of 58.1.

58.2 Let a be a linear transformation of V; let $L, \hat{L}$ be two bases of V; let $A = M(a; L)$ and $\hat{A} = M(a; \hat{L})$. Then $\hat{A} = P^{-1}AP$, where $P = M(L \to \hat{L})$.

The result 58.2 prompts the introduction of an important piece of terminology in matrix theory. For $A, B \in F_{n \times n}$, we say that B is **similar** to A if and only if $B = P^{-1}AP$ for some nonsingular matrix $P \in F_{n \times n}$.

It is not difficult to show that the relation of similarity so defined has what we call the reflexive, symmetric and transitive properties; that is:

(1) every matrix in $F_{n \times n}$ is similar to itself;

(2) "A is similar to B" is synonymous with "B is similar to A" ($A, B \in F_{n \times n}$);

(3) if A is similar to B and B is similar to C, then A is similar to C $(A, B, C \in F_{n\times n})$.
(Students familiar with the idea of an equivalence relation will recognize that similarity is an equivalence relation on $F_{n\times n}$.)

From 58.2 and 57.3, it follows that:

58.3 If a is a linear transformation of V and A is the matrix of a w.r.t. some particular basis of V, then the set of all matrices arising as the matrices of a w.r.t. the different possible bases of V is the set of matrices similar to A.

From 58.1 and 57.3, it is apparent that the corresponding result for a linear mapping is as follows.

58.4 Let $a: V \to W$ be a linear mapping, and let m, n be the dimensions of W, V, respectively. If A is the matrix of a w.r.t. some particular bases of V and W, then the set of all matrices arising as the matrices of a w.r.t. the different possible bases of V and W is the set of all matrices expressible in the form QAP, with Q a nonsingular matrix in $F_{m\times m}$ and P a nonsingular matrix in $F_{n\times n}$.

At first sight, the multiplicity of matrices that can arise as "the matrix of a" may seem a tiresome complication. In fact, however, it is a complication that we can exploit to our advantage. For example, among the many matrices representing a given linear transformation a of V (w.r.t. the different bases of V) it may be possible to find a particularly simple matrix A_1—simple in the sense that calculations involving A_1 are easy: then facts about A_1 are readily obtainable, and (cf. §56) these can be translated into facts about a that might otherwise remain obscure. It would be especially convenient if the matrix A_1 is diagonal—a theme illustrated in the following worked example, and a theme that motivates much of chapter 8.

Worked example. The linear transformation a of $\mathbb{R}^2$ is defined by

$$a(x_1, x_2) = (5x_1 - 4x_2, 3x_1 - 2x_2).$$

Use 58.2 to find the matrix $\hat{A}$ of a w.r.t. the basis $\hat{L} = (\mathbf{f}_1, \mathbf{f}_2)$ of $\mathbb{R}^2$, where $\mathbf{f}_1 = (1, 1)$ and $\mathbf{f}_2 = (4, 3)$. Taking advantage of the simple form of the answer, find a formula for $a^n(x_1, x_2)$.

Solution. The matrix of a w.r.t. the standard basis $L = (\mathbf{e}_1, \mathbf{e}_2)$ of $\mathbb{R}^2$ is $A = \begin{bmatrix} 5 & -4 \\ 3 & -2 \end{bmatrix}$ (cf. example 1 in §55).

$M(L \to \hat{L})$ is $P = \begin{bmatrix} 1 & 4 \\ 1 & 3 \end{bmatrix}$. (Its columns are the component columns of $\mathbf{f}_1$ and

$\mathbf{f}_2$ w.r.t. L.) Hence, by 58.2,

$$\hat{A} = P^{-1}AP = (-1)\begin{bmatrix} 3 & -4 \\ -1 & 1 \end{bmatrix}\begin{bmatrix} 5 & -4 \\ 3 & -2 \end{bmatrix}\begin{bmatrix} 1 & 4 \\ 1 & 3 \end{bmatrix} (P^{-1} \text{ obtained by 19.4})$$

$$= \begin{bmatrix} -3 & 4 \\ 1 & -1 \end{bmatrix}\begin{bmatrix} 1 & 8 \\ 1 & 6 \end{bmatrix} = \begin{bmatrix} 1 & 0 \\ 0 & 2 \end{bmatrix} = \operatorname{diag}(1,2).$$

Clearly, $\hat{A}^n = \operatorname{diag}(1, 2^n)$; and (cf. §56, 56.3 in particular) this is the matrix of a^n w.r.t. $\hat{L}$. By 58.2, the matrix of a^n w.r.t. L is

$$P \times \operatorname{diag}(1, 2^n) \times P^{-1} = \begin{bmatrix} 1 & 4 \\ 1 & 3 \end{bmatrix}\begin{bmatrix} 1 & 0 \\ 0 & 2^n \end{bmatrix}\begin{bmatrix} -3 & 4 \\ 1 & -1 \end{bmatrix}$$

$$= \begin{bmatrix} 1 & 4 \\ 1 & 3 \end{bmatrix}\begin{bmatrix} -3 & 4 \\ 2^n & -2^n \end{bmatrix}$$

$$= \begin{bmatrix} -3+2^{n+2} & 4(1-2^n) \\ 3(2^n-1) & 4-3.2^n \end{bmatrix}.$$

Hence the transformation a^n is given by

$$a^n(x_1, x_2) = ((-3+2^{n+2})x_1 + 4(1-2^n)x_2, 3(2^n-1)x_1 + (4-3.2^n)x_2).$$

59. The simplest matrix representing a linear mapping

We shall denote by $(K_r)_{m\times n}$ (or simply by K_r when the type is clear from the context) the $m \times n$ matrix

$$\begin{bmatrix} I_r & O \\ O & O \end{bmatrix}.$$

This is to be interpreted in the obvious way if $r = 0$ or if $r = m$ or n: $(K_0)_{m\times n}$ means $O_{m,n}$, and (for example) $(K_3)_{3\times 5}$ means $[I_3 \quad O_{3,2}]$.

The title of the section is a reference to the following fact.

59.1 Let $a: V \to W$ be a linear mapping with rank r; and let $n = \dim V$, $m = \dim W$. Then there are bases of V and W w.r.t. which a has matrix $(K_r)_{m\times n}$.

Proof. We introduce first a basis of ker a. Since

$$\dim(\ker a) = s(a) = \dim V - r(a) = n - r \qquad \text{(cf. 49.2)},$$

this basis must have length $n-r$, and the vectors in it may be numbered $r+1$, $r+2, \ldots, n$. So let $(\mathbf{e}_{r+1}, \mathbf{e}_{r+2}, \ldots, \mathbf{e}_n)$ be a basis of ker a. Then (cf. 41.4(iv)) let this basis of ker a be extended to a basis

$$L_V = (\mathbf{e}_1, \ldots, \mathbf{e}_r, \mathbf{e}_{r+1}, \ldots, \mathbf{e}_n)$$

of V. Next let vectors $\mathbf{f}_1, \ldots, \mathbf{f}_r$ in W be defined by $\mathbf{f}_k = a\mathbf{e}_k$ $(k = 1, 2, \ldots, r)$. By a small adaptation of the core of the proof of 49.2, it can be shown that $(\mathbf{f}_1, \ldots, \mathbf{f}_r)$ is a basis of im a; and (cf. 41.4(iv) again) it may be extended to a basis

$$L_W = (\mathbf{f}_1, \ldots, \mathbf{f}_r, \mathbf{f}_{r+1}, \ldots, \mathbf{f}_m)$$

of W.

Since $a\mathbf{e}_k = \mathbf{f}_k$ for $1 \leqslant k \leqslant r$ and $a\mathbf{e}_k = \mathbf{0}$ for $r+1 \leqslant k \leqslant n$, it is apparent that

$$M(a; L_V, L_W) = \left[\begin{array}{cccc|c} 1 & 0 & 0 & \ldots & \\ 0 & 1 & 0 & \ldots & O \\ 0 & 0 & 1 & \ldots & \\ & \ldots & & & \end{array}\right] = (K_r)_{m \times n},$$
$$\underbrace{\hphantom{1\ 0\ 0\ \ldots}}_{r} \quad \underbrace{\hphantom{O\ O}}_{n-r}$$

and the stated result is established.

The two remaining results in this section are among the interesting consequences of 59.1.

59.2 Let $a: V \to W$ be a linear mapping, and let A be the matrix of a w.r.t. any bases of V and W. Then $r(A) = r(a)$.

Proof. W.r.t. one choice of bases of V and W, a has matrix A; and (by 59.1) w.r.t. another choice of bases of V and W, a has matrix K_r, where $r = r(a)$. By 58.4, it follows that $A = QK_rP$ for some nonsingular matrices P, Q. Hence

$$\begin{aligned} r(A) &= r(K_r) && \text{(by 52.3M, } P \text{ and } Q \text{ being nonsingular)} \\ &= r && (K_r \text{ being an echelon matrix with } r \text{ nonzero rows}) \\ &= r(a), \end{aligned}$$

as claimed.

Remarks. (*a*) One of the themes of this chapter is the niceness of the correspondence $a \leftrightarrow M_a$ between linear mappings and their matrices w.r.t. prescribed bases. The result 59.2 reveals another nice feature of this correspondence—that it is, as we say, rank-preserving: i.e. mappings of rank r correspond to matrices of rank r.

(*b*) It should be realized that 59.2 does apply in the obvious way to a linear transformation of V: i.e. it is true that if a is such a transformation and $A = M(a; L)$ for some basis L of V, then $r(A) = r(a)$. An alert critic will note that, if our proof of 59.2 is to cover this case, we must be prepared to consider simultaneously two bases of V—one used when V is regarded as domain and the other used when V is thought of as codomain (playing the role of W in the general proof); but, even if this seems artificial, there is no logical objection to it.

(*c*) There is, however, no simple version of 59.1 for a linear transformation: if a is a linear transformation of V with rank r, it is *not* in general true that there is a basis L of V such that $M(a;L) = K_r$. (Roughly speaking, one needs the freedom to make two changes of basis, one in the domain and one in the codomain, to arrange for the matrix of a mapping to be K_r.) As already hinted in §58, we shall consider in chapter 8 the question, for a given linear transformation a of V, of whether and how one can find a basis L of V such that $M(a;L)$ is diagonal.

59.3 Let A be a matrix of rank r in $F_{m\times n}$. Then A can be expressed as BK_rC for some nonsingular matrices $B \in F_{m\times m}$ and $C \in F_{n\times n}$. (Here K_r stands for $(K_r)_{m\times n}$.)

Proof. Consider matrices representing the linear mapping $m_A: F_{n\times 1} \to F_{m\times 1}$, which has rank r (by 50.5). By 55.4, the matrix of m_A w.r.t. the standard bases of $F_{n\times 1}$ and $F_{m\times 1}$ is A; and, by 59.1, the matrix of m_A w.r.t. some other bases of $F_{n\times 1}$ and $F_{m\times 1}$ is K_r. Hence, by 58.4, $A = BK_rC$ for some nonsingular matrices B, C, as stated.

60. Vector space isomorphisms

Suppose that the dimensions of V and W are n and m, respectively, and that bases L_V and L_W of V and W have been chosen. Then, as we have seen (cf. concluding remarks in §56), working with linear mappings $a, b, \ldots \in \hom(V \to W)$ is exactly parallel to (one might say effectively the same as) working with the corresponding matrices $M_a, M_b, \ldots$ in $F_{m\times n}$. Thus, in some sense, the algebraic systems $\hom(V \to W)$ and $F_{m\times n}$ are the same, and the mapping $a \mapsto M_a$ is significant in expressing that sameness.

The ideas coming to the surface in the last sentence are what the words "isomorphic" and "isomorphism" refer to, and they are ideas much more widely and generally important than our current preoccupation with vector space theory might suggest. Taking temporarily a more general view, consider two algebraic systems A and B. (They might be two vector spaces or two fields, etc., etc.) We describe A and B as *isomorphic* (to each other) if there is a bijection f from the set A to the set B such that, whenever elements $x, y, z, \ldots$ of A are related in some way by the algebraic operations defined on A, the "corresponding" elements $f(x), f(y), f(z), \ldots$ of B are related in the corresponding way in B. (For example if $x+y = z$ in A, then, in B, $f(x)+f(y) = f(z)$.) It can be seen that when such a bijection f exists, the systems A and B are in some sense the same: indeed they would become formally indistinguishable if, for every $x \in A$, we relabelled the element $f(x)$ of B as x.

In general, a bijective mapping $f: A \to B$ as described above is called an *isomorphism* from A to B.

For each species of algebraic system (e.g. vector spaces over F) one can give a more explicit formulation of these concepts. After a little thought one comes to the conclusion that in the case of vector spaces over F the relevant type of bijection is a bijective *linear* mapping. Accordingly, the following definitions are made.

For vector spaces V, W over F, we say that V is **isomorphic to** W (written $V \cong W$) if and only if there exists a bijective linear mapping from V to W. And any bijective linear mapping from V to W is termed an **isomorphism** from V to W.

In harmony with the intuitive idea that when system A is isomorphic to system B, the systems A and B are in a certain sense the same, it is not difficult to prove from the above definitions for vector spaces that:

(1) for every vector space V, $V \cong V$ (the reflexive property);

(2) $V \cong W$ and $W \cong V$ mean the same thing (the symmetric property);

(3) if $U \cong V$ and $V \cong W$, then $U \cong W$ (the transitive property).

There is little to say about zero vector spaces in relation to isomorphisms: rather trivially, any two zero spaces are isomorphic to each other, and no zero space is isomorphic to a nonzero space.

The next two theorems reveal a further very simple story telling us what is isomorphic to what among nonzero f.d. vector spaces.

60.1 Let $\dim V = n$. Then $V \cong F^n$.

Proof. Introduce a basis $L = (\mathbf{e}_1, \ldots, \mathbf{e}_n)$ of V. Then consider the mapping $a: V \to F^n$ defined by

$$\mathbf{x} \mapsto \text{component vector of } \mathbf{x} \text{ w.r.t. } L \qquad (\mathbf{x} \in V),$$
$$\text{i.e. } \lambda_1\mathbf{e}_1 + \lambda_2\mathbf{e}_2 + \ldots + \lambda_n\mathbf{e}_n \mapsto (\lambda_1, \lambda_2, \ldots, \lambda_n).$$

It can be seen mentally that a is linear. And a is clearly surjective, so that $r(a) = \dim(F^n) = n$. Hence (by 49.2) $s(a) = 0$, so that (cf. 49.3) a is also injective. It is now apparent that a is a bijective linear mapping from V to F^n; and therefore $V \cong F^n$, as asserted.

A corollary is:

60.2 $V \cong W \Leftrightarrow \dim V = \dim W$.

Proof. (1) If $\dim V = \dim W = n$, then (by 60.1) $V \cong F^n$ and $W \cong F^n$, and it follows (by the symmetric and transitive properties mentioned above) that $V \cong W$.

(2) Conversely, if $V \cong W$, then there is a bijective linear mapping $a: V \to W$, and hence we have

$$\begin{aligned} \dim W &= \dim(aV) \qquad \text{(since } a \text{ is surjective)} \\ &= r(a) = \dim V - s(a) \qquad \text{(by 49.2)} \\ &= \dim V \qquad (s(a) \text{ being 0 since } a \text{ is injective}). \end{aligned}$$

Paragraphs (1) and (2) prove the stated result.

We are now able to re-express the ideas that sparked off our discussion of isomorphic algebraic systems at the beginning of the section.

60.3 Let $\dim V = n$ and $\dim W = m$; and let L_V, L_W be specific bases of V and W. Then the mapping $f: \hom(V \to W) \to F_{m\times n}$ given by

$$f(a) = M(a; L_V, L_W)$$

is an isomorphism and, therefore, the vector spaces $\hom(V \to W)$ and $F_{m\times n}$ are isomorphic to each other.

(That f is a bijection is seen from 55.3; and the linearity of f is proved by 56.1 and 56.2.)

In view of 60.2, an immediate corollary is:

60.4 $\dim[\hom(V \to W)] = (\dim V) \times (\dim W)$; and, in particular, the dimension of the dual space of V equals $\dim V$.

Suppose that $\dim V = n$. A special case of "$\hom(V \to W) \cong F_{m\times n}$" (the case $W = V$) tells us that end V and $F_{n\times n}$ are isomorphic vector spaces. However, there is more to say in this special case. As explained in §53, both end V and $F_{n\times n}$ are associative algebras; and the fact is that they are isomorphic associative algebras. Informally, this fact can be regarded as saying that the two associative algebras are, in some important senses, the same. The formal and precise meaning of the fact is that there exists a bijection $f: \text{end } V \to F_{n\times n}$ with the properties

(1) $f(a+b) = f(a)+f(b)$ for all $a, b \in \text{end } V$,

(2) $f(\lambda a) = \lambda f(a)$ for all $a \in \text{end } V$ and all $\lambda \in F$ and

(3) $f(ab) = f(a)f(b)$ for all $a, b \in \text{end } V$.

Such a mapping f is the mapping at $a \mapsto M(a; L)$, where L is any particular basis of V.

EXERCISES ON CHAPTER SEVEN

1. Let V, W be vector spaces of dimensions 3, 4, respectively, and let $L_V = (\mathbf{e}_1, \mathbf{e}_2, \mathbf{e}_3)$, $L_W = (\mathbf{f}_1, \mathbf{f}_2, \mathbf{f}_3, \mathbf{f}_4)$ be bases of V, W, respectively. Given that $M(a; L_V, L_W)$ $= \begin{bmatrix} 1 & 2 & 3 \\ 4 & 5 & 6 \\ 7 & 8 & 9 \\ 10 & 11 & 12 \end{bmatrix}$,

(i) write down $a\mathbf{e}_2$ as a linear combination of $\mathbf{f}_1, \mathbf{f}_2, \mathbf{f}_3, \mathbf{f}_4$, and

(ii) *by evaluating one matrix product*, obtain $a(2\mathbf{e}_1 + \mathbf{e}_2 - \mathbf{e}_3)$ as a linear combination of $\mathbf{f}_1, \mathbf{f}_2, \mathbf{f}_3, \mathbf{f}_4$.

2. (This question is about transformations of the x, y-coordinate plane that can be interpreted as linear transformations of E_2.)

Let l be the line making an angle $\theta\,(0 \leqslant \theta < \pi)$ with the positive direction along the x-axis; let b, c be, respectively, reflection in l and perpendicular projection on to l (both of which we may regard as linear transformations of E_2); and let B, C be the respective matrices of b, c w.r.t. the basis $(\mathbf{i}, \mathbf{j})$ of E_2.

(i) Show that

$$B = \begin{bmatrix} \cos 2\theta & \sin 2\theta \\ \sin 2\theta & -\cos 2\theta \end{bmatrix}.$$

Verify that $B^2 = I$, and explain why that is to be expected on geometrical grounds.

(ii) Explain why $C = \frac{1}{2}(B + I)$, and hence prove that

$$C = \begin{bmatrix} \cos^2 \theta & \sin \theta \cos \theta \\ \sin \theta \cos \theta & \sin^2 \theta \end{bmatrix}.$$

Verify that C is idempotent (i.e. $C^2 = C$), and explain why that is to be expected on geometrical grounds.

(iii) By considering the product of two matrices, prove the theorem in 2-dimensional geometry that the composition of the reflections in two non-parallel mirror lines intersecting at O is equal to a rotation about O through twice the angle between the mirror lines.

3. The linear mapping $a: \mathbb{R}^3 \to \mathbb{R}^2$ is such that $a(1,1,1) = (3,3)$, $a(1,0,1) = (0,3)$, and $a(1,-1,0) = (-2,2)$. Find the matrix of a w.r.t. the standard bases of $\mathbb{R}^3$ and $\mathbb{R}^2$.

4. The linear mapping $a: \mathbb{R}^3 \to \mathbb{R}^2$ is defined by

$$a(x_1, x_2, x_3) = (x_1 - x_2 + 2x_3, x_1 - x_3).$$

In $\mathbb{R}^2 L'$ is the basis $((1,1),(1,-1))$, and in $\mathbb{R}^3 L''$ is the basis $((1,1,0),(0,1,1),(1,0,1))$. Obtain: (i) the matrix of a w.r.t. the standard bases of $\mathbb{R}^3$ and $\mathbb{R}^2$; (ii) the matrix of a w.r.t. the standard basis of $\mathbb{R}^3$ and the basis L' of $\mathbb{R}^2$; (iii) the matrix of a w.r.t. the basis L'' of $\mathbb{R}^3$ and the standard basis of $\mathbb{R}^2$; (iv) $M(a; L'', L')$.

5. The linear transformation a of $\mathbb{R}^2$ is defined by

$$a(x_1, x_2) = (-5x_1 + 9x_2, -4x_1 + 7x_2),$$

and L is the basis $((3,2),(1,1))$ of $\mathbb{R}^2$. Find $M(a; L)$. Hence show that, for every positive integer n, $a^n - i = n(a - i)$, where i stands for the identity transformation of $\mathbb{R}^2$. Deduce a general formula for $a^n(x_1, x_2)$.

6. Let $A, B \in F_{n \times n}$. Show that, if A is nonsingular, then AB and BA are similar to each other.

7.* Let a be a linear transformation of the nonzero f.d. vector space V such that $r(a) = r(a^2) = r$. By the worked example in §49, $V = aV \oplus \ker a$. Using this fact along with 43.4(i), show that there is a basis of V w.r.t. which a has matrix of the form $\begin{bmatrix} A_1 & O \\ O & O \end{bmatrix}$, where A_1 is a nonsingular $r \times r$ matrix. Deduce that there exists a linear transformation b of V such that $r(b)$ is also r and $ab^2 = b^2a = b$.

8. Let V be a nonzero f.d. vector space, and let a, b, c denote linear transformations of V.

(i) As a particular case of question 7, prove that if a is idempotent with rank r, then there is a basis of V w.r.t. which a has matrix K_r. Prove also the converse.

(ii) Suppose that b satisfies $b^2 = \lambda b$, where λ is a nonzero scalar. Show that there is a basis of V w.r.t. which b has matrix λK_r, where r is the rank of b. (Use part (i)!)

(iii)* Suppose that c satisfies $(c - \alpha i_V)(c - \beta i_V) = O$, where α, β are unequal scalars. Prove that there is a basis of V w.r.t. which c has matrix of the form diag $(\alpha, \alpha, \dots, \alpha, \beta, \beta, \dots, \beta)$. (Use part (ii)!) Show also that $r(c - \alpha i_V) + r(c - \beta i_V) = \dim V$.

9. Suppose that V is a 3-dimensional vector space and a a linear transformation of V such that $a \neq O$ but $a^2 = O$. Prove that $r(a) = 1$ and that, for any $\mathbf{e}_1$ in V that is not in $\ker a$, $V = \mathrm{sp}(\mathbf{e}_1) \oplus \ker a$. Deduce that there is a basis of V w.r.t. which a has matrix $\begin{bmatrix} 0 & 0 & 0 \\ 1 & 0 & 0 \\ 0 & 0 & 0 \end{bmatrix}$.

Find such a basis in the case where $V = \mathbb{R}^3$ and a is the linear transformation defined by

$$a(x_1, x_2, x_3) = (x_1 - x_2 + x_3, 2x_1 - 2x_2 + 2x_3, x_1 - x_2 + x_3).$$

10. Let A be an arbitrary matrix in $F_{m \times n}$. By using 59.3, show that there is a matrix $Q \in F_{n \times m}$ such that $AQA = A$, $QAQ = Q$ and $r(Q) = r(A)$. (Such a matrix Q is a sort of generalization of an inverse of A.)

11. Let V be a nonzero f.d. vector space over F, V' the dual space of V, and V'' the dual space of V'.

(i) Prove that, for each $\mathbf{x} \in V$, the mapping $\theta_{\mathbf{x}} : V' \to F$ defined by $\theta_{\mathbf{x}}(a) = a\mathbf{x}$ $(a \in V')$ is a member of V''.

(ii) Show that the mapping $\theta : V \to V''$ defined by $\theta(\mathbf{x}) = \theta_{\mathbf{x}}$ $(\mathbf{x} \in V)$ is an isomorphism from the vector space V to the vector space V''.

12. Let V be a nonzero f.d. vector space and suppose that a is a nonzero nilpotent linear transformation of V. (So some positive power of a is the zero transformation of V.) Show that there is no basis of V w.r.t. which the matrix of a is diagonal.

13.* Two linear transformations a, b of a nonzero f.d. vector space V are such that $ab = b$ and $r(a) = r(b)$. Prove that $aV = bV$, and deduce that a is idempotent. Using question 8(i), deduce further that there exists a nonsingular linear transformation c of V such that $b = ac$.

14. Suppose that V is a vector space over F and that $V = A \oplus B$ for certain subspaces A and B of V. Let W be the set of all ordered pairs of the form $(\mathbf{x}, \mathbf{y})$ with $\mathbf{x} \in A$ and $\mathbf{y} \in B$. Then, by checking through all the vector space axioms, one can show that W becomes a vector space when addition of elements of W and their multiplication by scalars are defined, respectively, by

$$(\mathbf{x}_1, \mathbf{y}_1) + (\mathbf{x}_2, \mathbf{y}_2) = (\mathbf{x}_1 + \mathbf{x}_2, \mathbf{y}_1 + \mathbf{y}_2) \qquad (\mathbf{x}_1, \mathbf{x}_2 \in A;\, \mathbf{y}_1, \mathbf{y}_2 \in B),$$
$$\lambda(\mathbf{x}, \mathbf{y}) = (\lambda\mathbf{x}, \lambda\mathbf{y}) \qquad (\mathbf{x} \in A,\ \mathbf{y} \in B,\ \lambda \in F).$$

The zero in the vector space W is $(\mathbf{0}, \mathbf{0})$. Show that the mapping $f\colon W \to V$ defined by

$$f(\mathbf{x}, \mathbf{y}) = \mathbf{x} + \mathbf{y} \qquad (\mathbf{x} \in A,\ \mathbf{y} \in B)$$

is an isomorphism.

CHAPTER EIGHT

EIGENVALUES, EIGENVECTORS AND DIAGONALIZATION

61. Introduction

The main definitions of this chapter (those of the terms "eigenvalue" and "eigenvector") come in §63. Once the basic theory associated with these terms has been developed, we shall be equipped to tackle effectively some questions whose importance was glimpsed in the course of chapter 7: e.g., for a given linear transformation a of a nonzero f.d. vector space V, is there a basis L of V such that $M(a; L)$ is a diagonal matrix, and, if so, how can we find such a basis of V? These questions and related questions about matrices come to the forefront of our attention in §§65 and 66. In all this work, we shall make extensive use of what are called characteristic polynomials—a topic discussed as a preliminary in §62. The last two sections of the chapter (§§67, 68) break new ground in the theory of complex matrices.

In §§62 and 63 we shall work over an arbitrary field of scalars F, but from §64 onwards we shall narrow the discussion to the case $F = \mathbb{C}$. This crucially simplifies the problems to be considered by ensuring that, within our system of scalars, all polynomials can be completely factorized into linear factors (a fact which follows from the so-called "fundamental theorem of algebra").

62. Characteristic polynomials

For $A \in F_{n \times n}$, the **characteristic polynomial** of A is defined to be $\det(tI_n - A)$, where t denotes a scalar indeterminate. As the name suggests, this turns out to be a polynomial in t. We denote it by $\chi_A(t)$.

For example, in the case

$$A = \begin{bmatrix} 3 & 3 & 2 \\ 2 & 4 & 2 \\ -1 & -3 & 0 \end{bmatrix},$$

$$\chi_A(t) = \begin{vmatrix} t-3 & -3 & -2 \\ -2 & t-4 & -2 \\ 1 & 3 & t \end{vmatrix};$$

and hence (as the student should verify by expanding this determinant by any row or column and then simplifying)

$$\chi_A(t) = t^3 - 7t^2 + 14t - 8 = (t-1)(t-2)(t-4).$$

There are alternative clever methods of arriving at the final version by spotting row/column operations on the original determinant that will produce factors: e.g. the e.r.o. $R_1 \to R_1 - R_2$ discloses the factor $(t-1)$.

More generally, for an arbitrary 3×3 matrix $A = [\alpha_{ik}]_{3\times 3}$,

$$\chi_A(t) = \begin{vmatrix} t-\alpha_{11} & -\alpha_{12} & -\alpha_{13} \\ -\alpha_{21} & t-\alpha_{22} & -\alpha_{23} \\ -\alpha_{31} & -\alpha_{32} & t-\alpha_{33} \end{vmatrix};$$

and, on expansion of this determinant by the first row, one sees that terms involving t to a higher power of the first come only from the product $(t-\alpha_{11})(t-\alpha_{22})(t-\alpha_{33})$ of the entries on the main diagonal. Thus

$$\begin{aligned}\chi_A(t) &= (t-\alpha_{11})(t-\alpha_{22})(t-\alpha_{33}) + (\text{terms of degree} \leqslant 1 \text{ in } t) \\ &= t^3 - (\alpha_{11}+\alpha_{22}+\alpha_{33})t^2 + \gamma t + \delta, \qquad \text{for some scalars } \gamma, \delta.\end{aligned}$$

The corresponding story for an arbitrary $n \times n$ matrix $A = [\alpha_{ik}]_{n\times n}$ is that

$$\det(tI_n - A) = \begin{pmatrix}\text{product of entries on main} \\ \text{diagonal of } tI_n - A\end{pmatrix} + \begin{pmatrix}\text{terms of degree} \\ \leqslant (n-2) \text{ in } t\end{pmatrix},$$

i.e. $\chi_A(t) = t^n - (\alpha_{11}+\alpha_{22}+\ldots+\alpha_{nn})t^{n-1} + (\text{terms of lower degree})$.

The sum $\alpha_{11}+\alpha_{22}+\ldots+\alpha_{nn}$ of the entries on the main diagonal of an $n \times n$ matrix A is called the **trace** of A, which we denote by $\operatorname{tr}(A)$. Thus our information so far about the general shape of $\chi_A(t)$ may be summarized as follows.

62.1 For an $n \times n$ matrix A, $\chi_A(t)$ is a polynomial of degree n in which the coefficient of t^n is 1 and the coefficient of t^{n-1} is $-\operatorname{tr}(A)$.

It is also easy to identify the constant term.

62.2 For an $n \times n$ matrix A, the constant term in $\chi_A(t)$ is $(-1)^n \det A$.

(For: constant term $= \chi_A(0) = \det(0I - A) = \det((-1)A)$
$= (-1)^n \det A$ (by 31.6).)

In the important case of a diagonal matrix the characteristic polynomial is easily obtained in fully factorized form.

62.3 For a diagonal matrix $D = \text{diag}(\alpha_1, \alpha_2, \ldots, \alpha_n)$,

$$\chi_D(t) = (t-\alpha_1)(t-\alpha_2)\ldots(t-\alpha_n).$$

$$\begin{aligned}(\text{For: } \chi_D(t) &= \det(tI-D) = \det[\text{diag}(t-\alpha_1, t-\alpha_2, \ldots, t-\alpha_n)] \\ &= (t-\alpha_1)(t-\alpha_2)\ldots(t-\alpha_n) \qquad (\text{by } 30.1).)\end{aligned}$$

Consider now two similar matrices $A, B \in F_{n\times n}$. By the definition of "similar", $B = P^{-1}AP$ for some nonsingular $P \in F_{n\times n}$. Exercise 5 on chapter 4 dealt with this situation with a result that is now relevant. To review the main details, one easily verifies that

$$P^{-1}(tI-A)P = tI-B;$$

and hence, on taking determinants and using the multiplicative property of determinants and the fact that $\det(P^{-1}) = 1/(\det P)$ (cf. 32.5), one obtains

$$\det(tI-B) = \det(tI-A); \qquad \text{i.e. } \chi_B(t) = \chi_A(t).$$

In view of 62.1 and 62.2, it follows further that $\text{tr}(A) = \text{tr}(B)$ and $\det A = \det B$. It should be pointed out that these two conclusions are also obtainable directly (without reference to characteristic polynomials): for the trace use the first part of exercise 5 on chapter 2 with A replaced by P^{-1} and B replaced by AP; for the determinant see exercise 5 on chapter 4.

Summing up the facts that have come to light about similar matrices, we have:

62.4 Similar matrices have the same characteristic polynomial, the same trace, and the same determinant.

Consider now a linear transformation a of a nonzero f.d. vector space V. All the matrices representing a (i.e. the matrices of a w.r.t. the different possible bases of V) are similar to one another (by 58.3). So, from 62.4, we deduce the significant fact that:

62.5 For a linear transformation a as described above, all the matrices representing a have the same characteristic polynomial, the same trace, and the same determinant.

Accordingly it is possible to make the following definitions for such a linear transformation a:

(i) the characteristic polynomial $\chi_a(t)$ of a is the characteristic polynomial of each and every matrix representing a;

(ii) the trace of a, $\text{tr}(a)$, is the trace of each and every matrix representing a;

(iii) the determinant of a, $\det a$, is the determinant of each and every matrix representing a.

For example, for the linear transformation a of $\mathbb{C}^2$ defined by

$$a(x_1, x_2) = (2x_1 + x_2, 2x_1 + 3x_2),$$

the matrix of a w.r.t. the standard basis of $\mathbb{C}^2$ is $A = \begin{bmatrix} 2 & 1 \\ 2 & 3 \end{bmatrix}$, and so

$$\chi_a(t) = \chi_A(t) = \begin{vmatrix} t-2 & -1 \\ -2 & t-3 \end{vmatrix} = t^2 - 5t + 4 = (t-1)(t-4);$$

$\operatorname{tr}(a) = \operatorname{tr}(A) = 5$; and $\det a = \det A = 4$.

Further to these definitions, the last three propositions of the section draw attention to some simple but important points.

62.6 For a linear transformation a of a nonzero f.d. vector space V, $\chi_a(t) = \det(ti_V - a)$.

Proof. Let L be any basis of V, and let $A = M(a; L)$. Then (cf. §56) $M(ti_V - a; L) = tI - A$; and so

$$\begin{aligned} \det(ti_V - a) &= \det(tI - A) && \text{(by meaning of determinant of a linear transformation)} \\ &= \chi_a(t) && \text{(by the definition of } \chi_a(t)\text{).} \end{aligned}$$

62.7 Let a be a linear transformation of a nonzero f.d. vector space V. Then a is singular if and only if $\det a = 0$.

Proof. Let A be the matrix of a w.r.t. some basis of V. Then:

$$\begin{aligned} a \text{ is singular} &\Leftrightarrow A \text{ is singular} && \text{(by 56.4)} \\ &\Leftrightarrow \det A = 0 && \text{(by 32.4)} \\ &\Leftrightarrow \det a = 0 && \text{(by the meaning of } \det a\text{);} \end{aligned}$$

and the result is proved.

62.8 For $A \in F_{n \times n}$, $\chi_{m_A}(t) = \chi_A(t)$.

(This follows from the definition of the characteristic polynomial of a linear transformation and the fact (cf. 55.4) that A is the matrix of m_A w.r.t. the standard basis of $F_{n \times 1}$.)

63. Eigenvalues and eigenvectors

Let a denote a linear transformation of a vector space V.

It is our stated intention to investigate the possibility of finding a basis of V w.r.t. which a has diagonal matrix. From 55.1(ii) it is apparent that a basis

$(\mathbf{e}_1, \mathbf{e}_2, \ldots, \mathbf{e}_n)$ of V has the property referred to if and only if each $\mathbf{e}_k$ is mapped to a scalar multiple of itself by a.

This observation motivates us to take an interest in nonzero vectors mapped to scalar multiples of themselves by a. (We stipulate "nonzero" because the fact that $a\mathbf{0}$ is a scalar multiple of $\mathbf{0}$ is too trivial to be interesting, quite apart from the irrelevance of $\mathbf{0}$ to the search for or selection of bases.) It will also benefit us to consider which scalars λ can appear in true equations of the form $a\mathbf{x} = \lambda\mathbf{x}$ with $\mathbf{x}$ a nonzero vector. All this leads naturally towards the major definitions which now follow.

Suppose that a is a linear transformation of V, a vector space over F. A scalar λ $(\in F)$ is called an **eigenvalue** of a if and only if there exists a *nonzero* vector $\mathbf{x} \in V$ such that $a\mathbf{x} = \lambda\mathbf{x}$. And when λ is such an eigenvalue of a, each *nonzero* vector $\mathbf{x} \in V$ satisfying $a\mathbf{x} = \lambda\mathbf{x}$ is called an **eigenvector** of a corresponding to the eigenvalue λ.

For example, for the linear transformation a of $\mathbb{C}^2$ used as an illustration towards the end of §62, it can be verified that

$$a(1,2) = (4,8); \qquad \text{i.e. } a(1,2) = 4(1,2).$$

This shows that 4 is an eigenvalue of this transformation and that $(1,2)$ is an eigenvector of the transformation corresponding to the eigenvalue 4.

Eigenvalues and eigenvectors of a square matrix $A \in F_{n \times n}$ are defined as the eigenvalues and eigenvectors of the transformation m_A of $F_{n \times 1}$. That succinct definition can be helpfully re-expressed as follows: a scalar λ $(\in F)$ is an eigenvalue of A if and only if there exists a nonzero column $X \in F_{n \times 1}$ such that $AX = \lambda X$; and when λ is such an eigenvalue of A, each nonzero $X \in F_{n \times 1}$ satisfying $AX = \lambda X$ is called an eigenvector of A corresponding to the eigenvalue λ.

For example, for the matrix $A = \begin{bmatrix} 3 & -2 \\ 2 & -2 \end{bmatrix}$, one can check that

$$A \times \operatorname{col}(2,1) = \operatorname{col}(4,2) = 2 \times \operatorname{col}(2,1).$$

This shows that 2 is an eigenvalue of A and that $\operatorname{col}(2,1)$ is an eigenvector of A corresponding to the eigenvalue 2.

It should be noted that the equations $a\mathbf{x} = \lambda\mathbf{x}$ and $AX = \lambda X$ that occur in the course of the above definitions can be rewritten in the forms

$$(\lambda i_V - a)\mathbf{x} = \mathbf{0} \quad \text{and} \quad (\lambda I - A)X = O.$$

This observation is often useful and, in particular, leads us quickly to the following result.

63.1 (i) Let a be a linear transformation of the nonzero f.d. vector space V. Then (a) the eigenvalues of a are the roots of $\chi_a(t)$ (i.e. the values of t in F that make $\chi_a(t)$ equal to 0); and (*b*) for any eigenvalue λ of a, the eigenvectors of a corresponding to the eigenvalue λ are the nonzero vectors in $\ker(\lambda i_V - a)$.

(ii) Let $A \in F_{n\times n}$. Then (*a*) the eigenvalues of A are the roots of $\chi_A(t)$; and (*b*) for any eigenvalue λ of A, the eigenvectors of A corresponding to the eigenvalue λ are the nonzero columns in the set

$$\{X \in F_{n\times 1} : (\lambda I - A)X = O\},$$

i.e. the nonzero columns in $\ker(\lambda i - m_A)$, where i stands for the identity transformation of $F_{n\times 1}$.

Proof. (i) (*a*) For $\lambda \in F$,

$$\begin{aligned}
\lambda \text{ is an eigenvalue of } a &\Leftrightarrow \text{there exists nonzero } \mathbf{x} \in V \text{ such that } a\mathbf{x} = \lambda\mathbf{x}\\
&\Leftrightarrow \text{there exists nonzero } \mathbf{x} \in V \text{ such that } (\lambda i_V - a)\mathbf{x} = \mathbf{0}\\
&\Leftrightarrow s(\lambda i_V - a) > 0\\
&\Leftrightarrow r(\lambda i_V - a) < \dim V \qquad \text{(by 49.2)}\\
&\Leftrightarrow \lambda i_V - a \text{ is singular} \qquad \text{(by 49.6)}\\
&\Leftrightarrow \det(\lambda i_V - a) = 0 \qquad \text{(by 62.7)}\\
&\Leftrightarrow \chi_a(\lambda) = 0 \qquad \text{(by 62.6).}
\end{aligned}$$

Therefore, the eigenvalues of a are the roots of $\chi_a(t)$.

(*b*) is straightforward. If λ is an eigenvalue of a, then, for $\mathbf{x} \in V$,

$$\begin{aligned}
\mathbf{x} \text{ is an eigenvector of } a \text{ corresponding to the eigenvalue } \lambda &\\
&\Leftrightarrow \mathbf{x} \neq \mathbf{0} \text{ and } a\mathbf{x} = \lambda\mathbf{x}\\
&\Leftrightarrow \mathbf{x} \neq \mathbf{0} \text{ and } (\lambda i_V - a)\mathbf{x} = \mathbf{0}\\
&\Leftrightarrow \mathbf{x} \neq \mathbf{0} \text{ and } \mathbf{x} \in \ker(\lambda i_V - a);
\end{aligned}$$

and thus the set of all such eigenvectors is as stated.

(ii) (*a*) follows from (i) (*a*) by taking $a = m_A$: the eigenvalues of A are the eigenvalues of m_A; these are, by (i) (*a*), the roots of $\chi_{m_A}(t)$, which (by 62.8) is the same as $\chi_A(t)$.

And (ii) (*b*) follows at once from (i) (*b*) on taking $a = m_A$.

Remarks. (*a*) The proposition 63.1 illustrates the fact that, in this area of work, many results come in two parallel versions—one for a linear transformation a, one for a square matrix A. In general, the two versions can be proved by virtually parallel proofs; or (more slickly, as demonstrated above) the matrix version can be deduced from the linear transformation version by taking $a = m_A$.

(*b*) In each case the kernel referred to in the (*b*) parts of 63.1 is called an **eigenspace**. In the linear transformation case, when λ is an eigenvalue of a, $\ker(\lambda i_V - a)$ is called the eigenspace of a corresponding to λ and it may be denoted by $E(\lambda, a)$. In the matrix case, when λ is an eigenvalue of A, $\ker(\lambda i - m_A)$ is called the eigenspace of A corresponding to λ and it may be denoted by $E(\lambda, A)$. In either case the eigenspace corresponding to λ consists of zero along with all the eigenvectors corresponding to λ. It must be stressed that in each case (by the definition of "eigenvector") zero itself is *not* an eigenvector.

(*c*) The dimension of the eigenspace corresponding to eigenvalue λ (whether of a linear transformation or of a matrix) is called the **geometric multiplicity** of λ as an eigenvalue. This geometric multiplicity is always at least 1 : for in every case an eigenspace corresponding to eigenvalue λ contains all eigenvectors corresponding to λ and therefore contains at least one nonzero vector.

(*d*) As remarked in 62.1, the characteristic polynomial of an $n \times n$ matrix has degree n. So, therefore, has the characteristic polynomial of a linear transformation of an n-dimensional vector space. Since a polynomial of degree n over F cannot have more than n roots in F, it follows from the (*a*) parts of 63.1 that:

63.2 A matrix in $F_{n \times n}$ cannot have more than n different eigenvalues; and a linear transformation of an n-dimensional vector space cannot have more than n different eigenvalues.

(*e*) It is left as an easy exercise for the student to show that:

63.3 For a linear transformation a of a nonzero f.d. vector space (or for a square matrix A), 0 is an eigenvalue if and only if a (or A) is singular.

There follow two worked examples on the finding of eigenvalues and eigenvectors.

1. As in §62 (near the beginning) let A be the matrix $\begin{bmatrix} 3 & 3 & 2 \\ 2 & 4 & 2 \\ -1 & -3 & 0 \end{bmatrix}$,

which we shall regard as a matrix in $\mathbb{C}_{3 \times 3}$. In §62 we found that

$$\chi_A(t) = (t-1)(t-2)(t-4).$$

So, by 63.1(ii, *a*), the eigenvalues of A are 1, 2, 4.

Let us find an eigenvector of A corresponding to the eigenvalue 2.

The nonzero column $X = \text{col}(x_1, x_2, x_3)$ is such an eigenvector

$$\Leftrightarrow AX = 2X \Leftrightarrow (2I - A)X = O$$

$$\Leftrightarrow \left\{\begin{aligned} -x_1 - 3x_2 - 2x_3 &= 0 \\ -2x_1 - 2x_2 - 2x_3 &= 0 \\ x_1 + 3x_2 + 2x_3 &= 0 \end{aligned}\right\}$$ (easy to write down because the entries of $tI - A$ were arrayed before us in the calculation of $\chi_A(t)$)

$$\Leftrightarrow \left\{\begin{aligned} x_1 \qquad + \tfrac{1}{2}x_3 &= 0 \\ x_2 + \tfrac{1}{2}x_3 &= 0 \end{aligned}\right\}$$ (on transformation of the system to reduced echelon form)

$\Leftrightarrow x_3 = -2\alpha, x_2 = \alpha, x_1 = \alpha$ for some nonzero $\alpha \in \mathbb{C}$

$\Leftrightarrow X = \alpha \, \text{col}(1, 1, -2)$ for some nonzero $\alpha \in \mathbb{C}$.

Hence *an* eigenvector of A corresponding to the eigenvalue 2 is col $(1, 1, -2)$. [It is also apparent that the geometric multiplicity of 2 as an eigenvalue of A is 1, the eigenspace $E(2, A)$ being the 1-dimensional space comprising all scalar multiples of col $(1, 1, -2)$.]

Similarly, as the student should work out in detail, one can find that eigenvectors corresponding to the eigenvalues 1 and 4 of A are, respectively, col $(1, 0, -1)$ and col $(1, 1, -1)$.

2. Consider the linear transformation a of $\mathbb{C}^2$ defined by

$$a(x_1, x_2) = (x_1 + x_2, 4x_1 + x_2).$$

This has matrix $A = \begin{bmatrix} 1 & 1 \\ 4 & 1 \end{bmatrix}$ w.r.t. the standard basis of $\mathbb{C}^2$, and so

$$\begin{aligned} \chi_a(t) = \chi_A(t) = \det(tI - A) &= \begin{vmatrix} t-1 & -1 \\ -4 & t-1 \end{vmatrix} \\ &= t^2 - 2t - 3 = (t+1)(t-3). \end{aligned}$$

Hence, by 63.1(i, a), the eigenvalues of a are -1 and 3.

Let us find an eigenvector of a corresponding to the eigenvalue 3.

The nonzero vector (x_1, x_2) is such an eigenvector

$$\begin{aligned} &\Leftrightarrow a(x_1, x_2) = 3(x_1, x_2) \\ &\Leftrightarrow (3i - a)(x_1, x_2) = (0, 0) \qquad (i \text{ the identity transformation of } \mathbb{C}^2) \\ &\Leftrightarrow \left\{ \begin{array}{r} 2x_1 - x_2 = 0 \\ -4x_1 + 2x_2 = 0 \end{array} \right\} \\ &\Leftrightarrow x_2 = 2x_1 \\ &\Leftrightarrow (x_1, x_2) = \alpha(1, 2) \text{ for some nonzero } \alpha \in \mathbb{C}. \end{aligned}$$

Hence an eigenvector of a corresponding to the eigenvalue 3 is $(1, 2)$.

Similarly, one can obtain $(1, -2)$ as an eigenvector of a corresponding to the eigenvalue -1.

Note. Examples like the above are designed to help the student by providing first-hand acquaintance with eigenvalues and eigenvectors in specific numerical problems and thereby giving a "feel" for these relatively subtle concepts. However, real-life computations tend to be rather harder—the matrices bigger, perhaps, with entries and eigenvalues which are not exact integers; and in that light the methods demonstrated above are not ideal for real-life applications. There do exist other methods, outside the scope of this text-book, for finding eigenvalues and eigenvectors—in particular iterative methods, which are suitable for electronic computation.

This section concludes with two useful technical theorems about eigenvectors (and eigenspaces). In each case there is a version of the theorem for a linear transformation and an equally valid version for a matrix (cf. remark (*a*) earlier in this section). Here the linear transformation versions only will be stated and proved.

63.4 Let a be a linear transformation of the vector space V. Any finite sum of eigenspaces of a corresponding to different eigenvalues of a is a direct sum.

Proof. The proof is by induction on the number of eigenspaces.

First consider the case of 2 eigenspaces $E(\lambda, a)$, $E(\mu, a)$ corresponding to unequal eigenvalues λ, μ of a.

Let $\mathbf{x} \in E(\lambda, a) \cap E(\mu, a)$. Then $a\mathbf{x} = \lambda\mathbf{x}$ because $\mathbf{x} \in E(\lambda, a)$; and $a\mathbf{x} = \mu\mathbf{x}$ because $\mathbf{x} \in E(\mu, a)$. Hence

$$(\lambda - \mu)\mathbf{x} = a\mathbf{x} - a\mathbf{x} = \mathbf{0},$$

and so, since $\lambda \neq \mu$, $\mathbf{x} = \mathbf{0}$. This proves that $E(\lambda, a) \cap E(\mu, a) = \{\mathbf{0}\}$; and hence (by 43.3) the sum $E(\lambda, a) + E(\mu, a)$ is a direct sum.

Thus the theorem is true for two eigenspaces. It will, therefore, suffice to prove it true for $(n+1)$ eigenspaces under the inductive assumption that it is true for n eigenspaces $(n \geqslant 2)$.

We now make that inductive assumption, and consider $(n+1)$ eigenspaces $E(\lambda_1, a)$, $E(\lambda_2, a), \ldots, E(\lambda_{n+1}, a)$ of a, corresponding to eigenvalues $\lambda_1, \lambda_2, \ldots, \lambda_{n+1}$ of a that are all different.

Suppose that

$$\mathbf{0} = \mathbf{u}_1 + \mathbf{u}_2 + \ldots + \mathbf{u}_n + \mathbf{u}_{n+1}, \tag{1}$$

where, for each i, $\mathbf{u}_i \in E(\lambda_i, a)$. We shall show that this forces every $\mathbf{u}_i$ to be $\mathbf{0}$: it will then follow by 43.2 that the sum of the subspaces $E(\lambda_1, a), \ldots, E(\lambda_{n+1}, a)$ is a direct sum.

Because $\mathbf{u}_i \in E(\lambda_i, a)$, $a\mathbf{u}_i = \lambda_i\mathbf{u}_i$ $(i = 1, 2, \ldots, n+1)$. Hence, applying the linear transformation a to both sides of (1), we deduce

$$\mathbf{0} = \lambda_1\mathbf{u}_1 + \lambda_2\mathbf{u}_2 + \ldots + \lambda_n\mathbf{u}_n + \lambda_{n+1}\mathbf{u}_{n+1}. \tag{2}$$

Hence, from (2)—λ_{n+1}(1),

$$\mathbf{0} = (\lambda_1 - \lambda_{n+1})\mathbf{u}_1 + (\lambda_2 - \lambda_{n+1})\mathbf{u}_2 + \ldots + (\lambda_n - \lambda_{n+1})\mathbf{u}_n. \tag{3}$$

It is clear that the terms on the right-hand side of (3) belong to the subspaces $E(\lambda_1, a)$, $E(\lambda_2, a), \ldots, E(\lambda_n, a)$, respectively. But, by the inductive assumption, the sum of these n subspaces is a direct sum, and so there is just one way to express $\mathbf{0}$ as a sum of vectors in these subspaces (viz. $\mathbf{0} + \mathbf{0} + \ldots + \mathbf{0}$). Hence it follows from (3) that

$$(\lambda_i - \lambda_{n+1})\mathbf{u}_i = \mathbf{0} \qquad \text{for } i = 1, 2, \ldots, n;$$

and therefore (since λ_{n+1} is different from all of $\lambda_1, \ldots, \lambda_n$) $\mathbf{u}_i = \mathbf{0}$ for $i = 1, 2, \ldots, n$. By (1), $\mathbf{u}_{n+1}$ is also $\mathbf{0}$; and thus all of $\mathbf{u}_1, \mathbf{u}_2, \ldots, \mathbf{u}_{n+1}$ are $\mathbf{0}$.

As explained in the earlier reference to 43.2, this proves that the sum of $E(\lambda_1, a), \ldots, E(\lambda_{n+1}, a)$ is a direct sum; and this completes proof by induction of the stated result.

Note. In the above proof we took 2 as the minimum number of eigenspaces in a "sum of eigenspaces". One could regard the theorem as true for trivial reasons for the degenerate case of the "sum of 1 eigenspace".

We deduce as a corollary:

63.5 Let a be a linear transformation of the vector space V, let $\lambda_1, \lambda_2, \ldots, \lambda_n$ be n different eigenvalues of a, and let $\mathbf{x}_1, \mathbf{x}_2, \ldots, \mathbf{x}_n$ be eigenvectors of a corresponding to the eigenvalues $\lambda_1, \lambda_2, \ldots, \lambda_n$, respectively. Then the sequence $(\mathbf{x}_1, \mathbf{x}_2, \ldots, \mathbf{x}_n)$ is L.I.

Proof. In the case $n = 1$ the result is trivial, by 39.8 and the fact that an eigenvector must be nonzero. Henceforth suppose $n \geqslant 2$.

Suppose that

$$\beta_1\mathbf{x}_1 + \beta_2\mathbf{x}_2 + \ldots + \beta_n\mathbf{x}_n = \mathbf{0} \qquad (\beta_1, \ldots, \beta_n \text{ scalars}). \qquad (*)$$

For each i, $\mathbf{x}_i$ is in the subspace $E(\lambda_i, a)$; and therefore so is $\beta_i\mathbf{x}_i$. Thus (*) expresses $\mathbf{0}$ as a sum of vectors in the subspaces $E(\lambda_1, a), \ldots, E(\lambda_n, a)$. But because (by 63.4) the sum of these subspaces is a direct sum, there is just one way to express $\mathbf{0}$ as a sum of vectors in these subspaces, viz. $\mathbf{0}+\mathbf{0}+ \ldots +\mathbf{0}$. Hence every term $\beta_i\mathbf{x}_i$ on the left-hand side of (*) is $\mathbf{0}$. But in every case $\mathbf{x}_i$, being an eigenvector, is nonzero. Therefore all of $\beta_1, \beta_2, \ldots, \beta_n$ are zero.

This proves the linear independence of the sequence $(\mathbf{x}_1, \ldots, \mathbf{x}_n)$.

64. Eigenvalues in the case $F = \mathbb{C}$

An instructive lesson can be learned from the matrix $A = \begin{bmatrix} 0 & -1 \\ 1 & 0 \end{bmatrix}$, which could enter the discussion either when $\mathbb{R}$ is the "scalar universe" or when $\mathbb{C}$ fills that role. The characteristic polynomial of A is easily found to be t^2+1. Therefore, if we are using $\mathbb{R}$ as "scalar universe", we shall conclude that A has no eigenvalues; but if instead we are working over $\mathbb{C}$, we shall conclude that A has two eigenvalues, namely i and $-i$.

The moral is clear: the phrase "the eigenvalues of A" assumes a precise meaning only when there is a precise understanding as to what field of scalars is in use.

Having noted that subtle point, let us adopt the standing convention that, for the remainder of this chapter, the field of scalars in use will be $\mathbb{C}$. One immediate advantage of this convention is that, for any given $n \times n$ matrix A,

$\chi_A(t)$ can be factorized completely into linear factors: say

$$\chi_A(t) = (t-\alpha_1)(t-\alpha_2)\dots(t-\alpha_n) \qquad (*)$$

with each $\alpha_j \in \mathbb{C}$. By 63.1(ii, *a*), the eigenvalues of A are the complex numbers appearing in the list $\alpha_1, \alpha_2, \dots, \alpha_n$. But, of course, this list may contain repetitions. If eigenvalue λ appears m times in the list $\alpha_1, \alpha_2, \dots, \alpha_n$, i.e. if $(t-\lambda)^m$ is the highest power of $(t-\lambda)$ dividing $\chi_A(t)$, then we describe λ as an eigenvalue with **algebraic multiplicity** m. Moreover, we use the slightly loose phraseology "$\alpha_1, \alpha_2, \dots, \alpha_n$ are the eigenvalues of A (with algebraic multiplicities)", meaning that $\chi_A(t)$ factorizes as given by (*).

From 62.3 it is clear that:

64.1 In a diagonal matrix, the entries on the main diagonal are the eigenvalues (with algebraic multiplicities).

And from 62.4 it follows that:

64.2 Similar matrices have the same eigenvalues with the same algebraic multiplicities.

We can now connect the trace and determinant of a square matrix with its eigenvalues.

64.3 Let $A \in \mathbb{C}_{n \times n}$ and let $\alpha_1, \alpha_2, \dots, \alpha_n$ be the eigenvalues of A (with algebraic multiplicities). Then:

(i) $\operatorname{tr}(A) = \alpha_1 + \alpha_2 + \dots + \alpha_n$;

(ii) $\det A = \alpha_1 \alpha_2 \dots \alpha_n$.

Proof. The given information about $\alpha_1, \dots, \alpha_n$ tells us that

$$\begin{aligned}\chi_A(t) &= (t-\alpha_1)(t-\alpha_2)\dots(t-\alpha_n)\\ &= t^n - (\alpha_1 + \alpha_2 + \dots + \alpha_n)t^{n-1} + \dots + (-1)^n \alpha_1 \alpha_2 \dots \alpha_n.\end{aligned}$$

Hence, by 62.1 and 62.2,

$$-\operatorname{tr}(A) = -(\alpha_1 + \alpha_2 + \dots + \alpha_n) \quad \text{and} \quad (-1)^n \det A = (-1)^n \alpha_1 \alpha_2 \dots \alpha_n,$$

from which the stated results follow.

So far we have introduced the algebraic multiplicity of an eigenvalue of a square matrix. In an exactly parallel way, we can introduce the algebraic multiplicity of an eigenvalue λ of a linear transformation a of a nonzero f.d. vector space: in particular, this multiplicity is m if $(t-\lambda)^m$ is the highest power of $(t-\lambda)$ dividing $\chi_a(t)$.

Useful for deducing results about matrices from the corresponding results about linear transformations is the fact that:

64.4 For $A \in \mathbb{C}_{n \times n}$, the matrix A and the linear transformation m_A have the same eigenvalues with the same algebraic multiplicities and the same geometric multiplicities.

Proof. That A and m_A have the same eigenvalues with the same algebraic multiplicities is clear from the fact that $\chi_A(t) = \chi_{m_A}(t)$ (cf. 62.8). Moreover, for any eigenvalue λ of A (and m_A), the geometric multiplicities of λ as an eigenvalue of A and of m_A are equal: for these are the dimensions of the eigenspaces $E(\lambda, A)$ and $E(\lambda, m_A)$, which coincide, both being $\ker(\lambda i - m_A)$, where i is the identity transformation of $\mathbb{C}_{n \times 1}$ (cf. remark (*b*) in §63, where eigenspaces were introduced).

A simple general fact about algebraic and geometric multiplicities is:

64.5 For any eigenvalue λ of a square matrix or of a linear transformation of a nonzero f.d. vector space,

$$(\text{geometric multiplicity of } \lambda) \leqslant (\text{algebraic multiplicity of } \lambda).$$

Proof. Consider a linear transformation a of an n-dimensional vector space V $(n \geqslant 1)$. Let λ be any eigenvalue of a, and let l be its geometric multiplicity, so that $\dim(E(\lambda, a)) = l$. Introduce a basis $(\mathbf{e}_1, \ldots, \mathbf{e}_l)$ of $E(\lambda, a)$, and (cf. 41.4(iv)) let this basis be extended to a basis $L = (\mathbf{e}_1, \ldots, \mathbf{e}_n)$ of V. Since $a\mathbf{e}_k = \lambda \mathbf{e}_k$ for $1 \leqslant k \leqslant l$, it is clear (cf. definition of $M(a; L)$) that $M(a; L)$ has the form

$$\begin{bmatrix} \lambda I_l & B \\ O & C \end{bmatrix}.$$

Hence:

$$\chi_a(t) = \det[tI_n - M(a; L)] = \det\begin{bmatrix} (t-\lambda)I_l & -B \\ O & tI_{n-l} - C \end{bmatrix}$$
$$= (t-\lambda)^l \chi_C(t) \qquad \text{(by repeated expansion by 1st column).}$$

Therefore the algebraic multiplicity of λ as an eigenvalue of a is at least l.

This proves the theorem for the linear transformation case. In view of 64.4, its truth for a square matrix A follows by taking $a = m_A$ in the above.

We classify an eigenvalue as **simple** or **repeated** according as its algebraic multiplicity is 1 or greater than 1. Since the geometric multiplicity of an eigenvalue is always at least 1 (cf. remark (*c*) in §63), it is apparent from 64.5 that:

64.6 In the case of a simple eigenvalue, the algebraic and geometric multiplicities coincide.

As specific examples reveal, the algebraic and geometric multiplicities may be equal or may be unequal in the case of a repeated eigenvalue.

Postscript on real matrices (in the light of our $F = \mathbb{C}$ convention). A real $n \times n$ matrix A is a special case of a matrix in $\mathbb{C}_{n \times n}$. The eigenvalues of A may be all real, or all non-real, or a mixture of both; and the phrase "an eigenvector of A" refers to a column in $\mathbb{C}_{n \times 1}$ which certainly need not have all its entries real. However, when we are seeking an eigenvector of A corresponding to a real eigenvalue, it is open to us temporarily to regard $\mathbb{R}$ as the system of scalars in use and hence to produce a real eigenvector. Thus:

64.7 Corresponding to a real eigenvalue of a real square matrix, there exists a real eigenvector.

65. Diagonalization of linear transformations

Throughout this section, a will denote a linear transformation of V, a nonzero f.d. vector space over $\mathbb{C}$. We denote dim V by n.

If L is a basis $(\mathbf{e}_1, \dots, \mathbf{e}_n)$ of V, the statement that $M(a; L) = \operatorname{diag}(\alpha_1, \alpha_2, \dots, \alpha_n)$ is (by the definition of $M(a; L)$) equivalent to the statement that $a\mathbf{e}_k = \alpha_k \mathbf{e}_k$ for $k = 1, 2, \dots, n$. Hence (bearing in mind that a basis vector must certainly be nonzero) we see that:

65.1 For a basis $L = (\mathbf{e}_1, \dots, \mathbf{e}_n)$ of V, the following two statements are equivalent:

(1) $M(a; L) = \operatorname{diag}(\alpha_1, \alpha_2, \dots, \alpha_n)$;

(2) $\alpha_1, \alpha_2, \dots, \alpha_n$ are eigenvalues of a, and, for each k, $\mathbf{e}_k$ is an eigenvector of a corresponding to the eigenvalue α_k.

The result 65.1 formally records what briefly came to our attention at the beginning of §63 as motivation for consideration of eigenvectors.

We describe a as **diagonable** if and only if there exists a basis L of V such that $M(a; L)$ is a diagonal matrix. The question of when exactly a is diagonable is one that interests us, and from 65.1 we can immediately give one answer, namely:

65.2 The linear transformation a is diagonable if and only if there exists a basis of V in which every vector is an eigenvector of a.

The next theorem gives one important fact about the diagonable case.

65.3 Suppose that a is diagonable and that, w.r.t. some particular basis of V, the matrix of a is $D = \operatorname{diag}(\alpha_1, \alpha_2, \dots, \alpha_n)$. Then $\alpha_1, \alpha_2, \dots, \alpha_n$ are the eigenvalues of a (with algebraic multiplicities).

Proof. By the definition of $\chi_a(t)$, $\chi_a(t)$ is the same as $\chi_D(t)$; and hence, by 62.3,

$$\chi_a(t) = (t - \alpha_1)(t - \alpha_2) \dots (t - \alpha_n),$$

which proves the stated result.

Now we come to an interesting criterion for diagonability.

65.4 The linear transformation a is diagonable if and only if, for every eigenvalue λ of a, the algebraic and geometric multiplicities of λ coincide.

Proof. (1) (The "$\Rightarrow$" half) Suppose that a is diagonable, so that there is a basis $L = (\mathbf{e}_1, \ldots, \mathbf{e}_n)$ of V w.r.t. which a has a diagonal matrix—say $D = \text{diag}(\alpha_1, \alpha_2, \ldots, \alpha_n)$, where (by 65.3) $\alpha_1, \alpha_2, \ldots, \alpha_n$ are the eigenvalues of a (with algebraic multiplicities). Let λ be an arbitrary eigenvalue of a, and let its algebraic multiplicity be m, so that m of $\alpha_1, \alpha_2, \ldots, \alpha_n$ are equal to λ: say that $\alpha_{j_1}, \alpha_{j_2}, \ldots, \alpha_{j_m}$ are equal to λ ($j_1, j_2, \ldots, j_m$ being m different integers in the range 1 to n). Then (cf. 65.1) $(\mathbf{e}_{j_1}, \mathbf{e}_{j_2}, \ldots, \mathbf{e}_{j_m})$ is a sequence of m eigenvectors of a corresponding to the eigenvalue λ and thus is a sequence of m vectors in $E(\lambda, a)$. And this sequence is L.I. (being a subsequence of the basis L). Hence (by 40.8(ii)) $\dim(E(\lambda, a)) \geqslant m$:

i.e. (geometric multiplicity of λ) $\geqslant$ (algebraic multiplicity of λ).

The reverse inequality is given by 64.5, and hence the equality of the two multiplicities of an arbitrary eigenvalue follows.

(2) (The "$\Leftarrow$" half) Suppose that each eigenvalue of a has coincident algebraic and geometric multiplicities. Let $\lambda_1, \ldots, \lambda_s$ be the different eigenvalues of a, and let m_j be the common value of the multiplicities of λ_j. From the fact that $m_1, \ldots, m_s$ are the algebraic multiplicities, we have

$$\chi_a(t) = (t-\lambda_1)^{m_1}(t-\lambda_2)^{m_2}\ldots(t-\lambda_s)^{m_s}.$$

Hence, since $\chi_a(t)$ has degree n, it is apparent that $\sum_{j=1}^{s} m_j = n$.

For each j, introduce a basis $(\mathbf{e}_{j1}, \ldots, \mathbf{e}_{jm_j})$ of $E(\lambda_j, a)$. [The length of the basis is m_j because m_j is the geometric multiplicity of λ_j, i.e. $\dim(E(\lambda_j, a))$.] Since the sum S of the eigenspaces $E(\lambda_1, a), \ldots, E(\lambda_S, a)$ is a direct sum (by 63.4), it follows by an (easily proved) generalization of 43.4(i) that

$$(\mathbf{e}_{11}, \ldots, \mathbf{e}_{1m_1}, \mathbf{e}_{21}, \ldots, \mathbf{e}_{2m_2}, \ldots, \mathbf{e}_{s1}, \ldots, \mathbf{e}_{sm_s})$$

is a basis of S and, in particular, is a L.I. sequence. The length of this L.I. sequence is

$$\sum_{j=1}^{s} m_j = n = \dim V,$$

and so (by 41.2) it is a basis of V—and a basis of V in which every vector is an eigenvector. Hence, by 65.2, a is diagonable.

Parts (1) and (2) of the proof together establish the result.

From 64.6 and 65.4 it is clear that a is diagonable if every eigenvalue of a is a simple eigenvalue (i.e. if a has n different eigenvalues). In this special case, as we now demonstrate, it is extremely easy to prove the diagonability of a (without reference to 65.4) and to produce a basis L such that $M(a; L)$ is diagonal.

65.5 Suppose that a ($\in$ end V) has n different simple eigenvalues $\alpha_1, \alpha_2, \ldots, \alpha_n$ ($n = \dim V$). For each k, let $\mathbf{e}_k$ be an eigenvector of a corresponding to α_k, and let $L = (\mathbf{e}_1, \ldots, \mathbf{e}_n)$. Then L is a basis of V, and $M(a; L) = \text{diag}(\alpha_1, \alpha_2, \ldots, \alpha_n)$.

Proof. The sequence L is L.I. by 63.5, and so is a basis of V by 41.2. And $M(a; L) = \text{diag}(\alpha_1, \ldots, \alpha_n)$ by 65.1.

In cases where every eigenvalue of a has coincident algebraic and geometric multiplicities but at least one of the eigenvalues is a repeated eigenvalue, it is less easy to produce an explicit basis L such that $M(a; L)$ is diagonal. The method of doing so is revealed by part (2) of the proof of 65.4: one obtains a basis of V of the required kind by sticking together bases of the different eigenspaces $E(\lambda, a)$ (cf. example 3 below).

The following numerical examples illustrate the main ideas that have emerged in this section—on discovering whether a given a is diagonable and, if it is, finding a basis w.r.t. which it has diagonal matrix.

Example 1. Consider the linear transformation a of $\mathbb{C}^2$ defined by

$$a(x_1, x_2) = (x_1 + x_2, 4x_1 + x_2).$$

We discussed the same linear transformation in worked example 2 in §63, and we noted there that a has 2 different eigenvalues, -1 and 3, and that corresponding eigenvectors are respectively $(1, -2)$ and $(1, 2)$. This is the easy case covered by 65.5: we can immediately say that

$$((1, -2), (1, 2))$$

is a basis of $\mathbb{C}^2$ w.r.t. which the matrix of a is $\text{diag}(-1, 3)$.

Example 2. Consider the linear transformation b of $\mathbb{C}^2$ defined by

$$b(x_1, x_2) = (x_1 + x_2, x_2).$$

By the usual method, we find that $\chi_b(t) = (t-1)^2$, so that 1 is the only eigenvalue of b and is an eigenvalue of algebraic multiplicity 2. But, when we look for corresponding eigenvectors, we discover that the eigenspace $E(1, b)$ is the 1-dimensional space spanned by $(1, 0)$. So here is a case with an eigenvalue whose algebraic and geometric multiplicities are unequal. By 65.4, the linear transformation b is not diagonable.

(*Note*: the question of the simplest form of matrix representing a non-diagonable linear transformation is one which has been well explored but which lies outside the scope of this textbook.)

Example 3. Consider the linear transformation c of $\mathbb{C}^3$ defined by

$$c(x_1, x_2, x_3) = (3x_1 - 2x_3, x_2, x_1).$$

We soon find that $\chi_c(t) = (t-1)^2(t-2)$, so that c has two eigenvalues, viz. 1 (with algebraic multiplicity 2) and 2 (a simple eigenvalue).

Investigating the crucial eigenspace $E(1, c)$, we have:

$$(x_1, x_2, x_3) \in E(1,c) \Leftrightarrow c(x_1, x_2, x_3) = (x_1, x_2, x_3)$$

$$\Leftrightarrow \left\{\begin{array}{rcl} -2x_1 & +2x_3 & = 0 \\ & 0 & = 0 \\ -x_1 & +\ x_3 & = 0 \end{array}\right\}$$

$$\Leftrightarrow (x_1, x_2, x_3) = (\alpha, \beta, \alpha) = \alpha(1,0,1) + \beta(0,1,0)$$
$$\text{for some } \alpha, \beta \in \mathbb{C}.$$

Hence it is apparent that $E(1, c)$ is 2-dimensional, with basis

$$L_1 = ((1,0,1),(0,1,0)).$$

It is also now apparent (in view of 64.6 and 65.4) that c is diagonable.

Through an analysis that follows a more familiar pattern, we find that $L_2 = ((2,0,1))$ is a basis of the 1-dimensional $E(2, c)$.

As shown in part (2) of the proof of 65.4, the sticking together of L_1 and L_2 gives a basis L of V, namely

$$L = ((1,0,1),(0,1,0),(2,0,1));$$

and (cf. 65.1) $M(c; L) = \text{diag}(1,1,2)$.

66. Diagonalization of square matrices

Throughout this section A will denote a matrix in $\mathbb{C}_{n \times n}$.

We describe the matrix A as **diagonable** if and only if the linear transformation m_A of $\mathbb{C}_{n \times 1}$ is diagonable. In view of 55.4 and 58.3, this can be re-expressed by saying that A is diagonable if and only if A is similar to a diagonal matrix.

For most purposes the latter explanation is the natural thing to bear in mind as the meaning of "A is diagonable". But, at this initial stage, the former explanation is useful because it enables us to obtain fundamental theorems about the diagonability of A by taking $a = m_A$ in the theorems in §65.

For example, from 65.2 and 65.4 we deduce:

66.1 The following three statements about A are equivalent to one another:

(1) A is diagonable;

(2) there is a basis of $\mathbb{C}_{n\times 1}$ in which every vector is an eigenvector of A;

(3) for every eigenvalue λ of A, the algebraic and geometric multiplicities of λ are equal.

And from 65.3 we could obtain the following result (easily obtainable otherwise from 64.1 and 64.2).

66.2 If A is similar to the diagonal matrix $\operatorname{diag}(\alpha_1, \alpha_2, \ldots, \alpha_n)$, then $\alpha_1, \alpha_2, \ldots, \alpha_n$ are the eigenvalues of A (with algebraic multiplicities).

A new problem arising in this section is that of finding, in the case when A is diagonable, a "diagonalizing matrix", i.e. a nonsingular matrix P such that $P^{-1}AP$ is a diagonal matrix. The next theorem tells how this may be done.

66.3 Suppose that A is diagonable, so that (cf. 66.1) there is a basis of $\mathbb{C}_{n\times 1}$ in which every vector is an eigenvector of A. Let $(X_1, X_2, \ldots, X_n)$ be any such basis; for each k, let α_k be the eigenvalue of A to which the eigenvector X_k corresponds; and let

$$P = [X_1 \ \ X_2 \ldots X_n] \qquad \text{(the matrix with columns } X_1, \ldots, X_n\text{)}.$$

Then P is nonsingular, and $P^{-1}AP = \operatorname{diag}(\alpha_1, \alpha_2, \ldots, \alpha_n)$.

Proof. Since the sequence $(X_1, \ldots, X_n)$ is a basis of $\mathbb{C}_{n\times 1}$ and is therefore L.I., it follows from 50.8 that P is nonsingular. Further:

$$\begin{aligned} AP &= A[X_1 \ \ X_2 \ldots X_n] = [AX_1 \ \ AX_2 \ldots AX_n] \\ &= [\alpha_1 X_1 \ \ \alpha_2 X_2 \ldots \alpha_n X_n] \qquad (\text{since } X_k \in E(\alpha_k, A)) \\ &= P \times \operatorname{diag}(\alpha_1, \alpha_2, \ldots, \alpha_n) \qquad (\text{by 17.3(ii)}); \end{aligned}$$

and so, P being nonsingular, $P^{-1}AP = \operatorname{diag}(\alpha_1, \alpha_2, \ldots, \alpha_n)$.

The stated result is now proved.

Thus in any case where A is diagonable, the finding of a "diagonalizing matrix" boils down to the finding of a basis of $\mathbb{C}_{n\times 1}$ in which every vector is an eigenvector of A. In the more difficult cases such a basis is obtained by sticking together bases of the different eigenspaces of A (cf. part (2) of the proof of 65.4 in the case $a = m_A$ and the last example in §65). The case where A has n different eigenvalues (all simple) is specially easy, as described by the following counterpart of 65.5.

66.4 Suppose that A $(\in \mathbb{C}_{n\times n})$ has n different simple eigenvalues $\alpha_1, \alpha_2, \ldots, \alpha_n$. For each k, let X_k be an eigenvector of A corresponding to α_k, and let $P = [X_1 \ \ X_2 \ldots X_n]$. Then P is nonsingular, and $P^{-1}AP = \operatorname{diag}(\alpha_1, \alpha_2, \ldots, \alpha_n)$.

(This follows from 66.3 and 65.5 with $a = m_A$.)

As an illustration of 66.4, let us return to the matrix $A = \begin{bmatrix} 3 & 3 & 2 \\ 2 & 4 & 2 \\ -1 & -3 & 0 \end{bmatrix}$.

We noted in §63 that this matrix has eigenvalues 1, 2, 4, corresponding eigenvectors being col $(1, 0, -1)$, col $(1, 1, -2)$, col $(1, 1, -1)$, respectively. Let P be the matrix with these columns; i.e. let $P = \begin{bmatrix} 1 & 1 & 1 \\ 0 & 1 & 1 \\ -1 & -2 & -1 \end{bmatrix}$. Then (in view of 66.4) P is nonsingular, and $P^{-1}AP = \text{diag}(1, 2, 4)$.

67. The hermitian conjugate of a complex matrix

This section will cover some new but fairly elementary work concerning complex matrices. We shall denote by $\bar{\alpha}$ the complex conjugate of the complex number α.

For an arbitrary complex matrix $A = [\alpha_{ik}]_{m \times n}$, we define $\bar{A}$ (the **complex conjugate** of A) to be the $m \times n$ matrix whose (i, k)th entry is $\bar{\alpha}_{ik}$—i.e. the matrix obtained from A by replacing every entry by its complex conjugate.

From the elementary properties of conjugates of complex numbers (i.e. $\overline{\alpha + \beta} = \bar{\alpha} + \bar{\beta}$ and $\overline{\alpha\beta} = \bar{\alpha}\bar{\beta}$ for all $\alpha, \beta \in \mathbb{C}$), it is easy to deduce the following basic properties of complex conjugates of matrices.

67.1 (i) If $A, B \in \mathbb{C}_{m \times n}$, $\overline{A + B} = \bar{A} + \bar{B}$.

(ii) If $A \in \mathbb{C}_{m \times n}$ and $\lambda \in \mathbb{C}$, $\overline{\lambda A} = \bar{\lambda}\bar{A}$.

(iii) If $A \in \mathbb{C}_{l \times m}$ and $B \in \mathbb{C}_{m \times n}$, $\overline{AB} = \bar{A}\bar{B}$.

Let $A = [\alpha_{ik}]_{m \times n}$ be an arbitrary complex matrix. Observe that $(\bar{A})^T = \overline{A^T}$ (since both are the $n \times m$ matrix whose (i, k)th entry is $\bar{\alpha}_{ki}$). We define the **hermitian conjugate** of A to be $(\bar{A})^T$ [which also equals $\overline{A^T}$]; and we denote the hermitian conjugate of A by A^*.

For example if $A = \begin{bmatrix} 2+i & 3-4i \\ 4 & 5i \end{bmatrix}$, then $A^* = \begin{bmatrix} 2-i & 4 \\ 3+4i & -5i \end{bmatrix}$.

The next proposition gives some basic properties of hermitian conjugates.

67.2 (i) $(A+B)^* = A^* + B^*$ for all $A, B \in \mathbb{C}_{m \times n}$.

(ii) $(A^*)^* = A$ for every complex matrix A.

(iii) $(\lambda A)^* = \bar{\lambda}A^*$ for every complex matrix A and every $\lambda \in \mathbb{C}$.

(iv) If $A \in \mathbb{C}_{l \times m}$ and $B \in \mathbb{C}_{m \times n}$, then $(AB)^* = B^*A^*$.

Parts (i), (ii), (iii) are almost trivial. There follows a proof of part (iv).

For $A \in \mathbb{C}_{l \times m}$, $B \in \mathbb{C}_{m \times n}$,

$$\begin{aligned} (AB)^* &= (\overline{AB})^T && \text{(by the definition of hermitian conjugate)} \\ &= (\bar{A}\bar{B})^T && \text{(by 67.1(iii))} \\ &= (\bar{B})^T(\bar{A})^T && \text{(by 18.4)} \\ &= B^*A^*. \end{aligned}$$

A complex matrix A that is equal to its own hermitian conjugate (i.e. is such that $A^* = A$) is termed a **hermitian matrix.** Observe that such a matrix must be square and that the entries on its main diagonal must be real. It should be noted that (clearly):

67.3 Every real symmetric matrix is a hermitian matrix.

More generally, it is true that, in many places where the transpose is used in the discussion of real matrices, the natural analogue in the theory of complex matrices involves the hermitian conjugate rather than the transpose. The following two-part theorem provides a key illustration of this.

67.4 (i) For a complex column X,

$$X^*X \text{ is real and nonnegative; and } X^*X = 0 \text{ if and only if } X = O.$$

(ii) For a real column $X, X^TX \geqslant 0$; and $X^TX = 0$ if and only if $X = O$.

Proof. (i) Let $X = \operatorname{col}(x_1, x_2, \ldots, x_n)$ be an arbitrary column in $\mathbb{C}_{n \times 1}$. Then

$$\begin{aligned} X^*X &= \operatorname{row}(\bar{x}_1, \bar{x}_2, \ldots, \bar{x}_n) \times \operatorname{col}(x_1, x_2, \ldots, x_n) \\ &= \sum_{j=1}^{n} \bar{x}_j x_j = \sum_{j=1}^{n} |x_j|^2; \end{aligned}$$

and from this it is clear that the claims made about X^*X are true.

We deal similarly with part (ii), by noting that if X is the real column $\operatorname{col}(x_1, x_2, \ldots, x_n)$, then $X^TX = \sum_{j=1}^{n} x_j^2$.

68. Eigenvalues of special types of matrices

Two important theorems come under this heading.

68.1 The eigenvalues of a hermitian matrix (in particular, the eigenvalues of a real symmetric matrix (cf. 67.3)) are real.

Proof. Let A be an arbitrary hermitian matrix and λ an arbitrary eigenvalue of A.

Let X be an eigenvector of A corresponding to the eigenvalue λ, so that

$$AX = \lambda X. \tag{1}$$

We find two different expressions for X^*AX.

From (1), we have $(AX)^* = (\lambda X)^*$; i.e. (cf. 67.2) $X^*A^* = \bar{\lambda}X^*$; i.e. (since A is hermitian) $X^*A = \bar{\lambda}X^*$. Hence (on postmultiplying by X)

$$X^*AX = \bar{\lambda}X^*X.$$

But also, by premultiplying both sides of (1) by X^*, we have X^*AX

$= X^*(\lambda X) = \lambda X^*X$. Hence, from the two versions of X^*AX,

$$(\bar{\lambda}-\lambda)X^*X = \bar{\lambda}X^*X - \lambda X^*X = X^*AX - X^*AX = 0.$$

But, since X is an eigenvector of A, X is a nonzero column, and so (by 67.4(i)) $X^*X \neq 0$. Hence

$$\bar{\lambda}-\lambda = 0; \qquad \text{i.e. } \bar{\lambda} = \lambda, \qquad \text{i.e. } \lambda \text{ is real.}$$

This proves the stated result.

The particular case of a real symmetric matrix is important. That special case of the result is often stated as: "The eigenvalues of a real symmetric matrix are real". In interpreting what this means, one must place it in the context where $\mathbb{C}$ (and not $\mathbb{R}$) is taken as the "scalar universe": the result is saying, for a real symmetric matrix A, that if we find all the eigenvalues in $\mathbb{C}$ of A, then it must turn out that these eigenvalues belong to the subset $\mathbb{R}$ of $\mathbb{C}$.

We come now to the second theorem.

68.2 All the eigenvalues (in $\mathbb{C}$) of a real orthogonal matrix have modulus 1.

Proof. Let A be a real orthogonal matrix, and let λ be an arbitrary eigenvalue of A.

Let X be an eigenvector of A corresponding to the eigenvalue λ, so that $AX = \lambda X$. Hence $(AX)^* = (\lambda X)^* = \bar{\lambda}X^*$ (cf. 67.2(iii)); and so

$$\begin{aligned} \bar{\lambda}X^* &= (AX)^* = X^*A^* && \text{(cf. 67.2(iv))} \\ &= X^*(\bar{A})^T = X^*A^T && (\bar{A} = A,\ A \text{ being real}). \end{aligned}$$

Since $AX = \lambda X$, it follows that

$$\begin{aligned} (\bar{\lambda}X^*)(\lambda X) &= (X^*A^T)(AX) = X^*(A^TA)X \\ &= X^*X \qquad (\text{since } A^TA = I,\ A \text{ being orthogonal}). \end{aligned}$$

So $\bar{\lambda}\lambda(X^*X) = X^*X$; i.e. $|\lambda|^2X^*X = X^*X$.

But $X \neq O$ (X being an eigenvector), and so (by 67.4(i)) $X^*X \neq 0$. Therefore we can cancel the scalar X^*X from the last equation to deduce that $|\lambda|^2 = 1$, i.e. $|\lambda| = 1$.

The stated result follows.

Remarks. (*a*) By 68.2, an eigenvalue of a real orthogonal matrix A may be 1 or -1 or a non-real number of the form $e^{i\theta}$ (i.e. $\cos\theta + i\sin\theta$) with $\theta \in \mathbb{R}$. Because the eigenvalues of such a real matrix A are the roots of the real polynomial $\chi_A(t)$, the non-real roots must (by a well-known theorem on the roots of real polynomials) occur in complex conjugate pairs (pairs of the form $\{e^{i\theta}, e^{-i\theta}\}$).

(*b*) The result 68.2 generalizes to all complex matrices A satisfying $A^*A = I$. Such complex matrices are termed **unitary**.

EXERCISES ON CHAPTER EIGHT

1. For the matrix $\begin{bmatrix} 1 & 1 & 3 \\ 0 & 2 & 1 \\ 4 & -4 & -1 \end{bmatrix}$, find (i) the characteristic polynomial, (ii) the eigenvalues, (iii) an eigenvector corresponding to each eigenvalue.

2. Show that the matrix $\begin{bmatrix} 1 & 1 & -1 \\ 4 & 1 & 1 \\ 5 & -2 & 4 \end{bmatrix}$ has just one eigenvalue. Find a basis of the corresponding eigenspace.

3. Find the eigenvalues of the linear transformation a of $\mathbb{C}^2$ defined by

$$a(x_1, x_2) = (-x_1 + 3x_2, 3x_1 - x_2),$$

and find an eigenvector corresponding to each eigenvalue.

4. Show that the linear transformation b of $\mathbb{C}^2$ defined by

$$b(x_1, x_2) = (2x_1 - x_2, x_1 + 4x_2)$$

has just one eigenvalue, and find a basis of the corresponding eigenspace.

5. Let a be rotation of 3-dimensional space through angle θ about an axis through the origin, this transformation being regarded as a linear transformation of E_3. Show that $\operatorname{tr}(a) = 1 + 2\cos\theta$.

6. Let c, d be the linear transformations of $\mathbb{C}^\infty$ defined by

$$c(x_1, x_2, x_3, x_4, x_5, \ldots) = (x_2, x_3, x_4, x_5, x_6, \ldots),$$
$$d(x_1, x_2, x_3, x_4, x_5, \ldots) = (0, x_1, x_2, x_3, x_4, \ldots).$$

Show that every complex number is an eigenvalue of c but that d has no eigenvalues. (This shows that the theory developed for the f.d. case breaks down totally in the infinite-dimensional case.)

7. Let A, B be arbitrary matrices in $F_{n \times n}$. Let M be the $(2n) \times (2n)$ matrix $\begin{bmatrix} tI_n & A \\ B & I_n \end{bmatrix}$.

By premultiplying M by $\begin{bmatrix} I & -A \\ O & I \end{bmatrix}$, prove that $\det M = \chi_{AB}(t)$. (Use the result of exercise 8 on chapter 4.) By postmultiplying M by a suitable matrix of the form $\begin{bmatrix} I & X \\ O & Y \end{bmatrix}$, deduce that AB and BA have the same characteristic polynomial.

8. Let A be a matrix in $\mathbb{C}_{n \times n}$, and let λ be an eigenvalue of A. By introducing an eigenvector, show that λ^2 is an eigenvalue of A^2 and, more generally that, for any positive integer s, λ^s is an eigenvalue of A^s.

9. Deduce from the result in question 8 that:

(i) if the complex square matrix A satisfies $A^s = I$, then every eigenvalue of A is an sth root of unity;

(ii) if A is a nilpotent matrix in $\mathbb{C}_{n \times n}$ (i.e. some positive power of A is O), then $\chi_A(t) = t^n$ and $\operatorname{tr}(A) = 0$.

10. Let A be a matrix in $\mathbb{C}_{n\times n}$ such that $A^2 = I$. By question 9(i), every eigenvalue of A is 1 or -1. Deduce that if $\operatorname{tr}(A) = n$, then $A = I$.

11. Let A be a nonsingular matrix in $\mathbb{C}_{n\times n}$. Show that

$$\chi_{A^{-1}}(t) = \frac{(-t)^n}{\det A}\chi_A\left(\frac{1}{t}\right) \qquad (t \neq 0).$$

Deduce that if $\alpha_1, \alpha_2, \ldots, \alpha_n$ are the eigenvalues of A (with algebraic multiplicities), then $1/\alpha_1, 1/\alpha_2, \ldots, 1/\alpha_n$ are the eigenvalues of A^{-1} (with algebraic multiplicities). Using the result of question 9(i), deduce further that if some positive power of A is equal to I, then $\operatorname{tr}(A^{-1}) = \overline{\operatorname{tr}(A)}$.

12. Find bases of $\mathbb{C}^2$ w.r.t. which (i) the linear transformation a of question 3, (ii) the linear transformation c of $\mathbb{C}^2$ defined by $c(x_1, x_2) = (-x_2, x_1)$, have diagonal matrices. Explain why the linear transformation b of question 4 is not diagonable.

13. Let A be the matrix in question 1. Find a nonsingular matrix P such that $P^{-1}AP$ is diagonal.

14. Suppose that a matrix (or linear transformation of a nonzero f.d. vector space) over $\mathbb{C}$ has just one eigenvalue but is not a scalar multiple of the identity. Prove that the matrix (or linear transformation) is not diagonable.

15. Let $A_1 = \begin{bmatrix} -3 & -4 & 2 \\ 4 & 5 & -2 \\ 4 & 4 & -1 \end{bmatrix}$ and $A_2 = \begin{bmatrix} 1 & 1 & -1 \\ 4 & 3 & -4 \\ 4 & 2 & -3 \end{bmatrix}$. Show that both A_1 and A_2 have 1 as a repeated eigenvalue, and in each case find a basis of the corresponding eigenspace. Deduce that exactly one of A_1, A_2 is diagonable; and if B is the diagonable one, find a nonsingular matrix P such that $P^{-1}BP$ is diagonal.

16. Let $A = \begin{bmatrix} 4 & 4 & 4 \\ 6 & 6 & 5 \\ -6 & -6 & -5 \end{bmatrix}$. Find a nonsingular matrix P such that $P^{-1}AP$ is diagonal. Hence find (i) a matrix B such that $B^2 = A$, (ii) a formula for A^n.

17. Suppose that A is an $n \times n$ diagonable matrix. Prove that (i) $r(A) = r(A^2)$, (ii) if A is singular, then $r(A) = n - m$, where m is the algebraic multiplicity of 0 as an eigenvalue. Give an example of non-diagonable matrix A for which (ii) does not hold.

18. Write down an example of a non-real 3×3 hermitian matrix.

19. Let S be an arbitrary real skew-symmetric matrix. Prove that every eigenvalue of S is of the form iy with y real (i.e. every eigenvalue of S is zero or purely imaginary).
(*) Deduce that $I+S$ is nonsingular, and prove further that $(I-S)(I+S)^{-1}$ is orthogonal.

20.* Show that every 3×3 real orthogonal matrix with determinant $+1$ has 1 as an eigenvalue.

21. Let A, B, C denote the angles of an arbitrary triangle, and a, b, c the sides opposite A, B, C, respectively. Let S be the symmetric matrix $\begin{bmatrix} 0 & \cos C & \cos B \\ \cos C & 0 & \cos A \\ \cos B & \cos A & 0 \end{bmatrix}$, and let X be $\operatorname{col}(a, b, c)$. By a geometrical argument, show that 1 is an eigenvalue of S, X being a corresponding eigenvector. Obtain the quadratic equation for the other two eigenvalues of S, and deduce from 68.1 that

$$\cos A \cos B \cos C \leqslant \tfrac{1}{8}.$$

22. Let A be an arbitrary $m \times n$ real matrix. Show that, for $X \in \mathbb{R}_{n \times 1}$,

$$X^T A^T A X = 0 \Rightarrow AX = O,$$

and hence prove that the mappings $m_{A^T A}$ and m_A have the same kernel. Deduce that $r(A^T A) = r(A)$, and deduce from this that $r(AA^T)$ is also equal to $r(A)$.

(*) Prove further that if B is a real symmetric $m \times m$ matrix and C is any real $m \times n$ matrix, then $r(B^2 C) = r(BC)$.

23. State and prove the complex analogue of the "$r(A^T A) = r(A)$" result in question 22.

24. Let A be an arbitrary $m \times n$ real matrix, and let λ be an arbitrary eigenvalue of the real symmetric matrix $I_n + A^T A$. By introducing an eigenvector, prove that $\lambda \geqslant 1$.

CHAPTER NINE

EUCLIDEAN SPACES

69. Introduction

In this chapter our interest is exclusively in real vector spaces (i.e. vector spaces over $\mathbb{R}$), and so throughout the chapter the field of scalars in use will be $\mathbb{R}$.

The main theme of the chapter is generalization of the scalar product operation on E_3 discussed in §9 of chapter 1. In the more general context we use the phrase "inner product" rather than "scalar product". The next paragraph gives the precise formal definition of an "inner product" on a real vector space. The student will immediately recognize that the definition says essentially that an inner product is an operation which is like the scalar product operation on E_3 and which, in particular, possesses basic properties akin to those of the scalar product operation on E_3.

(The formal definition) An **inner product** on the real vector space V is an operation which produces from each pair $\mathbf{x}, \mathbf{y}$ of vectors in V a corresponding real number $\mathbf{x} \cdot \mathbf{y}$ (called "the inner product of $\mathbf{x}, \mathbf{y}$") and which satisfies the following postulates:

(P1) $\mathbf{x} \cdot \mathbf{y} = \mathbf{y} \cdot \mathbf{x}$ for all $\mathbf{x}, \mathbf{y} \in V$;

(P2) "$\mathbf{x} \cdot \mathbf{y}$" is linear in both variables;

(P3) $\mathbf{x} \cdot \mathbf{x} \geqslant 0$ for all $\mathbf{x} \in V$; $\mathbf{0} \cdot \mathbf{0} = 0$; and $\mathbf{x} \cdot \mathbf{x} > 0$ for every nonzero vector $\mathbf{x}$ in V.

Remarks. (*a*) The meaning of (P2) is that both of the mappings $\mathbf{x} \mapsto \mathbf{x} \cdot \mathbf{y}$ (for fixed $\mathbf{y}$) and $\mathbf{y} \mapsto \mathbf{x} \cdot \mathbf{y}$ (for fixed $\mathbf{x}$) are linear mappings, i.e. that in every case

$$(\mathbf{x}_1 + \mathbf{x}_2) \cdot \mathbf{y} = \mathbf{x}_1 \cdot \mathbf{y} + \mathbf{x}_2 \cdot \mathbf{y}, \quad (\lambda \mathbf{x}) \cdot \mathbf{y} = \lambda(\mathbf{x} \cdot \mathbf{y}),$$
$$\mathbf{x} \cdot (\mathbf{y}_1 + \mathbf{y}_2) = \mathbf{x} \cdot \mathbf{y}_1 + \mathbf{x} \cdot \mathbf{y}_2, \quad \mathbf{x} \cdot (\lambda \mathbf{y}) = \lambda(\mathbf{x} \cdot \mathbf{y}),$$

where $\mathbf{x}, \mathbf{x}_1, \dots, \mathbf{y}_2$ denote vectors in V and λ denotes a real number.

(*b*) A real vector space on which an inner product is defined is termed a **euclidean space**.

(*c*) Suppose that $\mathbf{x}$ is a vector in a euclidean space. By (P3), $\sqrt{(\mathbf{x} \cdot \mathbf{x})}$ makes sense (as a nonnegative real number), and so (motivated by 9.6) we define the

length (or magnitude) of $\mathbf{x}$ to be $\sqrt{(\mathbf{x}\,.\,\mathbf{x})}$. In this general context we denote the length of $\mathbf{x}$ by $\|\mathbf{x}\|$. Notice that, by the detail of (P3):

69.1 In a euclidean space, $\|\mathbf{0}\| = 0$ and $\|\mathbf{x}\| > 0$ for every nonzero vector $\mathbf{x}$.

(*d*) A **unit vector** in a euclidean space means a vector whose length is 1.

(*e*) Obviously the scalar product operation discussed in §9 makes E_3 and E_2 into euclidean spaces. We regard these examples as the motivation for the general idea and for some side-issues that we shall explore. Other interesting simple examples of euclidean spaces will now be mentioned.

(1) Taking a lead from 9.3, let us define an operation on $\mathbb{R}^n$ (n an arbitrary positive integer) by

$$(x_1, x_2, \ldots, x_n)\,.\,(y_1, y_2, \ldots, y_n) = \sum_{j=1}^{n} x_j y_j.$$

It is readily seen that this operation is an inner product. We call it the standard inner product on $\mathbb{R}^n$; and henceforth we regard $\mathbb{R}^n$ as a euclidean space, the inner product being the standard inner product just defined. Note that in the euclidean space $\mathbb{R}^n$,

$$\|(x_1, x_2, \ldots, x_n)\| = \sqrt{\left(\sum_{j=1}^{n} x_j^2\right)}:$$

$$\text{e.g. in } \mathbb{R}^4,\ \|(1, -1, 3, 5)\| = \sqrt{(1^2 + (-1)^2 + 3^2 + 5^2)} = 6.$$

(2) Almost the same story is $\mathbb{R}_{n\times 1}$ as a euclidean space, the inner product being the operation defined by

$$\operatorname{col}(x_1, x_2, \ldots, x_n)\,.\,\operatorname{col}(y_1, y_2, \ldots, y_n) = \sum_{j=1}^{n} x_j y_j.$$

A useful feature here is that the inner product of two vectors can be expressed as a matrix product, since:

69.2 For $X, Y \in \mathbb{R}_{n\times 1}$, $X\,.\,Y = X^T Y$. Correspondingly, the length $\|X\|$ of a column X in $\mathbb{R}_{n\times 1}$ is $\sqrt{(X^T X)}$.

Needless to add, $\mathbb{R}_{1\times n}$ is made into a euclidean space in the same way—by defining

$$\operatorname{row}(x_1, x_2, \ldots, x_n)\,.\,\operatorname{row}(y_1, y_2, \ldots, y_n) = \sum_{j=1}^{n} x_j y_j.$$

(3) An example of an infinite-dimensional euclidean space is obtained through considering the vector space $C[a, b]$ of all real-valued continuous functions on some given closed interval $[a, b]$: this real vector space becomes a euclidean space when we define an inner product on it by

$$f\,.\,g = \int_a^b f(x) g(x)\, dx.$$

The verification that the operation so defined is an inner product is largely trivial, but there is a subtle point to negotiate in proving that $f.f > 0$ if $f \neq 0$.

In §70 we shall take note of some elementary results about euclidean spaces. In §71 our attention will be drawn to bases of a f.d. euclidean space which, like the basis $(\mathbf{i}, \mathbf{j}, \mathbf{k})$ of E_3, consist of unit vectors the inner product of any two of which is zero. This section will shed new light on real orthogonal matrices, as will the following section on the subject of length-preserving linear transformations of a euclidean space. Finally, §73 will reveal something remarkable about the diagonalization of symmetric matrices.

70. Some elementary results about euclidean spaces

Throughout this section V will denote an arbitrary euclidean space.

First comes a pair of results about lengths of vectors (generalizations of facts that are very familiar in the case of E_3).

70.1 Let $\mathbf{x} \in V$ and $\lambda \in \mathbb{R}$.

(i) $\|\lambda\mathbf{x}\| = |\lambda|\,\|\mathbf{x}\|$.

(ii) If $\mathbf{x} \neq \mathbf{0}$, $\dfrac{1}{\|\mathbf{x}\|}\mathbf{x}$ is a unit vector.

Proof. $\|\lambda\mathbf{x}\|^2 = (\lambda\mathbf{x}).(\lambda\mathbf{x})$ (by the definition of "length")
$= \lambda^2(\mathbf{x}.\mathbf{x})$ (by (P2)).

Hence, on taking square roots, $\|\lambda\mathbf{x}\| = \sqrt{(\lambda^2)}\sqrt{(\mathbf{x}.\mathbf{x})} = |\lambda|\,\|\mathbf{x}\|$.

This proves (i); and (ii) is deduced immediately by taking $\mathbf{x}$ to be an arbitrary nonzero vector and λ to be the positive scalar $1/\|\mathbf{x}\|$.

Worthy of mention at this stage is an inequality known as the "triangle inequality":

70.2 For all $\mathbf{x}, \mathbf{y} \in V$, $\|\mathbf{x}+\mathbf{y}\| \leqslant \|\mathbf{x}\| + \|\mathbf{y}\|$.

Guidance on how to prove this is supplied in exercise 2 at the end of the chapter.

We turn attention now to vanishing inner products. For $\mathbf{x}, \mathbf{y} \in V$, we describe $\mathbf{x}$ as **orthogonal** to $\mathbf{y}$ if and only if $\mathbf{x}.\mathbf{y} = 0$. In view of (P1), there is no difference between "$\mathbf{x}$ is orthogonal to $\mathbf{y}$" and "$\mathbf{y}$ is orthogonal to $\mathbf{x}$", and so we may rephrase either by saying "$\mathbf{x}, \mathbf{y}$ are orthogonal *to each other*". "Orthogonal" may be regarded as an alternative word for "perpendicular", more appropriate in a not explicitly geometrical context.

It is an immediate consequence of (P2) that:

70.3 $\mathbf{0}.\mathbf{x} = \mathbf{x}.\mathbf{0} = 0$ for all $\mathbf{x} \in V$.

(For example, for each $\mathbf{x}$, $\mathbf{0}.\mathbf{x} = (0\mathbf{x}).\mathbf{x} = 0(\mathbf{x}.\mathbf{x}) = 0$.)

So $\mathbf{0}$ is orthogonal to every vector in V.

Another obvious consequence of (P2) is:

70.4 If the vectors **x**, **y** in V are orthogonal to each other, then so are $\alpha\mathbf{x}$, $\beta\mathbf{y}$ for any $\alpha, \beta \in \mathbb{R}$.

The next proposition is less trivial and is fundamental to ideas developed in §71.

70.5 Let $L = (\mathbf{x}_1, \mathbf{x}_2, \ldots, \mathbf{x}_m)$ be a sequence of nonzero vectors in V that are mutually orthogonal (i.e. $\mathbf{x}_i \,.\, \mathbf{x}_j = 0$ if $i \neq j$). Then L is a L.I. sequence.

Proof. Suppose that $\lambda_1\mathbf{x}_1 + \lambda_2\mathbf{x}_2 + \ldots + \lambda_m\mathbf{x}_m = \mathbf{0}$ $(\lambda_1, \ldots, \lambda_m \in \mathbb{R})$.

Let j be an arbitrary integer in the range 1 to m. On taking the inner product of both sides of the above equation with $\mathbf{x}_j$, we obtain

$$\left(\sum_{i=1}^{m} \lambda_i\mathbf{x}_i\right) . \mathbf{x}_j = \mathbf{0} \,.\, \mathbf{x}_j;$$

i.e. (by (P2) and 70.3) $\sum_{i=1}^{m} \lambda_i(\mathbf{x}_i \,.\, \mathbf{x}_j) = 0.$

But, since $\mathbf{x}_i \,.\, \mathbf{x}_j = 0$ whenever $i \neq j$, $\sum_{i=1}^{m} \lambda_i(\mathbf{x}_i \,.\, \mathbf{x}_j)$ reduces to the single term $\lambda_j(\mathbf{x}_j \,.\, \mathbf{x}_j)$, which is $\lambda_j \|\mathbf{x}_j\|^2$. Therefore $\lambda_j \|\mathbf{x}_j\|^2 = 0$. Hence (since all the vectors in L are nonzero, so that $\|\mathbf{x}_j\| \neq 0$ (cf. 69.1)) λ_j must be 0. It follows (since j was arbitrary) that all of $\lambda_1, \lambda_2, \ldots, \lambda_m$ are 0.

This proves that L is L.I.

The final result in this section is a useful fact about eigenvectors of a real symmetric matrix.

70.6 Let A be a real $n \times n$ symmetric matrix, and let X, Y be real eigenvectors of A corresponding to two different eigenvalues. Then X and Y are orthogonal to each other (in $\mathbb{R}_{n \times 1}$).

Proof. Let λ, μ be the (unequal) eigenvalues of A to which the eigenvectors X, Y, respectively, correspond. Then $AX = \lambda X$ and $AY = \mu Y$. We obtain two expressions for X^TAY.

First, $X^TAY = X^TA^TY$ (since A is symmetric)
$= (AX)^TY = (\lambda X)^TY$ (since $AX = \lambda X$)
$= \lambda X^TY.$

Secondly, $X^TAY = X^T(\mu Y)$ (since $AY = \mu Y$)
$= \mu(X^TY).$

Hence $(\lambda - \mu)X^TY = X^TAY - X^TAY = 0.$

Therefore, since $\lambda \neq \mu$, $X^TY = 0$; i.e. $X \,.\, Y = 0$ (cf. 69.2); i.e. X and Y are orthogonal to each other, as asserted.

71. Orthonormal sequences and bases

Throughout this section V will again denote an arbitrary euclidean space.

A sequence of vectors $(\mathbf{x}_1, \mathbf{x}_2, \ldots, \mathbf{x}_m)$ in V is described as **orthonormal** if and only if (1) $\mathbf{x}_i$ and $\mathbf{x}_j$ are orthogonal to each other if $i \neq j$ $(1 \leqslant i, j \leqslant m)$ and (2) $\mathbf{x}_1, \mathbf{x}_2, \ldots, \mathbf{x}_m$ are all unit vectors. A sequence $(\mathbf{x}_1)$ of one unit vector counts as an orthonormal sequence.

Notice a concise alternative way of putting the definition: the sequence $(\mathbf{x}_1, \ldots, \mathbf{x}_m)$ is orthonormal if and only if $\mathbf{x}_i . \mathbf{x}_j = \delta_{ij}$ for all i, j in the range $1, 2, \ldots, m$.

From 70.5 it is immediately clear that:

71.1 Every orthonormal sequence is L.I.

An **orthonormal basis** of V simply means a basis of V that is an orthonormal sequence. The most familiar example is $(\mathbf{i}, \mathbf{j}, \mathbf{k})$, which is an orthonormal basis of E_3. Another obvious example of an orthonormal basis is the standard basis of $\mathbb{R}^n$.

In view of 71.1 and 41.2, it is clear that:

71.2 If V is f.d., then an orthonormal sequence $(\mathbf{x}_1, \ldots, \mathbf{x}_m)$ of vectors in V is an orthonormal basis of V if and only if $m = \dim V$.

As the following proposition shows, the story of components w.r.t. an orthonormal basis is particularly simple.

71.3 Suppose that $\dim V = n$ and that $L = (\mathbf{e}_1, \mathbf{e}_2, \ldots, \mathbf{e}_n)$ is an orthonormal basis of V. Then, for every $\mathbf{x} \in V$, the components of $\mathbf{x}$ w.r.t. L are $(\mathbf{e}_1 . \mathbf{x}), (\mathbf{e}_2 . \mathbf{x}), \ldots, (\mathbf{e}_n . \mathbf{x})$.

Proof. Let $\mathbf{x} \in V$, and let the components of $\mathbf{x}$ w.r.t. L be $\lambda_1, \lambda_2, \ldots, \lambda_m$, so that $\mathbf{x} = \sum_{i=1}^{n} \lambda_i \mathbf{e}_i$. Then, for each j in the range 1 to n,

$$\begin{aligned} \mathbf{e}_j . \mathbf{x} &= \sum_{i=1}^{n} \lambda_i (\mathbf{e}_j . \mathbf{e}_i) \qquad \text{(by (P2))} \\ &= \sum_{i=1}^{n} \lambda_i \delta_{ji} \qquad \text{(since } L \text{ is an orthonormal basis)} \\ &= \lambda_j \qquad \text{(cf. 17.1);} \end{aligned}$$

and from this the result follows.

Next comes a new characterization of real orthogonal matrices.

71.4 Let $A \in \mathbb{R}_{n \times n}$. Then A is orthogonal if and only if the columns of A form an orthonormal sequence in $\mathbb{R}_{n \times 1}$.

Proof. Let A have columns $C_1, C_2, \ldots, C_n$. Then, for all relevant i, k,

$$\begin{aligned}(i,k)\text{th entry of } A^T A &= (i\text{th row of } A^T) \times (k\text{th column of } A)\\ &= C_i^T C_k = C_i \,.\, C_k \qquad (\text{cf. } 69.2).\end{aligned}$$

Hence we have:

$$\begin{aligned}A \text{ is orthogonal} &\Leftrightarrow (i,k)\text{th entry of } A^T A = \delta_{ik} \text{ for all relevant}\\ &\qquad i,k \qquad (\text{cf. definition of "orthogonal"})\\ &\Leftrightarrow C_i \,.\, C_k = \delta_{ik} \text{ for all relevant } i,k\\ &\Leftrightarrow (C_1, \ldots, C_n) \text{ is an orthonormal sequence.}\end{aligned}$$

This establishes the result.

Remarks. (*a*) Since the number of columns in an $n \times n$ real orthogonal matrix equals $\dim(\mathbb{R}_{n \times 1})$, it is apparent from 71.4 and 41.2 that:

71.5 The columns of an $n \times n$ real orthogonal matrix form an orthonormal basis of $\mathbb{R}_{n \times 1}$.

(*b*) An obvious adaptation of the proof of 71.4 establishes a corresponding result for rows: an $n \times n$ real matrix is orthogonal if and only if its rows form an orthonormal sequence in $\mathbb{R}_{1 \times n}$.

We shall now consider a process by which an orthonormal sequence of vectors in V that does not span V can be extended—to a longer orthonormal sequence of vectors in V. As will be explained, the process enables one to produce an orthonormal basis of V if V is nonzero f.d. The process is known as the **Gram-Schmidt process**, and details of it are spelled out in the following lemma.

71.6 Suppose that we have an orthonormal sequence $L = (\mathbf{u}_1, \ldots, \mathbf{u}_m)$ of vectors in V ($m \geqslant 1$) such that L does not span V. Let the following three-stage programme be carried out to construct a vector $\mathbf{u}_{m+1}$:

(1) Take any vector $\mathbf{x}_{m+1}$ in V that is not in the subspace $\mathrm{sp}(\mathbf{u}_1, \ldots, \mathbf{u}_m)$.

(2) Form the vector $\mathbf{y}_{m+1} = \mathbf{x}_{m+1} - \sum_{j=1}^{m} (\mathbf{u}_j \,.\, \mathbf{x}_{m+1})\mathbf{u}_j$ (which is nonzero because $\mathbf{x}_{m+1} \notin \mathrm{sp}(\mathbf{u}_1, \ldots, \mathbf{u}_m)$).

(3) Take $\mathbf{u}_{m+1}$ to be $\dfrac{1}{\|\mathbf{y}_{m+1}\|}\mathbf{y}_{m+1}$.

Then the sequence $(\mathbf{u}_1, \ldots, \mathbf{u}_m, \mathbf{u}_{m+1})$ is an orthonormal sequence (of vectors in V) (and is, manifestly, an extension of the original sequence L).

Proof. Since L is an orthonormal sequence, it will suffice to check that

(*a*) $\mathbf{u}_{m+1}$ is a unit vector, and

(*b*) $\mathbf{u}_{m+1}$ is orthogonal to each of $\mathbf{u}_1, \mathbf{u}_2, \ldots, \mathbf{u}_m$.

Of these, (a) is clearly true from stage (3) of the construction of $\mathbf{u}_{m+1}$ (cf. 70.1(ii)). To verify (b), we need only show (since $\mathbf{u}_{m+1}$ is a scalar multiple of $\mathbf{y}_{m+1}$) that $\mathbf{y}_{m+1}$ is orthogonal to each of $\mathbf{u}_1, \mathbf{u}_2, \ldots, \mathbf{u}_m$ (cf. 70.4); and this is so because, for each i in the range 1 to m,

$$\begin{aligned}
\mathbf{u}_i \cdot \mathbf{y}_{m+1} &= \mathbf{u}_i \cdot \left(\mathbf{x}_{m+1} - \sum_{j=1}^{m} (\mathbf{u}_j \cdot \mathbf{x}_{m+1}) \mathbf{u}_j \right) \qquad \text{(see detail of stage (2))} \\
&= \mathbf{u}_i \cdot \mathbf{x}_{m+1} - \sum_{j=1}^{m} (\mathbf{u}_j \cdot \mathbf{x}_{m+1})(\mathbf{u}_i \cdot \mathbf{u}_j) \qquad \text{(by (P2))} \\
&= \mathbf{u}_i \cdot \mathbf{x}_{m+1} - \sum_{j=1}^{m} (\mathbf{u}_j \cdot \mathbf{x}_{m+1}) \delta_{ij} \qquad \text{(since } L \text{ is orthonormal)} \\
&= \mathbf{u}_i \cdot \mathbf{x}_{m+1} - \mathbf{u}_i \cdot \mathbf{x}_{m+1} \qquad \text{(cf. 17.1)} \\
&= 0.
\end{aligned}$$

The validity of the claim made for the process is thus established.

For a worked example, let us start with the vectors

$$\mathbf{u}_1 = \tfrac{1}{2}(1, 1, 1, 1), \qquad \mathbf{u}_2 = \frac{1}{2\sqrt{3}}(1, 1, 1, -3)$$

in $\mathbb{R}^4$. One can quickly verify that $\mathbf{u}_1$ and $\mathbf{u}_2$ are unit vectors and that $\mathbf{u}_1 \cdot \mathbf{u}_2 = 0$. So $(\mathbf{u}_1, \mathbf{u}_2)$ is an orthonormal sequence. Let us work through the Gram-Schmidt process (precisely as described in 71.6) to extend this given sequence to an orthonormal sequence of 3 vectors in $\mathbb{R}^4$.

Stage (1). We choose any vector $\mathbf{x}_3$ in $\mathbb{R}^4$ that is not in $\mathrm{sp}(\mathbf{u}_1, \mathbf{u}_2)$. Since all vectors in $\mathrm{sp}(\mathbf{u}_1, \mathbf{u}_2)$ have equal 1st and 2nd components, it will do to take $\mathbf{x}_3 = (1, 0, 0, 0)$.

Stage (2). As a preliminary, we evaluate $\mathbf{u}_1 \cdot \mathbf{x}_3$ and $\mathbf{u}_2 \cdot \mathbf{x}_3$:

$$\mathbf{u}_1 \cdot \mathbf{x}_3 = \tfrac{1}{2}, \quad \text{and} \quad \mathbf{u}_2 \cdot \mathbf{x}_3 = \frac{1}{2\sqrt{3}}.$$

And we work out

$$\begin{aligned}
\mathbf{y}_3 &= \mathbf{x}_3 - (\mathbf{u}_1 \cdot \mathbf{x}_3)\mathbf{u}_1 - (\mathbf{u}_2 \cdot \mathbf{x}_3)\mathbf{u}_2 \\
&= (1, 0, 0, 0) - \tfrac{1}{4}(1, 1, 1, 1) - \tfrac{1}{12}(1, 1, 1, -3) \\
&= \tfrac{1}{12}[(12, 0, 0, 0) - (3, 3, 3, 3) - (1, 1, 1, -3)] = \tfrac{1}{12}(8, -4, -4, 0) \\
&= \tfrac{1}{3}(2, -1, -1, 0).
\end{aligned}$$

Stage (3). $\|\mathbf{y}_3\| = \tfrac{1}{3}\|(2, -1, -1, 0)\| = \tfrac{1}{3}\sqrt{(2^2 + 1^2 + 1^2)} = \tfrac{1}{3}\sqrt{6}$.

So we arrive at:

$$\mathbf{u}_3 = \frac{3}{\sqrt{6}} \mathbf{y}_3 = \frac{1}{\sqrt{6}}(2, -1, -1, 0).$$

This completes the process, and the result is the extended orthonormal sequence

$$\left(\tfrac{1}{2}(1,1,1,1), \frac{1}{2\sqrt{3}}(1,1,1,-3), \frac{1}{\sqrt{6}}(2,-1,-1,0)\right).$$

Remarks. (*a*) Stage (3) can be carried out in a slightly more compact way. Having found $\mathbf{y}_3 = \tfrac{1}{3}(2,-1,-1,0)$, we can realize that $\mathbf{u}_3$ will be the unit vector of the form $\alpha(2,-1,-1,0)$ with $\alpha > 0$; and this is obviously

$$\frac{1}{\|(2,-1,-1,0)\|}(2,-1,-1,0), \quad \text{i.e.} \quad \frac{1}{\sqrt{6}}(2,-1,-1,0).$$

The process of calculating the unit vector which is a positive multiple of a given nonzero vector is often described as "normalizing".

(*b*) The above example could be continued. Since the sequence $(\mathbf{u}_1, \mathbf{u}_2, \mathbf{u}_3)$ is too short to span the 4-dimensional $\mathbb{R}^4$, the Gram-Schmidt process could be used to extend it to an orthonormal sequence $(\mathbf{u}_1, \mathbf{u}_2, \mathbf{u}_3, \mathbf{u}_4)$ of vectors in $\mathbb{R}^4$. This new sequence would (cf. 71.2) be an orthonormal basis of $\mathbb{R}^4$.

More generally, if V is f.d., then, given any orthonormal sequence of m vectors in V $(1 \leqslant m < \dim V)$, we can, by $(\dim V - m)$ applications of the Gram-Schmidt process, extend the given sequence to an orthonormal sequence of $(\dim V)$ vectors in V; and (cf. 71.2 again) this extended sequence will be an orthonormal basis of V. This makes apparent the truth of the following specialized version of 41.3.

71.7 If V is f.d. and L is an orthonormal sequence of one or more vectors in V, then L can be extended to an orthonormal basis of V. (Here it is to be understood that if L already is a basis, then the "extension" will be the "trivial extension" in which no extra vectors are appended.)

Provided that V is nonzero, the process of generating an orthonormal sequence of vectors in V can always be started: if $\mathbf{v}$ is any nonzero vector in V and we take $\mathbf{u}_1$ to be $(1/\|\mathbf{v}\|)\mathbf{v}$, then $(\mathbf{u}_1)$ is an orthonormal sequence. If V is f.d., this embryonic orthonormal sequence can (by 71.7) be extended to an orthonormal basis of V. Therefore:

71.8 Every nonzero f.d. euclidean space has an orthonormal basis.

The last result in this section shows that when an orthonormal basis is used in a nonzero f.d. euclidean space, the inner product of two vectors is given very straightforwardly in terms of their components; and this brings out a close resemblance between an arbitrary n-dimensional euclidean space and $\mathbb{R}_{n \times 1}$.

71.9 Suppose that V is nonzero and f.d. and that L is an orthonormal basis of V. Let $\mathbf{x}, \mathbf{y}$ be arbitrary vectors in V, and let their component columns w.r.t. L

be, respectively, $X = \mathrm{col}(x_1, x_2, \ldots, x_n)$ and $Y = \mathrm{col}(y_1, y_2, \ldots, y_n)$, n being dim V. Then

$$\mathbf{x}.\mathbf{y} = \sum_{i=1}^{n} x_i y_i = X^T Y \qquad (=\text{inner product } X.Y \text{ of } X, Y \text{ in } \mathbb{R}_{n\times 1});$$

and, in particular,

$$\|\mathbf{x}\| = \sqrt{\left(\sum_{i=1}^{n} x_i^2\right)} \qquad (=\text{length } \|X\| \text{ of } X \text{ in } \mathbb{R}_{n\times 1}).$$

Proof. Let $L = (\mathbf{e}_1, \mathbf{e}_2, \ldots, \mathbf{e}_n)$, so that

$$\mathbf{x} = x_1\mathbf{e}_1 + x_2\mathbf{e}_2 + \ldots + x_n\mathbf{e}_n \quad \text{and} \quad \mathbf{y} = y_1\mathbf{e}_1 + y_2\mathbf{e}_2 + \ldots + y_n\mathbf{e}_n.$$

$$\begin{aligned}\text{Then: } \mathbf{x}.\mathbf{y} &= \sum_{i=1}^{n} x_i(\mathbf{e}_i.\mathbf{y}) = \sum_{i=1}^{n} x_i\left(\sum_{j=1}^{n} y_j(\mathbf{e}_i.\mathbf{e}_j)\right) \qquad \text{(by (P2))}\\ &= \sum_{i=1}^{n} x_i\left(\sum_{j=1}^{n} y_j\delta_{ij}\right) \qquad \text{(since } L \text{ is orthonormal)}\\ &= \sum_{i=1}^{n} x_i y_i \qquad \text{(cf. 17.1);}\end{aligned}$$

and the whole stated result follows.

72. Length-preserving linear transformations of a euclidean space

The contents of this section can be motivated by an observation about linear transformations of E_2 and E_3 that arise naturally from geometry. Consider a familiar case—a linear transformation a of E_2 defined by a rotation of the x, y-plane about the origin. It is clear that, in addition to its linearity, a has the following two nice properties:

(1) it preserves magnitudes (lengths); i.e. $|a\mathbf{x}| = |\mathbf{x}|$ in every case;

(2) it preserves scalar (inner) products (because it preserves magnitudes and angles between vectors); i.e. $(a\mathbf{x}).(a\mathbf{y}) = \mathbf{x}.\mathbf{y}$ in every case. Other geometrically interesting linear transformations (of E_2 and E_3) have the same properties—e.g. 2- and 3-dimensional reflections in "mirrors" through the origin.

We introduce the corresponding notions for a linear transformation a of a general euclidean space V by the following two definitions:

(1) a is **length-preserving** if and only if $\|a\mathbf{x}\| = \|\mathbf{x}\|$ for all $\mathbf{x} \in V$;

(2) a is **inner-product-preserving** if and only if $(a\mathbf{x}).(a\mathbf{y}) = \mathbf{x}.\mathbf{y}$ for all $\mathbf{x}, \mathbf{y} \in V$.

Straight away, we can prove that the two notions are, in fact, exactly equivalent. Once this has been proved, we can drop the cumbersome phrase "inner-product-preserving" from our vocabulary, knowing that "length-preserving" means exactly the same thing.

72.1 Let a be a linear transformation of a euclidean space V. Then a is length-preserving if and only if a is inner-product-preserving.

Proof. (1) (The "⇒" half). Suppose that a is length-preserving.
Let $\mathbf{x}, \mathbf{y}$ be arbitrary vectors in V.
We observe that, for all $\mathbf{u}, \mathbf{v} \in V$

$$\begin{aligned}\|\mathbf{u}+\mathbf{v}\|^2 &= (\mathbf{u}+\mathbf{v}).(\mathbf{u}+\mathbf{v}) = \mathbf{u}.\mathbf{u}+2(\mathbf{u}.\mathbf{v})+\mathbf{v}.\mathbf{v} \qquad \text{(by (P2) and (P1))},\\ &= \|\mathbf{u}\|^2+2(\mathbf{u}.\mathbf{v})+\|\mathbf{v}\|^2,\end{aligned}$$

and therefore

$$\mathbf{u}.\mathbf{v} = \tfrac{1}{2}(\|\mathbf{u}+\mathbf{v}\|^2 - \|\mathbf{u}\|^2 - \|\mathbf{v}\|^2). \qquad (*)$$

Hence:

$$\begin{aligned}(a\mathbf{x}).(a\mathbf{y}) &= \tfrac{1}{2}(\|a\mathbf{x}+a\mathbf{y}\|^2 - \|a\mathbf{x}\|^2 - \|a\mathbf{y}\|^2) \qquad \text{(by (*) with } \mathbf{u} = a\mathbf{x},\ \mathbf{v} = a\mathbf{y})\\ &= \tfrac{1}{2}(\|a(\mathbf{x}+\mathbf{y})\|^2 - \|a\mathbf{x}\|^2 - \|a\mathbf{y}\|^2) \qquad \text{(since } a \text{ is linear)}\\ &= \tfrac{1}{2}(\|\mathbf{x}+\mathbf{y}\|^2 - \|\mathbf{x}\|^2 - \|\mathbf{y}\|^2) \qquad \text{(since } a \text{ is length-preserving)}\\ &= \mathbf{x}.\mathbf{y} \qquad \text{(by (*) with } \mathbf{u} = \mathbf{x},\ \mathbf{v} = \mathbf{y}).\end{aligned}$$

Since $\mathbf{x}, \mathbf{y}$ were arbitrary, this proves that a is inner-product-preserving.

(2) (The "⇐" half) If a is inner-product-preserving, then, for all $\mathbf{x} \in V$, $(a\mathbf{x}).(a\mathbf{x}) = \mathbf{x}.\mathbf{x}$ and, therefore,

$$\|a\mathbf{x}\| = \sqrt{[(a\mathbf{x}).(a\mathbf{x})]} = \sqrt{(\mathbf{x}.\mathbf{x})} = \|\mathbf{x}\|,$$

so that a is length-preserving.
Parts (1) and (2) of the proof together establish the stated result.

The other theorems in this section show how orthogonal matrices fit into the story of length-preserving linear transformations and orthonormal bases.

72.2 Let V be a nonzero f.d. euclidean space, let L be an orthonormal basis of V, and let a be a linear transformation of V. Then a is length-preserving if and only if $M(a; L)$ is orthogonal.

Proof. Let L be $(\mathbf{e}_1, \mathbf{e}_2, \ldots, \mathbf{e}_n)$, where $n = \dim V$. Let us write M_a for $M(a; L)$, which is a matrix in $\mathbb{R}_{n\times n}$.
(1) (The "⇒" half) Suppose that a is length-preserving.
Let $C_1, C_2, \ldots, C_n$ be the columns of M_a. These columns are (by the definition of M_a) the component columns of $a\mathbf{e}_1, a\mathbf{e}_2, \ldots, a\mathbf{e}_n$, respectively, w.r.t. L. Hence, for all i, k in the range 1 to n,

$$\begin{aligned}C_i^T C_k &= (a\mathbf{e}_i).(a\mathbf{e}_k) \qquad \text{(by 71.9 with } \mathbf{x} = a\mathbf{e}_i,\ \mathbf{y} = a\mathbf{e}_k)\\ &= \mathbf{e}_i.\mathbf{e}_k \qquad \text{(by 72.1, } a \text{ being length-preserving)}\\ &= \delta_{ik} \qquad \text{(since } L \text{ is orthonormal).}\end{aligned}$$

Therefore the sequence of columns $(C_1, C_2, \ldots, C_n)$ is orthonormal; and so (by 71.4) M_a is orthogonal.

(2) (The "$\Leftarrow$" half) Suppose that M_a is orthogonal.

Let $\mathbf{x}$ be an arbitrary vector in V, and let X be the component column of $\mathbf{x}$ w.r.t. L. By 71.9, $\|\mathbf{x}\| = \|X\|$. Moreover, by 55.1(iii), the component column of $a\mathbf{x}$ w.r.t. L is M_aX, and (by 71.9 again) $\|a\mathbf{x}\| = \|M_aX\|$. Hence

$$\begin{aligned}\|a\mathbf{x}\|^2 &= \|M_aX\|^2\\ &= (M_aX)^T(M_aX) \qquad \text{(by the definition of length in } \mathbb{R}_{n\times 1}\text{—cf. 69.2)}\\ &= X^TM_a^TM_aX\\ &= X^TX \qquad (M_a \text{ being orthogonal})\\ &= \|X\|^2 = \|\mathbf{x}\|^2;\end{aligned}$$

and therefore $\|a\mathbf{x}\| = \|\mathbf{x}\|$.

Since $\mathbf{x}$ was arbitrary in V, this proves that a is length-preserving.

Parts (1) and (2) of the proof together establish the stated result.

The following corollary provides another interesting characterization of orthogonal matrices.

72.3 Let $A \in \mathbb{R}_{n\times n}$. Then A is orthogonal if and only if $\|AX\| = \|X\|$ for all $X \in \mathbb{R}_{n\times 1}$.

Proof. Since the standard basis of $\mathbb{R}_{n\times 1}$ is orthonormal and since the matrix of the transformation m_A w.r.t. that basis is A (cf. 55.4), we deduce from 72.2 that m_A is length-preserving if and only if A is orthogonal. The stated result follows, because the condition "$\|AX\| = \|X\|$ for all $X \in \mathbb{R}_{n\times 1}$" is clearly just another way of putting the condition "m_A is length-preserving".

Last in this sequence of results is one about change of basis.

72.4 In a nonzero f.d. euclidean space V, let L be an orthonormal basis, let $\hat{L}$ be a second basis, and let $P = M(L \to \hat{L})$. Then $\hat{L}$ is also orthonormal if and only if P is orthogonal.

Proof. Let $\hat{L} = (\hat{\mathbf{e}}_1, \hat{\mathbf{e}}_2, \ldots, \hat{\mathbf{e}}_n)$, n being dim V, and let $C_1, C_2, \ldots, C_n$ be the columns of the $n \times n$ matrix P.

For each k in the range 1 to n, the component column of $\hat{\mathbf{e}}_k$ w.r.t. $\hat{L}$ is the kth column E_k of the identity matrix I_n, and hence

$$\begin{aligned}\text{component column of } \hat{\mathbf{e}}_k \text{ w.r.t. } L &= PE_k \qquad \text{(by 57.2)}\\ &= C_k \qquad \text{(by 21.1 with } A = P \text{ and } B = I_n).\end{aligned}$$

Since L is an orthonormal basis, it follows by 71.9 that

$$\hat{\mathbf{e}}_i \cdot \hat{\mathbf{e}}_j = C_i^TC_j$$

for all relevant i, j. Hence we have:

$$\begin{aligned}\hat{L} \text{ is an orthonormal basis} &\Leftrightarrow C_i^T C_j = \delta_{ij} \text{ for all relevant } i, j \\ &\Leftrightarrow (C_1, C_2, \ldots, C_n) \text{ is an orthonormal sequence} \\ &\Leftrightarrow P \text{ is orthogonal} \quad \text{(cf. 71.4);}\end{aligned}$$

and the stated result is now proved.

Remarks. (*a*) There is an alternative way of proving 72.4 that brings out its connection with 72.2. Suppose that the basis L is $(\mathbf{e}_1, \mathbf{e}_2, \ldots, \mathbf{e}_n)$; and let p be the linear transformation of V that maps $\mathbf{e}_k$ to $\hat{\mathbf{e}}_k$ $(k = 1, 2, \ldots, n)$, so that $M(p; L) = P$ (cf. discussion prior to 57.1). The skeleton of the alternative proof of 72.4 is:

$$\begin{aligned}\hat{L} \text{ is orthonormal} &\Leftrightarrow p \text{ is length-preserving [this has to be proved} \\ &\qquad \text{with the aid of 72.1 and 71.9]} \\ &\Leftrightarrow P \text{ is orthogonal (by 72.2, since } L \text{ is} \\ &\qquad \text{orthonormal and } M(p; L) = P).\end{aligned}$$

(*b*) Theorem 72.2 is well illustrated by the case where V is E_2 and L is the orthonormal basis $(\mathbf{i}, \mathbf{j})$.

Geometrical argument shows that there are just two kinds of length-preserving linear transformations of E_2—rotations about O and reflections in lines through O (O being the origin). The matrices w.r.t. $L(=(\mathbf{i}, \mathbf{j}))$ of arbitrary examples of both kinds have been found (in a worked example in §55, and in exercise 2 on chapter 7, respectively); and in both cases the matrix found was orthogonal, in accordance with 72.2.

Conversely, given any real 2×2 orthogonal matrix A, we are in a position to identify a length-preserving linear transformation of E_2 whose matrix w.r.t. L is A. It is not difficult to show that A must be

$$\begin{bmatrix} \cos\beta & -\sin\beta \\ \sin\beta & \cos\beta \end{bmatrix} \quad \text{or} \quad \begin{bmatrix} \cos\beta & \sin\beta \\ \sin\beta & -\cos\beta \end{bmatrix}$$

for some β with $0 \leqslant \beta < 2\pi$ (the former if $\det A = +1$, the latter if $\det A = -1$). In the former case, A is the matrix (w.r.t. L) of rotation about O through angle β in the positive direction; in the latter case, A is the matrix (w.r.t. L) of reflection in the line through O making angle $\frac{1}{2}\beta$ with the positive half of the x-axis. (Cf. §55 and exercise 2 on chapter 7, respectively.)

Notice that rotations correspond to orthogonal matrices with determinant $+1$, while reflections correspond to orthogonal matrices with determinant -1.

(*c*) The familiar place where we encounter change from one orthonormal basis to another is a change of coordinate system (in 2 or 3 dimensions) by rotation of the coördinate axes (the origin being fixed). The original coordinates of any point are the components of its position vector w.r.t. one orthonormal basis L ($(\mathbf{i}, \mathbf{j})$ in 2 dimensions, $(\mathbf{i}, \mathbf{j}, \mathbf{k})$ in 3 dimensions); and the "new" coordinates will be the components of the position vector w.r.t. the orthonormal basis $\hat{L}$ of unit vectors whose directions are the positive directions along the new coordinate axes. By 72.4, $M(L \to \hat{L})$ is orthogonal, and so (cf. 57.2) the old and new coordinates of an arbitrary point are related by a system of equations with orthogonal coefficient matrix. For example, in 2 dimensions, it is easily proved that, when the axes are rotated through angle θ in the positive direction, the old coordinates (x, y) and the new coordinates $(\hat{x}, \hat{y})$ of an arbitrary point are related by

$$\left.\begin{cases} x = (\cos\,\theta)\hat{x} - (\sin\,\theta)\hat{y} \\ y = (\sin\,\theta)\hat{x} + (\cos\,\theta)\hat{y} \end{cases}\right\}.$$

It is useful to note explicitly the fact that

72.5 An arbitrary orthogonal matrix P of the appropriate size can be used to specify a change of coordinates: i.e. a new coordinate system (with the same origin as we start with and with mutually perpendicular coordinate axes) can be introduced so that the equations relating the original and new coordinates of a general point have coefficient matrix P.

Proof in the 3-dimensional case. Let P be an arbitrary 3×3 orthogonal matrix. Let $\mathbf{e}$, $\mathbf{f}$, $\mathbf{g}$ be the vectors in E_3 whose component columns w.r.t. $L = (\mathbf{i}, \mathbf{j}, \mathbf{k})$ are the columns of P, and let $\hat{L} = (\mathbf{e}, \mathbf{f}, \mathbf{g})$. Then (by 71.9 and 71.5) $\hat{L}$ is an orthonormal basis of E_3; and clearly $M(L \to \hat{L}) = P$. Now let a new coordinate system be introduced by taking the same origin as before and new coordinate axes pointing in the directions of the mutually perpendicular unit vectors $\mathbf{e}$, $\mathbf{f}$, $\mathbf{g}$. Then, just as the original coordinates (x_A, y_A, z_A) of a general point A are the components of $\mathbf{r}_A$ w.r.t. L (cf. 8.2), so the new coordinates $(\hat{x}_A, \hat{y}_A, \hat{z}_A)$ of A are the components of r_A w.r.t. $\hat{L}$; and, by 57.2, the relation between these two sets of coordinates is

$$\operatorname{col}(x_A, y_A, z_A) = P \times \operatorname{col}(\hat{x}_A, \hat{y}_A, \hat{z}_A)$$

which proves the stated result.

73. Orthogonal diagonalization of a real symmetric matrix

It should be emphasized at the outset that in this section (as throughout this chapter) the field of scalars in use is $\mathbb{R}$.

Because orthogonal matrices correspond to what we might call geometrically interesting changes of basis in a euclidean space (cf. 72.4), it is natural to introduce the relation of **orthogonal similarity** defined as follows.

For $A, B \in \mathbb{R}_{n \times n}$, A is orthogonally similar to B if and only if there exists an orthogonal matrix $P \in \mathbb{R}_{n \times n}$ such that $P^{-1}AP = B$. (Note that, of course, $P^{-1}AP$ is the same as P^TAP when P is orthogonal.)

The question of which real square matrices are orthogonally similar to diagonal matrices is answered by the following major theorem.

73.1 Let $A \in \mathbb{R}_{n \times n}$. Then A is orthogonally similar to a diagonal matrix if and only if A is symmetric.

Proof. (1) (The "$\Rightarrow$" half) Suppose that A is orthogonally similar to a diagonal matrix D, so that we have $P^{-1}AP = D$, for some orthogonal matrix P. Hence:

$$A = PDP^{-1} = PDP^T. \qquad (P^{-1} = P^T \text{ because } P \text{ is orthogonal.})$$

It follows that:

$$\begin{aligned} A^T &= (PDP^T)^T = PD^TP^T \qquad \text{(by 18.1 and 18.4)} \\ &= PDP^T \qquad \text{(cf. 18.5)} \\ &= A; \end{aligned}$$

and thus A is symmetric.

(2) (The "$\Leftarrow$" half) We prove by induction on n the universal truth of the statement: (α) if $A \in \mathbb{R}_{n \times n}$ and A is symmetric, then A is orthogonally similar to a diagonal matrix.

The statement (α) is certainly true for $n = 1$. (For, if A is a 1×1 real symmetric matrix, then (since A is diagonal and I_1 is orthogonal) the equation $I_1^{-1}AI_1 = A$ shows that A is orthogonally similar to a diagonal matrix.) So it will suffice to prove the truth of (α) under the inductive assumption that every real $(n-1) \times (n-1)$ symmetric matrix is orthogonally similar to a diagonal matrix ($n \geqslant 2$).

We now make that inductive assumption, and consider an arbitrary $n \times n$ real symmetric matrix A.

By 68.1, $\chi_A(t)$ factorizes completely into real linear factors; and certainly, therefore, we may introduce a real eigenvalue α_1 of A. We may also (cf. 64.7) introduce a corresponding real eigenvector of A and hence (by "normalizing") produce a unit vector U_1 ($\in \mathbb{R}_{n \times 1}$) which is also in the eigenspace $E(\alpha_1, A)$. The sequence (U_1) is an orthonormal sequence; and so, by 71.7, it may be extended to an orthonormal basis $(U_1, U_2, \ldots, U_n)$ of $\mathbb{R}_{n \times 1}$. Let Q be the $n \times n$ matrix

$$[U_1 \ \ U_2 \ldots U_n],$$

which is orthogonal, by 71.4. Then, for each i in the range 1 to n, $(i, 1)$th entry of Q^TAQ = (ith row of Q^T) $\times A \times$ (1st column of Q) (cf. 21.2)

$$\begin{aligned} &= U_i^TAU_1 \\ &= U_i^T(\alpha_1 U_1) \qquad (\text{since } U_1 \in E(\alpha_1, A)) \\ &= \alpha_1(U_i^TU_1) \\ &= \alpha_1\delta_{i1} \qquad ((U_1, \ldots, U_n) \text{ being an orthonormal sequence}); \end{aligned}$$

and so the 1st column of Q^TAQ is $\operatorname{col}(\alpha_1, 0, 0, \ldots, 0)$. But

$$(Q^TAQ)^T = Q^TA^T(Q^T)^T = Q^TAQ \qquad (A \text{ being symmetric}),$$

so that Q^TAQ is also symmetric. Hence Q^TAQ is of the form

$$\begin{bmatrix} \alpha_1 & O \\ O & B \end{bmatrix},$$

where B is a (real) $(n-1) \times (n-1)$ symmetric matrix.

By the inductive assumption, there is an $(n-1) \times (n-1)$ orthogonal matrix S such that S^TBS is diagonal: say $S^TBS = \operatorname{diag}(\alpha_2, \ldots, \alpha_n)$. Let R be the $n \times n$ matrix $\begin{bmatrix} 1 & O \\ O & S \end{bmatrix}$, which is clearly orthogonal (since S is orthogonal). Then

$$\begin{aligned}(QR)^TA(QR) = R^T(Q^TAQ)R &= \begin{bmatrix} 1 & O \\ O & S^T \end{bmatrix}\begin{bmatrix} \alpha_1 & O \\ O & B \end{bmatrix}\begin{bmatrix} 1 & O \\ O & S \end{bmatrix} \\ &= \begin{bmatrix} \alpha_1 & O \\ O & S^TBS \end{bmatrix} \\ &= \operatorname{diag}(\alpha_1, \alpha_2, \ldots, \alpha_n) \qquad (\text{since } S^TBS = \operatorname{diag}(\alpha_2, \ldots, \alpha_n)).\end{aligned}$$

Thus P^TAP is diagonal, where $P = QR$, which is orthogonal since Q and R are orthogonal (cf. exercise 11 on chapter 2).

This shows that A is orthogonally similar to a diagonal matrix, and it completes a proof by induction of the truth of (α) for every positive integer n.

This and part (1) of the proof together establish the stated result.

Remarks. (*a*) The fact that a symmetric matrix A is orthogonally similar to a diagonal matrix is often expressed by saying that A is "orthogonally diagonable".

(*b*) Notice that if the symmetric matrix A is orthogonally similar to the diagonal matrix $D = \operatorname{diag}(\alpha_1, \alpha_2, \ldots, \alpha_n)$, then (in particular) A is similar to D, and so (by 66.2) $\alpha_1, \alpha_2, \ldots, \alpha_n$ must be the eigenvalues of A (with algebraic multiplicities).

(*c*) Suppose that A is a real $n \times n$ symmetric matrix with n different (simple) eigenvalues. It turns out to be particularly easy in this case to find an orthogonal matrix P such that P^TAP is diagonal. This emerges from part (ii) of the following theorem, which, incidentally also shows that the existence of such an orthogonal matrix P can be established (in this n different eigenvalues case) without reference to 73.1 or its proof. (This may sound like an anticlimax after the complexity of the proof of 73.1; but it will be realized that 73.1 remains of great importance because it tells us that *every* real symmetric matrix is orthogonally diagonable, whether or not its eigenvalues are all simple.)

73.2 Let A be a symmetric matrix in $\mathbb{R}_{n\times n}$.

(i) Suppose that $(V_1, V_2, \ldots, V_n)$ is an orthonormal sequence of n columns in $\mathbb{R}_{n\times 1}$, each of which is an eigenvector of A. Let $\beta_1, \beta_2, \ldots, \beta_n$ be the eigenvalues of A to which $V_1, V_2, \ldots, V_n$, respectively correspond; and let P be the $n \times n$ matrix $[V_1 \ \ V_2 \ldots V_n]$. Then P is orthogonal, and $P^TAP = \text{diag}(\beta_1, \beta_2, \ldots, \beta_n)$.

(ii) Suppose that A has n different eigenvalues $\alpha_1, \alpha_2, \ldots, \alpha_n$. Let respective corresponding real *unit* eigenvectors $U_1, U_2, \ldots, U_n$ be introduced (i.e. eigenvectors of length 1, obtainable by normalizing arbitrary real eigenvectors); and let P be the $n \times n$ matrix $[U_1 \ \ U_2 \ldots U_n]$. Then P is orthogonal, and $P^TAP = \text{diag}(\alpha_1, \alpha_2, \ldots, \alpha_n)$.

Proof. (i) It follows immediately from 71.4 that P is orthogonal. Moreover, since (for each k) $AV_k = \beta_k V_k$, we can quickly prove (by the method seen in the proof of 66.3) that $AP = PD$, where $D = \text{diag}(\beta_1, \beta_2, \ldots, \beta_n)$. Hence ($P$ being orthogonal) $P^{-1}AP = D$, i.e. $P^TAP = D$. This completes proof of part (i).

(ii) Any two of $U_1, U_2, \ldots, U_n$ are orthogonal to each other (by 70.6). So, since each U_k is a unit vector, the sequence $(U_1, U_2, \ldots, U_n)$ is orthonormal. The whole of part (ii) is now immediately deducible from part (i) (by taking V_k to be U_k and β_k to be α_k).

Worked examples. 1. (This example illustrates the ease with which one can, in the light of 73.2(ii), find an orthogonal matrix P such that P^TAP is diagonal in the case where every eigenvalue of the given symmetric matrix A is simple.)

We consider the matrix $A = \begin{bmatrix} 1 & 3 \\ 3 & 1 \end{bmatrix}$.

$\chi_A(t)$ works out to be $t^2-2t-8 = (t-4)(t+2)$. So A has eigenvalues 4 and -2. By the procedure familiar from chapter 8, one finds that corresponding eigenvectors are, respectively, col(1, 1) and col(1, -1). Hence corresponding *unit* eigenvectors (obtained by normalizing) are $\frac{1}{\sqrt{2}}$ col (1, 1) and $\frac{1}{\sqrt{2}}$ col (1, -1).

Let P be the matrix with these columns; i.e. $P = \frac{1}{\sqrt{2}}\begin{bmatrix} 1 & 1 \\ 1 & -1 \end{bmatrix}$. Then (cf. 73.2(ii)) P is orthogonal, and $P^TAP = \text{diag}(4, -2)$.

2. Let A be an $n \times n$ real symmetric matrix whose eigenvalues are all positive. (Such a symmetric matrix is described as **positive-definite**.) Prove that there is a positive-definite symmetric matrix B such that $B^2 = A$.

Solution. By 73.1, there is an orthogonal matrix P such that P^TAP is a diagonal matrix D. Suppose that $D = \text{diag}(\alpha_1, \alpha_2, \ldots, \alpha_n)$. Then (cf. remark (*b*) following 73.1) $\alpha_1, \alpha_2, \ldots, \alpha_n$ are the eigenvalues of A (with algebraic multiplicities), and these are all positive.

For each k, let β_k be the (positive) square root of α_k. Then let $E = \text{diag}(\beta_1, \ldots, \beta_n)$, so that $E^2 = D$. Since $P^TAP = D$, we have

$$\begin{aligned} A &= PDP^T \qquad \text{(since } P \text{ is orthogonal)} \\ &= PE^2P \\ &= PEP^TPEP^T \qquad \text{(since } P \text{ is orthogonal)} \\ &= B^2, \qquad \text{where } B = PEP^T. \end{aligned}$$

We note that $B^T = (PEP^T)^T = (P^T)^T E^T P^T = PEP^T = B$ (cf. 18.1, 18.4, 18.5); and so B is symmetric. Moreover

$$B = Q^{-1}EQ,$$

where $Q = P^T$. So B is similar to E, and hence (cf. 66.2) the eigenvalues of B are $\beta_1, \beta_2, \ldots, \beta_n$, which are all positive.

Thus B is a positive-definite symmetric matrix whose square is A.

(We shall come to more about positive-definite matrices and the reasons why they are significant in §§79 and 80 of chapter 10.)

3. (An illustration of the case of a symmetric matrix with a repeated eigenvalue). Consider $A = \begin{bmatrix} 0 & 1 & 1 \\ 1 & 0 & 1 \\ 1 & 1 & 0 \end{bmatrix}$. We set ourselves the task of finding an orthogonal matrix P such that P^TAP is diagonal.

In this case, $\chi_A(t)$ works out to be $(t+1)^2(t-2)$. So the eigenvalues of A are -1 (a repeated eigenvalue with algebraic multiplicity 2) and 2 (a simple eigenvalue).

The eigenspace $E(-1, A)$ is soon found to be the 2-dimensional space of all columns in $\mathbb{R}_{3\times 1}$ with entries totalling zero. One obvious unit vector in this space is $V_1 = \dfrac{1}{\sqrt{2}}\operatorname{col}(1, -1, 0)$. The Gram-Schmidt process enables us to extend the embryonic orthonormal sequence (V_1) to an orthonormal basis of $E(-1, A)$; and one possible outcome is the orthonormal basis (V_1, V_2), where $V_2 = \dfrac{1}{\sqrt{6}}\operatorname{col}(1, 1, -2)$.

The simple eigenvalue 2 is much easier to deal with, and one finds that a corresponding unit eigenvector is $V_3 = \dfrac{1}{\sqrt{3}}\operatorname{col}(1, 1, 1)$.

We now have orthonormal bases (V_1, V_2) of $E(-1, A)$ and (V_3) of $E(2, A)$. Sticking these together gives the sequence (V_1, V_2, V_3), which is also orthonormal! (The point to be checked is that V_3 is orthogonal to V_1 and V_2: this is guaranteed by 70.6.) Now let P be the matrix with columns V_1, V_2, V_3; i.e.

$$P = \frac{1}{\sqrt{6}}\begin{bmatrix} \sqrt{3} & 1 & \sqrt{2} \\ -\sqrt{3} & 1 & \sqrt{2} \\ 0 & -2 & \sqrt{2} \end{bmatrix}.$$

Then (cf. 73.2(i)) P is orthogonal, and $P^TAP = \operatorname{diag}(-1, -1, 2)$.

EXERCISES ON CHAPTER NINE

1. In $\mathbb{R}^4$, $\mathbf{x}$ is the vector $(2, 2, -3, 8)$. (*a*) What is the length of $\mathbf{x}$? (*b*) What is the value of α if $(-1, 2, \alpha, 2)$ is orthogonal to $\mathbf{x}$?

2. Let $\mathbf{x}$, $\mathbf{y}$ be arbitrary vectors in a euclidean space V.

(i) Show that, for all $\lambda \in \mathbb{R}$,

$$\|\lambda\mathbf{x}+\mathbf{y}\|^2 = a\lambda^2+b\lambda+c,$$

where $a = \|\mathbf{x}\|^2$, $b = 2\mathbf{x}.\mathbf{y}$, and $c = \|\mathbf{y}\|^2$. It follows that $a\lambda^2+b\lambda+c$ is a quadratic in λ that is nonnegative for all $\lambda \in \mathbb{R}$. Use this to prove that

$$|\mathbf{x}.\mathbf{y}| \leqslant \|\mathbf{x}\|\,\|\mathbf{y}\|. \qquad (*)$$

(ii) Explain why (*) enables the angle between $\mathbf{x}$ and $\mathbf{y}$ to be defined when $\mathbf{x}$ and $\mathbf{y}$ are nonzero.

(iii) By considering the difference $(\|\mathbf{x}\|+\|\mathbf{y}\|)^2 - \|\mathbf{x}+\mathbf{y}\|^2$ and using (*), prove 70.2.

3. Deduce from (*) in question 2 an inequality concerning integrals of continuous functions over a closed interval $[a, b]$.

4. Let $\mathbf{x}_1 = \frac{1}{7}(2, 3, 6)$. Check that $\mathbf{x}_1$ is a unit vector in $\mathbb{R}^3$, and use the Gram-Schmidt process to find an orthonormal basis of $\mathbb{R}^3$ in which $\mathbf{x}_1$ is one of the vectors.

5. Find an orthonormal basis of the solution space of the equation

$$x_1+x_2+2x_3 = 0.$$

6. Let V be the space of all real polynomials with degrees not exceeding 2, and let the inner product on V be that defined by

$$f.g = \int_0^1 f(x)g(x)dx \qquad (f, g \in V).$$

Find an orthonormal basis of V.

7. Find an orthogonal matrix whose first column is $\dfrac{1}{\sqrt{11}}\operatorname{col}(1, 1, 3)$.

8. If $(\mathbf{e}_1, \mathbf{e}_2, \ldots, \mathbf{e}_n)$ is an orthonormal sequence of vectors in a euclidean space, what is the length of $\sum_{j=1}^{n} \mathbf{e}_j$?

9. Suppose that $(\mathbf{e}_1, \mathbf{e}_2, \ldots, \mathbf{e}_n)$ is an orthonormal basis of a euclidean space V, and let $\mathbf{x}$ denote a vector in V. Show that the sequence

$$(\mathbf{e}_1, \mathbf{e}_2, \ldots, \mathbf{e}_{n-1}, \mathbf{x})$$

is an orthonormal basis of V if and only if $\mathbf{x} = \mathbf{e}_n$ or $\mathbf{x} = -\mathbf{e}_n$.

10.* Let V be a nonzero f.d. euclidean space and S a subspace of V. Let $S^\perp$ be defined as the subset

$$\{\mathbf{x} \in \mathbf{V} : \mathbf{x} \text{ is orthogonal to every vector in } S\}$$

of V. Prove that S is a subspace of V and that $V = S \oplus S^\perp$.

11. Take A to be each of the following matrices in turn. In each case find an orthogonal matrix P such that P^TAP is diagonal.

(i) $\begin{bmatrix} 6 & 2 \\ 2 & 3 \end{bmatrix}$; (ii) $\begin{bmatrix} -2 & 1 & 2 \\ 1 & 1 & 1 \\ 2 & 1 & -2 \end{bmatrix}$; (iii) $\begin{bmatrix} 5 & 2 & 4 \\ 2 & 8 & -2 \\ 4 & -2 & 5 \end{bmatrix}$.

12. Show that, apart from I_n, there is no real symmetric matrix in with cube equal to I_n.

13. (i) State why it is true that every positive-definite symmetric matrix in $\mathbb{R}_{n \times n}$ is nonsingular. (See worked example 2 in §73 for the meaning of "positive-definite".)

(ii) Let A, S be symmetric matrices in $\mathbb{R}_{n \times n}$, A being positive-definite. Using the result of worked example 2 in §73, show that there is a nonsingular matrix Q in $\mathbb{R}_{n \times n}$ such that $Q^T A Q = I_n$. Verify that $Q^T S Q$ is symmetric, and deduce that there is a nonsingular matrix $P \in \mathbb{R}_{n \times n}$ such that $P^{\prime} A P = I_n$ and $P^{\prime} S P$ is diagonal. (This foreshadows theorem 79.5, which appears in the next chapter.)

14. Let A be a symmetric matrix in $\mathbb{R}_{n \times n}$. Show that A can be expressed as the difference $B - C$ of two symmetric matrices B and C, each having nonnegative eigenvalues, such that $BC = O$. Illustrate taking A to be the given matrix in part (ii) of exercise 11.

15.* Let M be an arbitrary nonsingular matrix in $\mathbb{R}_{n \times n}$. Prove that $M^T M$ is a positive-definite symmetric matrix. By worked example 2 in §73, there is a positive-definite symmetric matrix B such that $B^2 = M^T M$. Prove that MB^{-1} is orthogonal. Deduce that the arbitrary nonsingular M is expressible as a product PB, with P orthogonal and B positive-definite symmetric.

CHAPTER TEN

QUADRATIC FORMS

74. Introduction

As in earlier chapters, F will denote an arbitrary field of scalars, except that in this chapter, where virtually everything rests on the possibility of dividing by 2, we impose throughout the restriction char $F \neq 2$ (cf. the final remark in §11). Once again, readers for whom such technicalities are unexplored territory can be assured that they will lose nothing if they simply regard F as standing for a familiar number system like $\mathbb{R}$ or $\mathbb{C}$; and indeed some of the most interesting things in the chapter relate exclusively to the case $F = \mathbb{R}$.

A **quadratic form of order** n over F means a mapping—q, say—from F^n to F specifiable by a formula of the form

$$\left.\begin{aligned} q(x_1, x_2, \dots, x_n) = \alpha_{11}x_1^2 + \alpha_{22}x_2^2 + \dots + \alpha_{nn}x_n^2 \\ +2\alpha_{12}x_1x_2 + 2\alpha_{13}x_1x_3 + \dots + 2\alpha_{1n}x_1x_n \\ +2\alpha_{23}x_2x_3 + \dots + 2\alpha_{2n}x_2x_n \\ + \dots + 2\alpha_{n-1,n}x_{n-1}x_n \qquad (x_1, \dots, x_n \in F) \end{aligned}\right\} \tag{1}$$

or (the same thing tidied up by using Σ-notation)

$$q(x_1, x_2, \dots, x_n) = \sum_{i=1}^{n} \alpha_{ii}x_i^2 + \sum_{\substack{i,k=1 \\ i<k}}^{n} 2\alpha_{ik}x_ix_k \qquad (x_1, \dots, x_n \in F), \tag{2}$$

each of the coefficients α_{11}, etc. being a constant in F (constant in the sense of independent of $x_1, \dots, x_n$). The cardinal thing about the defining expression for $q(x_1, x_2, \dots, x_n)$ is that it is a polynomial in $x_1, x_2, \dots, x_n$ in which *every* term is of degree 2, being either a "square term" of the form $\alpha_{ii}x_i^2$ or a "mixed product term" of the form $2\alpha_{ik}x_ix_k$ with $i \neq k$. The *order* of the quadratic form q refers to the dimension of its domain, F^n.

To give a specific example, the mapping $q_1 : \mathbb{R}^3 \to \mathbb{R}$ defined by

$$q_1(x_1, x_2, x_3) = x_1^2 - 4x_2^2 + 5x_3^2 + 2x_1x_2 - 6x_1x_3 + 3x_2x_3 \quad (x_1, x_2, x_3 \in \mathbb{R}) \tag{3}$$

is a quadratic form of order 3 over $\mathbb{R}$.

Quadratic forms appear in many places in mathematics and its applications. At the end of the chapter some details will be given of how the theory of quadratic forms can be used in the discussion of stationary points of functions of two or more real variables. In physics the energy of a system may often be expressed as a quadratic form in the coordinates of the particles of the system and the components of their velocities.

The polynomial expressions used to specify quadratic forms (e.g. the right-hand sides of the above equations (1), (2) and (3)) have, in general, a complicated and unwieldy appearance. So let us straight away give attention to the following proposition, which enables such expressions to be re-written in an extremely simple way as matrix products.

74.1 Let $X = \text{col}(x_1, x_2, \ldots, x_n)$ and let $A = [\alpha_{ik}]_{n\times n}$ be a *symmetric* matrix (so that $\alpha_{ik} = \alpha_{ki}$ in all relevant cases). Then

$$X^TAX = \sum_{i=1}^{n} \alpha_{ii}x_i^2 + \sum_{\substack{i,k=1\\ i<k}}^{n} 2\alpha_{ik}x_ix_k$$

(= right-hand side of the above equations (1) and (2)).

Proof.

$$X^TAX = \sum_{i=1}^{n} (i\text{th entry of the row } X^T) \times (i\text{th entry of the column } AX)$$

$$= \sum_{i=1}^{n}\left(x_i\left(\sum_{k=1}^{n} \alpha_{ik}x_k\right)\right)$$

$$= \sum_{i=1}^{n}\sum_{k=1}^{n} \alpha_{ik}x_ix_k$$

$$= \sum_{i=1}^{n} \alpha_{ii}x_i^2 + \sum_{\substack{i,k=1\\ i\neq k}}^{n} \alpha_{ik}x_ix_k \quad \text{(after separation of the terms for which } i = k \text{ and those for which } i \neq k)$$

$$= \sum_{i=1}^{n} \alpha_{ii}x_i^2 + \sum_{\substack{i,k=1\\ i<k}}^{n} 2\alpha_{ik}x_ix_k \quad \text{(since } A \text{ is symmetric).}$$

As an illustration of how this works in practice in abbreviating the defining expression for a quadratic form, take the right-hand side of (3), namely

$$x_1^2 - 4x_2^2 + 5x_3^2 + 2x_1x_2 - 6x_1x_3 + 3x_2x_3.$$

By 74.1 this can be re-expressed as X^TAX, where $X = \text{col}(x_1, x_2, x_3)$ and A

is the symmetric matrix $\begin{bmatrix} 1 & 1 & -3 \\ 1 & -4 & \frac{3}{2} \\ -3 & \frac{3}{2} & 5 \end{bmatrix}$. All the details of how to write down A are apparent from a close look at 74.1: we simply take (in every case)

$$(i,i)\text{th entry of } A = \text{coefficient of } x_i^2, \quad \text{and}$$
$$(i,k)\text{th and } (k,i)\text{th entries of } A = \tfrac{1}{2}\times(\text{coefficient of } x_i x_k) \quad (i \neq k).$$

More generally, 74.1 reveals that an arbitrary quadratic form of order n over F can be specified by an equation of the simple form

$$q(x_1, x_2, \ldots, x_n) = X^T AX \qquad (x_1, x_2, \ldots, x_n \in F), \tag{4}$$

where X stands for col$(x_1, x_2, \ldots, x_n)$ and A is a symmetric matrix in $F_{n\times n}$. Furthermore (4) may be re-written in the even more compact form

$$q(X) = X^T AX \qquad (X \in F_{n\times 1}),$$

this being an area where it pays to ignore the difference between F^n and $F_{n\times 1}$ and to regard the n-tuple $(x_1, \ldots, x_n)$ and the column col$(x_1, \ldots, x_n)$ as identical (cf. remarks made when F^n was introduced as a vector space in §35).

Conversely, if we start with any symmetric matrix A in $F_{n\times n}$, then we can define a corresponding quadratic form, for which we shall always use the notation q_A, by

$$q_A(X) = X^T AX \qquad (X \in F_{n\times 1}).$$

For example, in the case $F = \mathbb{R}$ and $n = 2$, if $A = \begin{bmatrix} 1 & 2 \\ 2 & 3 \end{bmatrix}$, then q_A is the quadratic form whose unabbreviated specification is

$$q_A(x_1, x_2) = x_1^2 + 3x_2^2 + 4x_1x_2 \qquad (x_1, x_2 \in \mathbb{R}),$$

(the right-hand side being $X^T AX$ where $X = \text{col}(x_1, x_2)$).

A significant question arising is whether one quadratic form q could be equal both to q_A and to q_B for two different symmetric matrices A and B. That is, for a given quadratic form q of order n over F, could we possibly have

$$q(X) = X^T AX \quad \text{and} \quad q(X) = X^T BX \quad \text{for } \textit{every } X \in F_{n\times 1}$$

with A and B unequal symmetric matrices in $F_{n\times 1}$? The answer is "no", as is revealed by a result that we can also use elsewhere—namely

74.2 Suppose that A and B are symmetric matrices in $F_{n \times n}$ such that

$$X^T AX = X^T BX \qquad (*)$$

for *every* $X \in F_{n \times 1}$. Then $A = B$.

Proof. Let α_{ik} and β_{ik} denote the (i, k)th entries of A and B, respectively, and let $E_1, E_2, \ldots, E_n$ denote the columns of the identity matrix I_n. Then for all relevant i, k,

$$\begin{aligned} E_i^T A E_k &= (i\text{th row of } I_n) \times A \times (k\text{th column of } I_n) \\ &= (i, k)\text{th entry of } I_n A I_n (= A) \qquad (\text{by } 21.2) \\ &= \alpha_{ik}, \end{aligned}$$

and similarly $E_i^T BE_k = \beta_{ik}$.

Hence, for each relevant i, taking $X = E_i$ in (*) gives

$$E_i^T AE_i = E_i^T BE_i, \qquad \text{i.e. } \alpha_{ii} = \beta_{ii}$$

which shows that the main diagonals of A and B are identical.

Further, for all relevant unequal i, k, taking $X = E_i + E_k$ in (*) gives, after expansion,

$$\begin{aligned} &E_i^T AE_i + E_i^T AE_k + E_k^T AE_i + E_k^T AE_k = E_i^T BE_i + E_i^T BE_k + E_k^T BE_i + E_k^T BE_k, \\ &\text{i.e.} \quad \alpha_{ii} + \alpha_{ik} + \alpha_{ki} + \alpha_{kk} = \beta_{ii} + \beta_{ik} + \beta_{ki} + \beta_{kk}, \\ &\text{i.e.} \quad \alpha_{ik} + \alpha_{ki} = \beta_{ik} + \beta_{ki} \quad (\text{since } A, B \text{ have identical main diagonals}), \\ &\text{i.e.} \quad 2\alpha_{ik} = 2\beta_{ik} \quad (\text{since } A \text{ and } B \text{ are symmetric}), \\ &\text{i.e.} \quad \alpha_{ik} = \beta_{ik}. \end{aligned}$$

It is now apparent that $A = B$, as asserted.

As indicated in the preamble to 74.2, it follows, in particular, that, for any given quadratic form q of order n over F, there is a *unique* symmetric matrix $A \in F_{n \times n}$ such that $q = q_A$, i.e. such that $q(X) = X^T AX$ for all $X \in F_{n \times 1}$. Since it is uniquely determined, it makes sense to refer to this matrix A as "*the* matrix of the quadratic form q".

Having uncovered the connection between quadratic forms and symmetric matrices, we are ready to explore aspects of quadratic form theory that are interesting and relevant to applications. A helpful way to begin a preview of what lies ahead is to note that if D is a diagonal matrix (say $D = \operatorname{diag}(\gamma_1, \gamma_2, \ldots, \gamma_n)$), then the unabbreviated general formula for $q_D(x_1, \ldots, x_n)$ takes a nice simple form:

$$q_D(x_1, \ldots, x_n) = \gamma_1 x_1^2 + \gamma_2 x_2^2 + \ldots + \gamma_n x_n^2,$$

with no mixed product terms. This observation motivates what we call the *diagonalization* of an arbitrary given quadratic form q—by which we mean changing variables so that the general expression for $q(x_1, \ldots, x_n)$ in terms of the new variables ($\hat{x}_1, \ldots, \hat{x}_n$, say) takes the form

$$\gamma_1\hat{x}_1^2 + \gamma_2\hat{x}_2^2 + \ldots + \gamma_n\hat{x}_n^2 \qquad \text{(with no mixed product terms).}$$

Generalities concerning change of variable and diagonalization of a quadratic form will be dealt with in detail in §§75 and 76, and then in §77 it will be seen that diagonalization of a quadratic form brings to light some important intrinsic properties of the form. Not surprisingly, in the case $F = \mathbb{R}$ the orthogonal diagonalization of a symmetric matrix A can be applied to the diagonalization of the quadratic form q_A, and that will form the basic subject matter of §78. Moreover, in the case $F = \mathbb{R}$ diagonalization enables us to spot quadratic forms q for which (as in the case given by $q(x_1, \ldots, x_n) = x_1^2 + \ldots + x_n^2$) $q(x_1, \ldots, x_n)$ always takes a positive value except in the trivial case $x_1 = x_2 = \ldots = x_n = 0$. Such quadratic forms are described as *positive-definite*, and the chapter will reach a climax in §§79 and 80 with major theorems about positive-definite quadratic forms and their matrices.

Before we proceed, let us introduce a notational convention which will save us from repetitious explanations in the pages ahead. This convention is that, where columns are denoted by capital letters, their entries may (without the need for explanatory comment) be denoted by the corresponding small letters with subscripts: e.g. the entries of column X may be denoted by $x_1, x_2, \ldots$, the entries of column Y by $y_1, y_2, \ldots$, the entries of column $\hat{X}$ by $\hat{x}_1, \hat{x}_2, \ldots$, etc. And conversely, where small letters with subscripts are already in use for the entries of columns, the entire columns may (again without the need for explanatory comment) be denoted by the corresponding capital letters: e.g. col $(x_1, x_2, \ldots)$ may be denoted by X.

75. Change of basis and change of variable

An arbitrary quadratic form q of order n over F may, as was seen in §74, be defined by an equation

$$q(X) = X^T A X \qquad (X \in F_{n\times 1}), \qquad (\alpha)$$

A being a symmetric $n \times n$ matrix. Do not lose sight of the fact that the compact right-hand side of (α) is simply (cf. 74.1) a neat way of writing a certain quadratic polynomial in the entries $x_1, x_2, \ldots, x_n$ of X. Those entries are the components of X w.r.t. the standard basis of $F_{n\times 1}$ (for which standard basis we shall henceforth use the notation L), so that the equation (α) gives a general expression for $q(X)$ in terms of the components of X w.r.t. L. That remark might seem, at first sight, merely to be putting a gloss of sophistication on a simple situation. But it is a purposeful remark, as it points to the possibility of obtaining a general formula giving $q(X)$ in terms of the components ($\hat{x}_1, \hat{x}_2, \ldots, \hat{x}_n$, say) of X w.r.t. another basis $\hat{L}$ of $F_{n\times 1}$; and it takes little imagination to appreciate that by this means we might obtain a simpler polynomial expression for $q(X)$ than the one we started with. (We shall return to the simplification idea in §76.)

It will be helpful first to gain a feel for what we are talking about by working through a numerical example. (Note: this will be the first place where we freely use the notational convention introduced in the final paragraph of §74—the entries of column X are denoted by x_1, x_2, and col $(\hat{x}_1, \hat{x}_2)$ is denoted by $\hat{X}$.) Taking $F = \mathbb{R}$, consider the quadratic form q of order 2 defined by

$$q(X) = x_1^2 + 3x_2^2 + 4x_1x_2$$
$$\left(= X^T A X, \quad \text{where } A = \begin{bmatrix} 1 & 2 \\ 2 & 3 \end{bmatrix}\right) \qquad (X \in \mathbb{R}_{2\times 1}).$$

Let $\hat{L}$ be the basis $\left(\begin{bmatrix} 1 \\ 1 \end{bmatrix}, \begin{bmatrix} -2 \\ 1 \end{bmatrix}\right)$ of $\mathbb{R}_{2\times 1}$, so that $M(L \to \hat{L}) = \begin{bmatrix} 1 & -2 \\ 1 & 1 \end{bmatrix}$ ($= P$, say). [See §57 for the theory and notations of change of basis.] For arbitrary $X \in \mathbb{R}_{2\times 1}$, the components $\hat{x}_1$, $\hat{x}_2$ of X w.r.t. $\hat{L}$ are given (cf. 57.2) by $X = P\hat{X}$, i.e. by

$$\begin{Bmatrix} x_1 = \hat{x}_1 - 2\hat{x}_2 \\ x_2 = \hat{x}_1 + \hat{x}_2 \end{Bmatrix}.$$

Hence, by substituting in the defining equation for $q(X)$, we obtain

$$\begin{aligned} q(X) &= (\hat{x}_1 - 2\hat{x}_2)^2 + 3(\hat{x}_1 + \hat{x}_2)^2 + 4(\hat{x}_1 - 2\hat{x}_2)(\hat{x}_1 + \hat{x}_2) \\ &= 8\hat{x}_1^2 - \hat{x}_2^2 - 2\hat{x}_1\hat{x}_2 \qquad \text{(after simplification).} \end{aligned}$$

This being true in all cases, we reach the conclusion that, for every $X \in \mathbb{R}_{2\times 1}$,

$$q(X) = \hat{X}^T B \hat{X},$$

where $B = \begin{bmatrix} 8 & -1 \\ -1 & -1 \end{bmatrix}$, $\hat{X}$ denoting the component column of X w.r.t. the basis $\hat{L}$.

The generalization of this re-expression of $q(X)$ in terms of $\hat{X}$ is described by:

75.1 Let q be the quadratic form of order n over F given by $q(X) = X^T AX$ ($X \in F_{n\times 1}$), A being a symmetric matrix in $F_{n\times n}$. Let $\hat{L}$ be an arbitrary basis of $F_{n\times 1}$, and let $P = M(L \to \hat{L})$, the (necessarily nonsingular) matrix of the change of basis from the standard basis L to the basis $\hat{L}$. Then, for every

$X \in F_{n\times 1}$,

$$q(X) = \hat{X}^T B\hat{X},$$

where B is the symmetric matrix P^TAP and $\hat{X}$ denotes the component column of X w.r.t. $\hat{L}$.

Proof. For every $X \in F_{n\times 1}$, $X = P\hat{X}$ (by 57.2) and hence

$$q(X) = X^TAX = (P\hat{X})^TA(P\hat{X}) = \hat{X}^TP^TAP\hat{X} = \hat{X}^TB\hat{X},$$

where $B = P^TAP$. And this matrix B is symmetric, since

$$\begin{aligned} B^T &= P^TA^TP && \text{(by 18.4)} \\ &= P^TAP && \text{(A being symmetric)} \\ &= B. \end{aligned}$$

Remarks. (*a*) In future we shall write things like

$$q(X) \equiv X^TAX \quad \text{and} \quad q(X) \equiv \hat{X}^TB\hat{X},$$

the '$\equiv$' indicating that we have an equation holding in all cases, i.e. for every relevant column X.

(*b*) In the notation of 75.1, observe that as X ranges through $F_{n\times 1}$, $\hat{X}$ also takes every possible value in $F_{n\times 1}$. (That is obvious from the fact that $\hat{X}$ is the component column of X w.r.t. a certain basis.) This observation enables us to see that, in 75.1, B $(= P^TAP)$ is the *only* symmetric matrix that can validly fill the blank in

$$q(X) \equiv \hat{X}^T(\ldots)\hat{X}:$$

for if $q(X) \equiv \hat{X}^TC\hat{X}$ (where C is a symmetric matrix), then we have $\hat{X}^TC\hat{X} = \hat{X}^TB\hat{X}$ *for all* $\hat{X} \in F_{n\times 1}$, and hence $B = C$ (by 74.2).

(*c*) In §74 we brought in the phrase "the matrix of the quadratic form q" for the unique symmetric matrix A such that $q(X) \equiv X^TAX$. In the context of change of basis and the result 75.1, we tighten our phraseology and call A the matrix of q w.r.t. the standard basis L of $F_{n\times 1}$, to distinguish it from B $(= P^TAP)$, which we naturally term the matrix of q w.r.t. the basis $\hat{L}$. The notation $M(q;\Lambda)$ will be used for the matrix of q w.r.t. the basis Λ, so that, in 75.1, $M(q;L) = A$ and $M(q;\hat{L}) = B$. Further, the content of 75.1 can be re-expressed as follows.

75.2 If q is a quadratic form of order n over F, then for any basis $\hat{L}$ of $F_{n\times 1}$,

$$M(q;\hat{L}) = P^TM(q;L)P$$

where $P = M(L \to \hat{L})$ (and where, as always in this chapter, L denotes the standard basis of $F_{n\times 1}$).

(*d*) The result 75.1 and the numerical example preceding it were about starting with the general value $q(X)$ of a quadratic form expressed in terms of $x_1, x_2, \ldots$ and re-expressing $q(X)$ in terms of $\hat{x}_1, \hat{x}_2, \ldots$. We shall adopt the phrase "change of variable" as the obvious informal description of the re-expression process. E.g. in the numerical example preceding 75.1, we could say that the re-expression of $q(X)$ is brought about by the change of variables defined by

$$\left\{\begin{aligned} x_1 &= \hat{x}_1 - 2\hat{x}_2 \\ x_2 &= \hat{x}_1 + \hat{x}_2 \end{aligned}\right\}.$$

In future we shall often want to carry out such changes of variable without digressing to mention the underlying change of basis. The main point to note is that a change of variables from "original variables" $x_1, \ldots, x_n$ to "new variables" $\hat{x}_1, \ldots, \hat{x}_n$ can always legitimately be made when the two sequences of variables are connected by an *invertible* system of linear equations, i.e. by $X = P\hat{X}$ for some *nonsingular* matrix P: for this change of variables is precisely what results from the change of basis with matrix P.

76. Diagonalization of a quadratic form

Let q be a quadratic form of order n over F. Suppose that, for a certain basis $\hat{L}$ of $F_{n\times 1}$, $M(q; \hat{L})$ is diagonal—say $M(q; \hat{L}) = D = \operatorname{diag}(\gamma_1, \gamma_2, \ldots, \gamma_n)$. Then, where $\hat{x}_1, \hat{x}_2, \ldots, \hat{x}_n$ denote the components of X w.r.t. $\hat{L}$, we have

$$q(X) \equiv \hat{X}^T D \hat{X},$$

$$\text{i.e.} \quad q(X) \equiv \gamma_1 \hat{x}_1^2 + \gamma_2 \hat{x}_2^2 + \ldots + \gamma_n \hat{x}_n^2.$$

The right-hand side of this last equation (with no mixed product terms) is called a *diagonal version* of $q(X)$ (corresponding to the fact that $M(q; \hat{L})$ is diagonal).

Diagonalization of a given quadratic form q means the making of a change of basis (or the corresponding change of variable—cf. remark (*d*) at the end of §75) so as to produce a diagonal version of $q(X)$. This section is devoted to demonstrating a systematic method by which an arbitrary quadratic form can be diagonalized. The method is often called the *rational reduction* of a quadratic form, reflecting the fact it involves no operations other than addition, subtraction, multiplication and division; and, essentially because of that fact, it can be applied to quadratic forms over any field F (char $F \neq 2$).

To follow the method, it is vital to appreciate that the expansion of

$(x_1+x_2+\ldots+x_n)^2$ is

$$\sum_{i=1}^{n} x_i^2 + \sum_{\substack{i,k=1\\ i<k}}^{n} 2x_ix_k,$$

consisting of every x_i^2 term and all the different mixed product terms $2x_ix_k$. For example,

$$(x_1+x_2+x_3)^2 = x_1^2+x_2^2+x_3^2+2x_1x_2+2x_1x_3+2x_2x_3.$$

It is vital also to be conversant with the process of completion of the square. This crops up in elementary mathematics, where, in the most straightforward case, one is given a quadratic in x (say x^2+bx+c) in which the coefficient of x^2 is 1, and the set task is to re-express the quadratic in the form

$$(x+\text{constant})^2+(\text{constant}),$$

so that all mentions of x are absorbed into the square term at the beginning. This is achieved through noticing that $(x+\frac{1}{2}b)^2$ reproduces the x^2+bx part of the given quadratic but contains also the "excess term" $\frac{1}{4}b^2$. Therefore, we have

$$x^2+bx+c = (x+\tfrac{1}{2}b)^2-\tfrac{1}{4}b^2+c$$

(excess term subtracted off: $-\tfrac{1}{4}b^2$; original constant term: c)

$$= (x+\tfrac{1}{2}b)^2+k,$$

where $k = c-\frac{1}{4}b^2$. For example,

$$x^2+8x+13 = (x+4)^2-16+13 = (x+4)^2-3.$$

The less straightforward case where the coefficient of x^2 (a, say) is not 1 can be handled in a similar way after taking the factor a out of the terms involving x. For example,

$$2x^2+8x+13 = 2[x^2+4x]+13 = 2[(x+2)^2-4]+13 = 2(x+2)^2+5.$$

This detailed review of the elementary process prepares the way for understanding how the completion of the square idea can be used in the diagonalization of a given quadratic form q of order n. Again let us begin with a simple case—the case where the coefficient of x_1^2 in the given defining expression for $q(X)$ is 1. Suppose that the total of the terms involving x_1 in that expression is

$$x_1^2+2\alpha_{12}x_1x_2+2\alpha_{13}x_1x_3+\ldots+2\alpha_{1n}x_1x_n \qquad (= E,\text{ say}).$$

We note that the expansion of $\hat{x}_1^2$, where

$$\hat{x}_1 = x_1+\alpha_{12}x_2+\alpha_{13}x_3+\ldots+\alpha_{1n}x_n,$$

contains all the terms in E, along with excess terms that do not involve x_1, namely the squares of $\alpha_{12}x_2, \ldots, \alpha_{1n}x_n$ and all the mixed products $2(\alpha_{1i}x_i)(\alpha_{1k}x_k)$. Therefore we have

$$q(X) \equiv E + (\text{original terms not involving } x_1)$$

$$\equiv \hat{x}_1^2 - \begin{pmatrix}\text{excess terms not}\\ \text{involving } x_1\end{pmatrix} + \begin{pmatrix}\text{original terms}\\ \text{not involving } x_1\end{pmatrix}$$

$$\equiv \hat{x}_1^2 + r_1(X),$$

where the residual expression $r_1(X)$ involves only $x_2, x_3, \ldots, x_n$ (and not x_1). If instead the original coefficient (β_1, say) of x_1^2 is a nonzero scalar different from 1, we could take out the factor β_1 from the total E of the terms involving x_1, then proceed as before. So, in any event, provided only that the coefficient of x_1^2 in $q(X)$ is nonzero, the manoeuvre just outlined gives us

$$q(X) \equiv \beta_1\hat{x}_1^2 + r_1(X),$$

where $r_1(X)$ involves only $x_2, \ldots, x_n$. We call this manoeuvre "completion of the square to absorb all terms involving x_1". Readers eager for an illustration of it should look ahead to the first stage of worked example 1 (it may be helpful to read the whole of this example in parallel with the present explanation).

As will now be indicated, repeated use of completion of the square offers a procedure for diagonalizing q that will work in most (but not all) cases. Provided the coefficient β_2 of x_2^2 in $r_1(X)$ is nonzero, we can complete the square within $r_1(X)$ so as to absorb all references to x_2: that will give us

$$r_1(X) \equiv \beta_2\hat{x}_2^2 + r_2(X)$$

and hence

$$q(X) \equiv \beta_1\hat{x}_1^2 + \beta_2\hat{x}_2^2 + r_2(X),$$

where $\hat{x}_2 = x_2 + \lambda_3x_3 + \ldots + \lambda_nx_n$ for some constants $\lambda_3, \ldots, \lambda_n$ and where $r_2(X)$ involves only $x_3, x_4, \ldots, x_n$ (but not x_1 and x_2). Then, if the coefficient β_3 of x_3^2 in $r_2(X)$ is nonzero, a further completion of the square will give

$$q(X) \equiv \beta_1\hat{x}_1^2 + \beta_2\hat{x}_2^2 + \beta_3\hat{x}_3^2 + r_3(X),$$

where $r_3(X)$ involves only $x_4, \ldots, x_n$; and so on. The basic aim is to produce an expression for $q(X)$ in which there are no mixed product terms. That will be achieved if at any stage the residual expression $r_k(X)$ consists only of square terms (including the case $r_k(X) = 0$); and this is bound to happen after, at most, $(n-1)$ completions of the square. (For, after $(n-1)$ completions of the square, the residual expression $r_{n-1}(X)$ would involve x_n only and so would be of the form $\beta_nx_n^2$.)

Quite a number of loose ends remain to be tied up. We shall return to deal with them after the following straightforward worked example in which the method outlined above can be executed smoothly to bring about diagonalization of the given quadratic form.

Worked example 1. Diagonalize the quadratic form q of order 4 defined by

$$q(X) \equiv 2x_1^2+3x_2^2+6x_3^2+x_4^2-4x_1x_2+8x_1x_3-4x_1x_4-6x_2x_3-2x_2x_4-2x_3x_4.$$

$$q(X) \equiv 2(x_1^2-2x_1x_2+4x_1x_3-2x_1x_4)+3x_2^2+6x_3^2+x_4^2-6x_2x_3 -2x_2x_4-2x_3x_4$$

$$\equiv 2(x_1-x_2+2x_3-x_4)^2$$

$$-2x_2^2-8x_3^2-2x_4^2+8x_2x_3-4x_2x_4+8x_3x_4 \quad \swarrow \text{excess terms subtracted off}$$

$$+3x_2^2+6x_3^2+x_4^2-6x_2x_3-2x_2x_4-2x_3x_4 \quad \leftarrow \text{original terms}$$

$$\equiv 2\hat{x}_1^2+\underbrace{x_2^2+2x_2x_3-6x_2x_4-2x_3^2-x_4^2+6x_3x_4}$$

"$r_1(X)$", in which we now complete the square to absorb all references to x_2

$$\equiv 2\hat{x}_1^2+(x_2+x_3-3x_4)^2-x_3^2-9x_4^2+6x_3x_4 -2x_3^2-x_4^2+6x_3x_4$$

$$\equiv 2\hat{x}_1^2+\hat{x}_2^2 \underbrace{-3x_3^2+12x_3x_4-10x_4^2}$$

"$r_2(X)$", in which we now complete the square to absorb all references to x_3

$$\equiv 2\hat{x}_1^2+\hat{x}_2^2-3(x_3^2-4x_3x_4)-10x_4^2$$

$$\equiv 2\hat{x}_1^2+\hat{x}_2^2-3(x_3-2x_4)^2+12x_4^2-10x_4^2$$

$$\equiv 2\hat{x}_1^2+\hat{x}_2^2-3\hat{x}_3^2+2x_4^2$$

$$\equiv 2\hat{x}_1^2+\hat{x}_2^2-3\hat{x}_3^2+2\hat{x}_4^2,$$

where

$$\left\{\begin{array}{rl} \hat{x}_1 = & x_1-x_2+2x_3-x_4 \\ \hat{x}_2 = & x_2+x_3-3x_4 \\ \hat{x}_3 = & x_3-2x_4 \\ \hat{x}_4 = & x_4 \end{array}\right\},$$

a system of equations which (clearly) defines an invertible linear change of variables.

There follows a series of remarks that deal with the loose ends referred to earlier, including the question of how to cope with the cases so far excluded—the cases where at some stage the residual expression $r_{k-1}(X)$ is nonzero but does not contain a nonzero $\beta_k x_k^2$ term (with the result that completion of the square to absorb all references to x_k cannot be started).

(*a*) In all cases it is important that the change of variables made in diagonalizing the given quadratic form be invertible. Fortunately (as worked example 1 illustrates), when the standard procedure that we have described works, it produces a change of variables of the form

$$\begin{cases} \hat{x}_1 = x_1 + \mu_{12}x_2 + \mu_{13}x_3 + \ldots + \mu_{1n}x_n \\ \hat{x}_2 = \qquad x_2 + \mu_{23}x_3 + \ldots + \mu_{2n}x_n \\ \hat{x}_3 = \qquad\qquad x_3 + \ldots + \mu_{3n}x_n \\ \qquad\qquad \ldots \end{cases},$$

which is easily seen to be invertible. (The coefficient matrix obviously has determinant 1.)

(*b*) When the procedure we have described terminates early through the obtaining of a zero $r_k(X)$, giving us

$$q(X) \equiv \beta_1\hat{x}_1^2 + \ldots + \beta_k\hat{x}_k^2$$

with k less than n, the order of q, it is important that we do not regard $\hat{x}_{k+1}, \ldots, \hat{x}_n$ as non-existent: for $\hat{x}_1, \ldots, \hat{x}_n$ denote the components of X w.r.t. a certain basis of the n-dimensional space $F_{n \times 1}$. When in such a case we wish to detail the change of variables that is being made, we can just define $\hat{x}_{k+1}, \ldots, \hat{x}_n$ in any way at all that gives $\hat{X} =$ (a *nonsingular* matrix) $\times X$. It is usually a triviality to achieve this in a simple way that makes the whole change of variables visibly invertible. An illustration of all this is the definition of $\hat{x}_4$ to be x_4 at the end of worked example 2 below.

(*c*) So far we have regarded x_1 as the "first" original variable and we have begun by completing the square to absorb all mentions of x_1. But we can take the original variables in any order: at any stage, if it suits us, we can regard any particular x_j as the "next variable" and complete the square to absorb all terms involving it. For example, given

$$q(X) = x_2^2 + 3x_3^2 - 2x_4^2 + (\text{mixed product terms in } x_1, \ldots, x_4),$$

we are not stuck at the beginning because x_1^2 has zero coefficient: for we can regard x_2 as the first variable and complete the square to absorb all terms involving x_2, so getting

$$q(X) = \hat{x}_1^2 + (\text{terms involving } x_1, x_3, x_4 \text{ only}).$$

(*d*) Taking stock after remark (*c*), we can see that our technique of repeated completion of the square will bring about the diagonalization of an arbitrary given quadratic form q except in cases where either the original $q(X)$ or the residual expression $r_k(X)$ obtained at some point is nonzero and consists exclusively of mixed product terms. A simple supplementary device enables us to cope with all such exceptional cases: if $\beta x_i x_k$ is a nonzero mixed product in the awkward expression (the expression that is nonzero but contains no square terms), we put $x_i = u+v$ and $x_k = u-v$. This device (illustrated at the first stage of worked example 2, below) creates square terms, as it turns $\beta x_i x_k$ into $\beta u^2 - \beta v^2$; and hence it enables us to continue as before, using completion of the square. Notice, incidentally, that the change of variables from x_i, x_k to u, v is invertible (being reversed by $u = \frac{1}{2}(x_i + x_k)$, $v = \frac{1}{2}(x_i - x_k)$), and for this reason it is compatible with the over-riding condition that the entire change of variables made in diagonalizing a quadratic form must be invertible.

Because the supplementary device just described covers all the previously outstanding cases, we are in the happy position of being able to record that:

76.1 (By the methods outlined in this section) every quadratic form over an arbitrary field F (char $F \neq 2$) can be diagonalized.

Worked example 2. Diagonalize the quadratic form q defined by

$$q(X) = x_1 x_2 + x_1 x_3 + 2x_1 x_4 + 2x_2 x_3 + x_2 x_4 + x_3 x_4,$$

recording at the end a change of variables by which the diagonalization is accomplished.

Since $q(X)$ contains no square terms, we begin by putting $x_1 = u+v$, $x_2 = u-v$. Hence

$$
\begin{aligned}
q(X) &\equiv u^2 - v^2 + (u+v)x_3 + 2(u+v)x_4 + 2(u-v)x_3 + (u-v)x_4 + x_3 x_4 \\
&\equiv u^2 + 3ux_3 + 3ux_4 - v^2 - vx_3 + vx_4 + x_3 x_4 \quad \leftarrow \left[\begin{array}{l} \text{from now on it's} \\ \text{completion of the} \\ \text{square as in} \\ \text{worked example 1} \end{array}\right] \\
&\equiv (u + \tfrac{3}{2}x_3 + \tfrac{3}{2}x_4)^2 - \tfrac{9}{4}x_3^2 - \tfrac{9}{4}x_4^2 - \tfrac{9}{2}x_3 x_4 \\
&\quad - v^2 - vx_3 + vx_4 + x_3 x_4 \\
&\equiv \hat{x}_1^2 - (v^2 + vx_3 - vx_4) - \tfrac{9}{4}x_3^2 - \tfrac{9}{4}x_4^2 - \tfrac{7}{2}x_3 x_4 \\
&\equiv \hat{x}_1^2 - (v + \tfrac{1}{2}x_3 - \tfrac{1}{2}x_4)^2 + \tfrac{1}{4}x_3^2 + \tfrac{1}{4}x_4^2 - \tfrac{1}{2}x_3 x_4 \\
&\quad - \tfrac{9}{4}x_3^2 - \tfrac{9}{4}x_4^2 - \tfrac{7}{2}x_3 x_4 \\
&\equiv \hat{x}_1^2 - \hat{x}_2^2 - 2x_3^2 - 2x_4^2 - 4x_3 x_4 \\
&\equiv \hat{x}_1^2 - \hat{x}_2^2 - 2(x_3^2 + 2x_3 x_4) - 2x_4^2 \\
&\equiv \hat{x}_1^2 - \hat{x}_2^2 - 2(x_3 + x_4)^2 + 2x_4^2 - 2x_4^2 \\
&\equiv \hat{x}_1^2 - \hat{x}_2^2 - 2\hat{x}_3^2 \\
&(\text{or } \hat{x}_1^2 - \hat{x}_2^2 - 2\hat{x}_3^2 + 0\hat{x}_4^2),
\end{aligned}
$$

where

$$
\left\{
\begin{aligned}
\hat{x}_1 &= u + \tfrac{3}{2}x_3 + \tfrac{3}{2}x_4 = \tfrac{1}{2}(x_1 + x_2 + 3x_3 + 3x_4) \\
\hat{x}_2 &= v + \tfrac{1}{2}x_3 - \tfrac{1}{2}x_4 = \tfrac{1}{2}(x_1 - x_2 + x_3 - x_4) \\
\hat{x}_3 &= \phantom{v + \tfrac{1}{2}x_3 - \tfrac{1}{2}x_4 = \tfrac{1}{2}(} x_3 + x_4 \\
\hat{x}_4 &= \phantom{v + \tfrac{1}{2}x_3 - \tfrac{1}{2}x_4 = \tfrac{1}{2}(x_3 + } x_4
\end{aligned}
\right\}.
$$

(*Note* (cf. earlier remark (*b*)). The definition of $\hat{x}_4$ here is to a large extent arbitrary. Taking $\hat{x}_4 = x_4$ has the advantage of being a simple way of making the entire change of variable visibly invertible.)

77. Invariants of a quadratic form

Let us begin by supposing that q is some particular quadratic form and that $\left\{\begin{array}{c}\text{change of basis}\\ \text{change of variables}\end{array}\right\}$ produces the diagonal version

$$q(X) \equiv \hat{x}_1^2 + 4\hat{x}_2^2 + 9\hat{x}_3^2$$

with coefficients 1, 4, 9. Then the further change of variables defined by

$$y_1 = \hat{x}_1, \quad y_2 = 2\hat{x}_2, \quad y_3 = 3\hat{x}_3$$

gives us

$$q(X) \equiv y_1^2 + y_2^2 + y_3^2,$$

in which the coefficients are 1, 1, 1. The moral is that the coefficients appearing in a diagonal version of a quadratic form q are *not* uniquely determined by q: for in fact they depend on which of several possible changes of variables is used to diagonalize q.

In contrast, it is possible to find, for an arbitrary quadratic form q, features of diagonal versions of $q(X)$ which are the same in all such diagonal versions and which are, therefore, independent of the particular method used to diagonalize q. Any such feature is called an **invariant** of q. An example of an invariant (as the next result, 77.1, shows) is the number of nonzero coefficients in a diagonal version of $q(X)$: two diagonal versions of $q(X)$ may have different coefficients, but they must, as 77.1 shows, contain the same number of nonzero coefficients. That is true over an arbitrary field F. The second result of this section (77.2) will reveal further invariants in the case $F = \mathbb{R}$.

It will be noticed that, in discussing these results, we often assume that the terms in a diagonal version of a quadratic form come in an order that is convenient to us (e.g. all the terms with positive coefficients first). This is perfectly all right: it corresponds to the fact that when we $\left\{\begin{array}{c}\text{change variables}\\ \text{change basis}\end{array}\right\}$, we can number the $\left\{\begin{array}{c}\text{new variables}\\ \text{new basis vectors}\end{array}\right\}$ in any order we please.

77.1 Let q be a quadratic form, of order n, say, over F. Then in any diagonal version of $q(X)$ the number of nonzero coefficients is equal to the rank of the matrix $M(q; L)$. This number of nonzero coefficients is, therefore, an invariant of q.

Proof. Let $M(q; L) = A$, and let

$$q(X) \equiv \alpha_1 \hat{x}_1^2 + \alpha_2 \hat{x}_2^2 + \ldots + \alpha_r \hat{x}_r^2 (+0\hat{x}_{r+1}^2 + \ldots + 0\hat{x}_n^2),$$

in which the coefficients $\alpha_1, \alpha_2, \ldots, \alpha_r$ are all nonzero, be an arbitrary diagonal version of $q(X)$. Here $\hat{x}_1, \hat{x}_2, \ldots, \hat{x}_n$ are the components of X w.r.t. some basis $\hat{L}$ of $F_{n \times 1}$, and we have

$$M(q; \hat{L}) = \text{diag}(\alpha_1, \alpha_2, \ldots, \alpha_r, 0, \ldots, 0) \qquad (= D, \text{say}).$$

By 75.2, $D = P^T A P$, where P is the nonsingular matrix $M(L \to \hat{L})$. Since P^T is also nonsingular (cf. 20.3), it follows by 52.3M that $r(D) = r(A)$. But clearly $r(D) = r$. So $r = r(A)$—which proves the result.

In view of 77.1, it makes sense to refer to the number of nonzero coefficients in any (and therefore every) diagonal version of $q(X)$ as the *rank of the quadratic form q*.

From now on in this chapter we are exclusively concerned with real quadratic forms—i.e. quadratic forms over $\mathbb{R}$. The following theorem, generally known as *Sylvester's Law of Inertia*, shows that, in the case of a real quadratic form, the numbers of positive and negative coefficients appearing in a diagonal version are invariants.

77.2 Let q be a real quadratic form of order n and rank r. Let

$$q(X) \equiv \alpha_1 \hat{x}_1^2 + \ldots + \alpha_s \hat{x}_s^2 - \alpha_{s+1} \hat{x}_{s+1}^2 - \ldots - \alpha_r \hat{x}_r^2 (+0\hat{x}_{r+1}^2 + \ldots + 0\hat{x}_n^2)$$

and

$$q(X) \equiv \beta_1 \tilde{x}_1^2 + \ldots + \beta_t \tilde{x}_t^2 - \beta_{t+1} \tilde{x}_{t+1}^2 - \ldots - \beta_r \tilde{x}_r^2 (+0\tilde{x}_{r+1}^2 + \ldots + 0\tilde{x}_n^2),$$

where $\alpha_1, \ldots, \alpha_r, \beta_1, \ldots, \beta_r$ are all positive, be two arbitrary diagonal versions of $q(X)$. Then $s = t$; i.e. the two diagonal versions contain the same number of positive coefficients (and therefore also the same number of negative coefficients).

Proof. Let $\hat{L} = (\hat{\mathbf{e}}_1, \ldots, \hat{\mathbf{e}}_n)$ be the basis of $\mathbb{R}_{n \times 1}$ w.r.t. which X has components $\hat{x}_1, \ldots, \hat{x}_n$, and let $\tilde{L} = (\tilde{\mathbf{e}}_1, \ldots, \tilde{\mathbf{e}}_n)$ be the basis of $\mathbb{R}_{n \times 1}$ w.r.t. which X has components $\tilde{x}_1, \ldots, \tilde{x}_n$.

With the aim of producing a contradiction, assume that s and t are unequal numbers. In view of the symmetric nature of the hypotheses, we may assume further (without loss of generality) that $s < t$.

Let

$$V = \{X \in \mathbb{R}_{n\times 1} : \hat{x}_1 = \hat{x}_2 = \ldots = \hat{x}_s = 0\}, \quad \text{i.e. } V = \mathrm{sp}(\hat{\mathbf{e}}_{s+1}, \ldots, \hat{\mathbf{e}}_n),$$

and

$$W = \{X \in \mathbb{R}_{n\times 1} : \tilde{x}_{t+1} = \tilde{x}_{t+2} = \ldots = \tilde{x}_n = 0\}, \quad \text{i.e. } W = \mathrm{sp}(\tilde{\mathbf{e}}_1, \ldots, \tilde{\mathbf{e}}_t),$$

so that V and W are subspaces of $\mathbb{R}_{n\times 1}$ with dimensions $n-s$ and t, respectively. Further, let X now denote an arbitrary column in $V \cap W$ (so that $\hat{x}_1 = \ldots = \hat{x}_s = \tilde{x}_{t+1} = \ldots = \tilde{x}_n = 0$). Then

$$q(X) = -\alpha_{s+1}\hat{x}_{s+1}^2 - \ldots - \alpha_r \hat{x}_r^2, \quad \text{which is } \leqslant 0 \text{ (each } \alpha_i \text{ being } > 0),$$

and

$$q(X) = \beta_1 \tilde{x}_1^2 + \ldots + \beta_t \tilde{x}_t^2, \quad \text{which is } \geqslant 0 \text{ (each } \beta_i \text{ being } > 0).$$

It follows that $q(X)$ must be zero, and thus

$$\beta_1 \tilde{x}_1^2 + \ldots + \beta_t \tilde{x}_t^2 = 0.$$

Hence (as $\beta_1, \ldots, \beta_t$ are all positive [and $\tilde{x}_1, \ldots, \tilde{x}_t$ denote real numbers]) $\tilde{x}_1 = \tilde{x}_2 = \ldots = \tilde{x}_t = 0$. Therefore, in fact, all of $\tilde{x}_1, \ldots, \tilde{x}_t, \tilde{x}_{t+1}, \ldots, \tilde{x}_n$ (the components of X w.r.t. the basis $\tilde{L}$) are zero, and hence $X = O$.

This proves that $V \cap W = \{O\}$. Hence

$$\begin{aligned} \dim(V+W) &= \dim V + \dim W \quad \text{(cf. 42.3)} \\ &= (n-s)+t \\ &> n \quad (\text{since } s < t) \end{aligned}$$

—a contradiction, since $V+W$ is a subspace of the n-dimensional space $\mathbb{R}_{n\times 1}$.

From this contradiction the stated result follows.

In the light of 77.2 it makes sense to define the **signature** of a real quadratic form q to be the difference

$$\begin{pmatrix}\text{number of positive} \\ \text{coefficients}\end{pmatrix} - \begin{pmatrix}\text{number of negative} \\ \text{coefficients}\end{pmatrix}$$

in any (and therefore every) diagonal version of $q(X)$. This also, obviously, is an invariant of q.

To illustrate the definition of signature, recall that the two worked examples of §76 both involved real and quadratic forms of order 4 and produced the respective diagonal versions

$$2\hat{x}_1^2 + \hat{x}_2^2 - 3\hat{x}_3^2 + 2\hat{x}_4^2 \qquad \text{and} \qquad \hat{x}_1^2 - \hat{x}_2^2 - 2\hat{x}_3^2 (+0\hat{x}_4^2).$$

It is apparent that the quadratic forms have ranks 4 and 3, respectively, while their respective signatures are

$$3-1 = 2 \qquad \text{and} \qquad 1-2 = -1.$$

Notice that if we know the values of the rank r and the signature σ of a real quadratic form, then we can easily deduce a and b, the number of positive and negative coefficients in a diagonal version of the quadratic form: for clearly $r = a+b$, while $\sigma = a-b$, so that

$$a = \tfrac{1}{2}(r+\sigma) \qquad \text{and} \qquad b = \tfrac{1}{2}(r-\sigma).$$

78. Orthogonal diagonalization of a real quadratic form

This section is exclusively about real quadratic forms. In this context L will denote the standard basis of $\mathbb{R}_{n\times 1}$, which, it should be noted, is clearly an orthonormal basis of $\mathbb{R}_{n\times 1}$. Orthonormal bases were discussed in general in chapter 9, where we went on to prove the major theorem (see 73.1) that every real symmetric matrix is orthogonally similar to a diagonal matrix. The pattern at the heart of this theorem

$$P^T \times (\text{symmetric matrix}) \times P = \text{diagonal matrix}$$

makes it obviously relevant to quadratic form theory. It almost immediately yields the following results, 78.1 and 78.2, which should be regarded as two versions of the same story—the first version told in terms of change of basis, the second in terms of the corresponding change of variables.

78.1 Let q be a real quadratic form of order n, and let $A = M(q; L)$ [so that A is a real symmetric $n \times n$ matrix and $q(X) \equiv X^T AX$]. Then there exists an orthonormal basis $\hat{L}$ of $\mathbb{R}_{n\times 1}$ such that

$$M(q; \hat{L}) = \text{diag}\,(\alpha_1, \alpha_2, \ldots, \alpha_n),$$

where $\alpha_1, \alpha_2, \ldots, \alpha_n$ are the eigenvalues of A (with algebraic multiplicities).

Proof. By 73.1, there is an orthogonal matrix P such that $P^T AP$ is a diagonal matrix—D, say. Let $\hat{L}$ be the basis of $\mathbb{R}_{n\times 1}$ such that $M(L \to \hat{L}) = P$. Then, by 72.4, since L is an orthonormal basis and P is an orthogonal matrix, $\hat{L}$ too is an orthonormal basis of $\mathbb{R}_{n\times 1}$. Further, by 75.2,

$$M(q; \hat{L}) = P^T AP = D;$$

and, finally, since the diagonal matrix D is (orthogonally) similar to A, 66.2 tells us that the entries on the main diagonal of D are the eigenvalues of A (with algebraic multiplicities). The stated result is now fully established.

Post-script. In the situation described in 78.1 there could be several orthonormal bases of $\mathbb{R}_{n\times 1}$ w.r.t. which q has diagonal matrix. It is easy to use the ideas in the above proof to show that every diagonal matrix arising in this way has on its main diagonal the eigenvalues of A (with algebraic multiplicities).

The change of variables version of 78.1 is:

78.2 Let q be an arbitrary real quadratic form. Then there is an orthogonal change of variables (i.e. a change of variables taking the form $X = P\hat{X}$ with P an orthogonal matrix) that produces

$$q(X) \equiv \alpha_1\hat{x}_1^2+\alpha_2\hat{x}_2^2+\ldots+\alpha_n\hat{x}_n^2,$$

where $\alpha_1, \alpha_2, \ldots, \alpha_n$ are the eigenvalues of $M(q; L)$ (with algebraic multiplicities).

Proof (continuing the notation of 78.1 and its proof). Since $M(L \rightarrow \hat{L}) = P$, the re-expressing of $q(X)$ in terms of the components $\hat{x}_1, \ldots, \hat{x}_n$ of X w.r.t. $\hat{L}$ is what we informally describe as making the change of variables $X = P\hat{X}$ (cf. final note in §75); and, since P is an orthogonal matrix, this is an orthogonal change of variables. Since $M(q; \hat{L}) = D$, it produces

$$q(X) \equiv \hat{X}^T D\hat{X}, \qquad \text{i.e. } q(X) \equiv \alpha_1\hat{x}_1^2+\ldots+\alpha_n\hat{x}_n^2,$$

where (as in 78.1) $\alpha_1, \ldots, \alpha_n$ are the eigenvalues of $M(q; L)$ (with algebraic multiplicities). All that was asserted has now been proved.

The twin results 78.1 and 78.2 represent a powerful contribution to the study of real quadratic forms, and in §79 we shall see the use of 78.2 as the key idea in the proof of an important and profound theorem about real quadratic forms (79.5). Meanwhile the rest of this section is devoted to an outline of how orthogonal diagonalization of quadratic forms can be used in the coordinate geometry of central conics.

There are two kinds of central conic—ellipses and hyperbolas. We call these curves "central" conics because each has an identifiable centre where two (perpendicular) axes of symmetry intersect. (Incidentally, in this discussion, a circle may be regarded as a special case of an ellipse.)

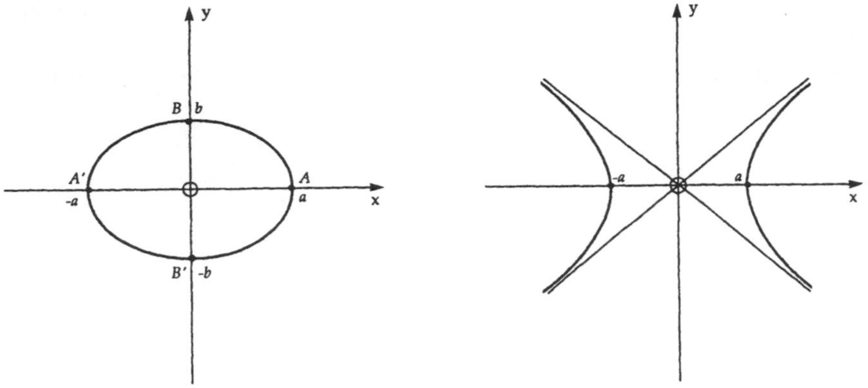

ELLIPSE HYPERBOLA

Figure XIX

With the origin at their centres and x- and y-axes along their axes of symmetry (cf. figure XIX), an ellipse and a hyperbola have equations of the forms

$$\frac{x^2}{a^2}+\frac{y^2}{b^2}=1 \qquad \text{and} \qquad \frac{x^2}{a^2}-\frac{y^2}{b^2}=1,$$

respectively (a and b denoting positive constants). In each case interesting geometrical properties of the curve can be deduced from a knowledge of the values of a and b: e.g. in the case of the ellipse with equation $(x^2/a^2)+(y^2/b^2)=1$, the axes ($A'A$ and $B'B$ in figure XIX) have lengths $2a$ and $2b$.

In general, however, the equation of a central conic will not present itself in so tidy a form, because the axes of symmetry of the curve will not usually lie along the originally chosen coordinate axes. At least, however, there are easy procedures for translating the coordinate axes so that the origin is made to coincide with the centre of the curve in question; and it can be shown that, once that is done, the equation of the conic takes the form

$$\alpha x^2+2\beta xy+\gamma y^2=1, \qquad \text{i.e. } q_A(x,y)=1, \tag{1}$$

where A is the symmetric matrix $\begin{bmatrix}\alpha & \beta\\ \beta & \gamma\end{bmatrix}$. (Recall that q_A means the quadratic form defined by $q_A(X)=X^TAX$.)

Obvious questions now arise concerning equations of the form (1). For example:

(a) Given such an equation out of the blue, how can we tell whether it defines an ellipse or a hyperbola or neither?

(*b*) If (1) does define an ellipse, how can we find geometrical details of the ellipse, e.g. the lengths of its axes?

In §72 it was explained that, in coordinate geometry, a change of coordinates by taking new perpendicular axes through the existing origin corresponds to an orthogonal change of variable. Accordingly 78.2 enables us to answer questions like the above in terms of the eigenvalues (λ_1 and λ_2, say) of A: for by 78.2 there is $\left\{\begin{matrix}\text{an orthogonal change of variables}\\ \text{a change of coordinates}\end{matrix}\right\}$ to new coordinate variables $\hat{x}$, $\hat{y}$ that transforms (1) into

$$\lambda_1\hat{x}^2+\lambda_2\hat{y}^2 = 1.$$

Hence it is apparent that (1) will define an ellipse if the eigenvalues λ_1 and λ_2 are both positive, while (1) will define a hyperbola in the case where λ_1 and λ_2 have opposite signs. (There are, of course, other cases, where (1) defines neither an ellipse nor a hyperbola.) Moreover, in the case where λ_1 and λ_2 are both positive and (1) defines an ellipse, the equation of the ellipse in terms of $\hat{x}$ and $\hat{y}$ can be written

$$\frac{\hat{x}^2}{1/\lambda_1}+\frac{\hat{y}^2}{1/\lambda_2} = 1,$$

from which it is evident that the lengths of the axes are $2/\sqrt{\lambda_1}$ and $2/\sqrt{\lambda_2}$.

In 3-dimensional coordinate geometry analogous questions arise concerning the equation

$$q_A(x, y, z) = 1, \tag{2}$$

where A is a 3×3 real symmetric matrix, and these questions can be handled in a very similar way to the corresponding 2-dimensional problems. In particular, coordinates can be changed to transform (2) into

$$\lambda_1\hat{x}^2+\lambda_2\hat{y}^2+\lambda_3\hat{z}^2 = 1,$$

where $\lambda_1, \lambda_2, \lambda_3$ are the eigenvalues of A. And, for example, if these eigenvalues are all positive, then the surface defined is an ellipsoid whose axes have lengths $2/\sqrt{\lambda_1}$, $2/\sqrt{\lambda_2}$, $2/\sqrt{\lambda_3}$.

79. Positive-definite real quadratic forms

Let q denote a real quadratic form of order n. We say that q is **positive-semidefinite** if and only if $q(X) \geqslant 0$ for every $X \in \mathbb{R}_{n\times 1}$. And we say that q is **positive-definite** if and only if $q(X)$ is strictly positive for *all nonzero* $X \in \mathbb{R}_{n\times 1}$, i.e. if and only if q is positive-semidefinite and $q(X)$ is zero only in the case $X = O$.

As simple illustrations, consider the quadratic forms q_1 and q_2, each of order 4, defined by

$$q_1(X) = x_1^2+x_2^2+x_3^2+0x_4^2 \qquad (X \in \mathbb{R}_{4\times 1}),$$
$$q_2(X) = x_1^2+x_2^2+x_3^2+x_4^2 \qquad (X \in \mathbb{R}_{4\times 1}).$$

Clearly q_1 and q_2 are both positive-semidefinite since, for every $X \in \mathbb{R}_{4\times 1}$, $q_1(X) \geqslant 0$ and $q_2(X) \geqslant 0$; and q_2 is positive-definite since $q_2(X)$ is strictly positive whenever $X \neq O$, i.e. whenever at least one of x_1, x_2, x_3, x_4 is nonzero. But q_1 is not positive-definite because q_1 takes the value zero in more cases than the case $X = O$ (e.g. the case $X = \text{col}\,(0, 0, 0, 1)$).

There are corresponding terminologies with "positive" replaced by "negative". We describe the real quadratic form q of order n as **negative-semidefinite** if and only if $q(X) \leqslant 0$ for all $X \in \mathbb{R}_{n\times 1}$, **negative-definite** if and only if $q(X) < 0$ for all nonzero $X \in \mathbb{R}_{n\times 1}$. Finally, we say that q is **indefinite** if q is neither positive-semidefinite nor negative-semidefinite, i.e. if there exist columns X_1, $X_2 \in \mathbb{R}_{n\times 1}$ such that $q(X_1) > 0$ and $q(X_2) < 0$.

As foreshadowed in the above simple illustrations, one can tell immediately from a diagonal version of $q(X)$ which of the descriptions we have introduced fit(s) a given real quadratic form q. In detail:

79.1 Let q be a real quadratic form of order n, and suppose that, for some basis $\hat{L}$ of $\mathbb{R}_{n\times 1}$, $M(q; \hat{L}) = \text{diag}\,(\alpha_1, \alpha_2, \ldots, \alpha_n)$, so that, in terms of the components $\hat{x}_1, \ldots, \hat{x}_n$ of X w.r.t. $\hat{L}$,

$$q(X) \equiv \alpha_1\hat{x}_1^2+\alpha_2\hat{x}_2^2+\ldots+\alpha_n\hat{x}_n^2.$$

Then q is

(i) positive-semidefinite if and only if every α_i is $\geqslant 0$,
(ii) positive-definite if and only if every α_i is positive,
(iii) negative-semidefinite if and only if every α_i is $\leqslant 0$,
(iv) negative-definite if and only if every α_i is negative,
(v) indefinite if and only if the sequence $\alpha_1, \ldots, \alpha_n$ includes at least one positive number and at least one negative number.

It is fairly easy to see the truth of all these assertions. As a sample of the thinking involved, here is a full proof of part (ii).

If any of the α_is (say α_j) is $\leqslant 0$, then $q(X)$ takes non-positive value in the case $\left\{\begin{matrix} \hat{x}_j = 1 \\ \hat{x}_i = 0 \text{ for all } i \neq j \end{matrix}\right\}$—which ($\hat{x}_1, \ldots, \hat{x}_n$ being the components of X w.r.t. the basis $\hat{L}$) is a case other than $X = O$; and that reveals that q is not positive-definite. Thus

$$q \text{ is positive-definite} \Rightarrow \text{all of } \alpha_1, \ldots, \alpha_n \text{ are positive.}$$

And, conversely, if all of $\alpha_1, \ldots, \alpha_n$ are positive, then clearly $q(X) \left[\equiv \sum_{i=1}^{n} \alpha_i \hat{x}_i^2 \right] > 0$ in all cases except the case $\hat{x}_1 = \ldots = \hat{x}_n = 0$, i.e. the case $X = O$, and therefore q is positive-definite.

Notice further that, as a consequence of 79.1(ii):

79.2 If q is a positive-definite real quadratic form of order n, there is a basis $\hat{L}$ of $\mathbb{R}_{n \times 1}$ such that $M(q; \hat{L}) = I$, i.e. there is a change of variables (to new variables $\hat{x}_1, \ldots, \hat{x}_n$, say) that produces

$$q(X) \equiv \hat{x}_1^2 + \hat{x}_2^2 + \ldots + \hat{x}_n^2.$$

(For, if q is a positive-definite quadratic form of order n, diagonalization of q will give, say,

$$q(X) \equiv \alpha_1 y_1^2 + \alpha_2 y_2^2 + \ldots + \alpha_n y_n^2,$$

with every α_i positive (by 79.1(ii)), and then the further change of variables

$$\hat{x}_1 = \alpha_1^{1/2} y_1, \quad \hat{x}_2 = \alpha_2^{1/2} y_2, \ldots, \hat{x}_n = \alpha_n^{1/2} y_n$$

produces $q(X) \equiv \hat{x}_1^2 + \hat{x}_2^2 + \ldots + \hat{x}_n^2 \; (= \hat{X}^T I \hat{X})$.)

Next, putting together 78.2 and 79.1, we can characterize positive-definite quadratic forms, positive-semidefinite quadratic forms, etc. in terms of the eigenvalues of their matrices.

79.3 Let A be a real symmetric matrix. Then the quadratic form q_A [defined by $q_A(X) \equiv X^T A X$] is

(i) positive-semidefinite if and only if all the eigenvalues of A are $\geqslant 0$,

(ii) positive-definite if and only if all the eigenvalues of A are positive;

and there are corresponding results for negative-semidefinite, negative-definite, and indefinite.

(In the case of part (ii), for example, 78.2 tells us that there is a change of variables giving

$$q_A(X) \equiv \alpha_1 \hat{x}_1^2 + \alpha_2 \hat{x}_2^2 + \ldots + \alpha_n \hat{x}_n^2,$$

with $\alpha_1, \alpha_2, \ldots, \alpha_n$ the eigenvalues of A, and, by 79.1, q_A is positive-definite if and only if all these eigenvalues are positive.)

In view of the close connection between quadratic forms and symmetric matrices, it is natural to define a **positive-definite** (real) **matrix** to be a (real) symmetric matrix (A, say) with the property that the quadratic form q_A is positive-definite. However, in worked example 2 of §73, before we encountered quadratic forms, we defined a positive-definite matrix to be a symmetric matrix whose eigenvalues are all positive. Fortunately, as 79.3(ii) shows, the two definitions give exactly the same meaning to the phrase "positive-definite real matrix". A handy summary of all this is that, for a real symmetric matrix A, the three statements

(1) A is a positive-definite matrix,

(2) q_A is a positive-definite quadratic form,

and (3) all the eigenvalues of A are positive

are all equivalent. One detail which should not be overlooked is that describing a real matrix as positive-definite *presupposes that it is symmetric.*

As all the eigenvalues of a positive-definite matrix are positive, it is clear from 64.3(ii) that:

79.4 Every positive-definite matrix has positive determinant.

Positive-definite matrices and quadratic forms have a number of interesting properties. They also appear quite widely. For example, as readers with some knowledge of physics will understand, the quadratic form giving the kinetic energy T of a system in terms of the velocity components of its particles is positive-definite (since T cannot be negative and is zero only when all the particles are at rest). That is the beginning of the reason why the following theorem (79.5) finds significant application in physics. The theorem is called the *simultaneous diagonability theorem*, because it tells us that, by the same $\left\{\begin{array}{c}\text{change of basis}\\ \text{change of variables}\end{array}\right\}$, one can diagonalize both of two given quadratic forms if one of them is positive-definite.

79.5 Let A and B be symmetric matrices in $\mathbb{R}_{n\times n}$, A being positive-definite. Then there is a change of variables (to new variables $\hat{x}_1, \dots, \hat{x}_n$, say) that produces

$$q_A(X) \equiv \hat{x}_1^2 + \hat{x}_2^2 + \dots + \hat{x}_n^2 \quad \text{and} \quad q_B(X) = \gamma_1 \hat{x}_1^2 + \gamma_2 \hat{x}_2^2 + \dots + \gamma_n \hat{x}_n^2,$$

where $\gamma_1, \gamma_2, \dots, \gamma_n$ are the roots of the equation

$$\det(tA - B) = 0$$

(with algebraic multiplicities). [Equivalently, there is a basis $\hat{L}$ of $\mathbb{R}_{n\times 1}$ such that $M(q_A; \hat{L}) = I$ and $M(q_B; \hat{L}) = \operatorname{diag}(\gamma_1, \dots, \gamma_n)$, where $\gamma_1, \dots, \gamma_n$ are as specified above.]

Proof. By 79.2, since q_A is positive-definite, there is a change of variables ($X = QY$, say, Q being nonsingular) that gives

$$q_A(X) \equiv y_1^2 + y_2^2 + \dots + y_n^2 \; (= Y^T Y).$$

The same change of variables also gives (cf. 75.1)

$$q_B(X) \equiv Y^T C Y,$$

where $C = Q^T B Q$. By 78.2, there is an orthogonal change of variables ($Y = P\hat{X}$, say, with P orthogonal) that transforms $Y^T C Y$ into $\gamma_1 \hat{x}_1^2 + \ldots + \gamma_n \hat{x}_n^2$, where $\gamma_1, \ldots, \gamma_n$ are the eigenvalues of C. So we have

$$q_B(X)\,(\equiv Y^T C Y) \equiv \gamma_1 \hat{x}_1^2 + \gamma_2 \hat{x}_2^2 + \ldots + \gamma_n \hat{x}_n^2,$$

while at the same time

$$\begin{aligned} q_A(X)\,(\equiv Y^T Y) &= (P\hat{X})^T (P\hat{X}) \equiv \hat{X}^T (P^T P) \hat{X} \equiv \hat{X}^T I \hat{X} \quad (P \text{ being orthogonal}) \\ &\equiv \hat{x}_1^2 + \hat{x}_2^2 + \ldots + \hat{x}_n^2. \end{aligned}$$

It remains to prove that $\gamma_1, \ldots, \gamma_n$ are as stated in the theorem. For this purpose, note that, since the change of variables $X = QY$ transformed $q_A(X)$ [i.e. $X^T A X$] into $Y^T I Y$, we have (cf. 75.1) $Q^T A Q = I$; and hence

$$\begin{aligned} \det(tI - C) &= \det(tQ^T A Q - C) \\ &= \det(tQ^T A Q - Q^T B Q) \quad (\text{since } C = Q^T B Q, \text{ as recorded above}) \\ &= \det[Q^T (tA - B) Q] \\ &= (\det Q)^2 \det(tA - B) \quad (\text{by 32.6 and 31.1}). \end{aligned}$$

So, since the nonsingular matrix Q has nonzero determinant, it is apparent that the equation $\det(tA - B) = 0$ has the same roots as the equation $\det(tI - C) = 0$, namely the eigenvalues $\gamma_1, \ldots, \gamma_n$ of C. Thus $\gamma_1, \ldots, \gamma_n$ are indeed as asserted, and the theorem is fully proved.

80. The leading minors theorem

Consider an arbitrary real symmetric matrix $A = [\alpha_{ik}]_{n \times n}$. For each p in the range 1 to n, inclusive, the **leading $p \times p$ submatrix** of A (which we shall denote by A_p) means the top left-hand $p \times p$ corner of A, i.e. the $p \times p$ submatrix formed by the entries of A that lie both in its first p rows and in its first p columns. The smallest and largest cases are, respectively, $A_1 = \alpha_{11}$ and $A_n = A$.

We introduce the further notation $\Delta_p(A)$ for the determinant of A_p ($1 \leqslant p \leqslant n$); and we call $\Delta_1(A), \Delta_2(A), \ldots, \Delta_p(A)$ the **leading minors** of A. Note that $\Delta_1(A) = \alpha_{11}$ and that $\Delta_n(A) = \det A$.

Since A is symmetric, it is clear that the leading submatrices $A_1, \ldots, A_n$ are all symmetric. So, for each p ($1 \leqslant p \leqslant n$), we can introduce the quadratic form q_{A_p} with matrix A_p: q_{A_p} is a quadratic form of order p and is given by

$$q_{A_p}(x_1, x_2, \ldots, x_p) = \sum_{i=1}^{p} \alpha_{ii} x_i^2 + \sum_{\substack{i,k=1 \\ i<k}}^{p} 2\alpha_{ik} x_i x_k.$$

Notice that the right-hand side here is obtainable from the corresponding defining expression for $q_A(x_1, x_2, \ldots, x_n)$ simply by suppressing all the terms involving $x_{p+1}, \ldots, x_n$—an observation that enables us to see the truth of two equations which are recorded in our next result (80.1) and which will be very helpful to us in what follows thereafter.

80.1 (With notation as above)

(i) for each p in the range 1 to $n-1$ (inclusive)
$$q_{A_p}(x_1, x_2, \ldots, x_p) = q_A(x_1, x_2, \ldots, x_p, 0, 0, \ldots, 0);$$
(ii) $$q_A(x_1, x_2, \ldots, x_{n-1}, x_n) = q_{A_{n-1}}(x_1, \ldots, x_{n-1}) + \sum_{i=1}^{n-1} 2\alpha_{in}x_i x_n + \alpha_{nn}x_n^2.$$

As an easy step towards the main theorem of this section, we now show that:

80.2 With notation as above, if A is positive-definite, then so are all of $A_1, A_2, \ldots, A_n$.

Proof. Suppose that A is positive-definite (so that the quadratic form q_A is positive-definite). Then, for each p with $1 \leqslant p \leqslant n-1$, the quadratic form q_{A_p} is also positive-definite since

(1) for all $x_1, \ldots, x_p \in \mathbb{R}$,
$$\begin{aligned} q_{A_p}(x_1, \ldots, x_p) &= q_A(x_1, \ldots, x_p, 0, \ldots, 0) \qquad \text{(cf. 80.1(i))} \\ &\geqslant 0 \qquad (q_A \text{ being positive-definite}), \end{aligned}$$

and (2) for $x_1, \ldots, x_p \in \mathbb{R}$,
$$\begin{aligned} q_{A_p}(x_1, \ldots, x_p) = 0 &\Rightarrow q_A(x_1, \ldots, x_p, 0, \ldots, 0) = 0 \qquad \text{(by 80.1(i))} \\ &\Rightarrow (x_1, \ldots, x_p, 0, \ldots, 0) = (0, 0, \ldots, 0) \qquad (q_A \text{ being positive-definite}) \\ &\Rightarrow \text{all of } x_1, \ldots, x_p \text{ are zero.} \end{aligned}$$

So $A_1, A_2, \ldots, A_{n-1}$ are all positive-definite, while A_n $(= A)$ is positive-definite by hypothesis. The stated result follows.

Having attended to all the above preliminaries, we are ready for the big theorem of this section—the *leading minors theorem*, which gives a useful (and strikingly simple) necessary and sufficient condition for a real symmetric matrix to be positive-definite.

80.3 (The leading minors theorem) Let A be a symmetric matrix in $\mathbb{R}_{n \times n}$. Then A is positive-definite if and only if the leading minors $\Delta_1(A)$, $\Delta_2(A), \ldots, \Delta_n(A)$ of A are all positive.

Proof. (1) (The "$\Rightarrow$" half) If A is positive-definite, then (by 80.2) so are all the leading submatrices $A_1, \dots, A_n$, and therefore (by 79.4) $\det A_1, \dots, \det A_n$ are all positive, i.e. the leading minors $\Delta_1(A), \dots, \Delta_n(A)$ are all positive.

(2) (The "$\Leftarrow$" half) We prove by induction on n the universal truth of the statement: (α) if A is a symmetric matrix in $\mathbb{R}_{n\times n}$ and $\Delta_1(A), \dots, \Delta_n(A)$ are all positive, then A is positive-definite.

In the case $n = 1$ (α) boils down to the clearly true assertion that if a is a positive number, then the 1×1 matrix $[a]$ is positive-definite. So it will suffice to prove the truth of (α) under the inductive assumption that every real $(n-1)\times(n-1)$ symmetric matrix whose leading minors are all positive is positive-definite.

We now make that inductive assumption and consider an arbitrary real $n \times n$ symmetric matrix $A = [\alpha_{ik}]_{n\times n}$ with $\Delta_1(A), \dots, \Delta_n(A)$ all positive.

A key observation is that the leading minors of the $(n-1)\times(n-1)$ leading submatrix A_{n-1} are simply $\Delta_1(A), \dots, \Delta_{n-1}(A)$, which are all positive. So, by the induction hypothesis, A_{n-1} is positive-definite; i.e. the quadratic form $q_{A_{n-1}}$ is positive-definite. Hence, by 79.2, there is an invertible change of variables—from original variables $x_1, \dots, x_{n-1}$ to new variables $y_1, \dots, y_{n-1}$, say, that yields

$$q_{A_{n-1}}(x_1, \dots, x_{n-1}) \equiv y_1^2 + y_2^2 + \dots + y_{n-1}^2.$$

This change of variables is given, let us suppose, by

$$\operatorname{col}(x_1, \dots, x_{n-1}) = P \times \operatorname{col}(y_1, \dots, y_{n-1}),$$

P being a nonsingular $(n-1)\times(n-1)$ matrix. We now refer to 80.1(ii), which gives the connection relevant here between $q_{A_{n-1}}$ and q_A and which can be re-written as

$$q_A(x_1, \dots, x_{n-1}, x_n) \equiv q_{A_{n-1}}(x_1, \dots, x_{n-1}) + 2x_n \sum_{i=1}^{n-1} \alpha_{in} x_i + \alpha_{nn} x_n^2. \qquad (*)$$

In this equation we make the change of variables defined by

$$\operatorname{col}(x_1, \dots, x_{n-1}, x_n) = \underbrace{\begin{bmatrix} P & O \\ O & 1 \end{bmatrix}} \times \operatorname{col}(y_1, \dots, y_{n-1}, y_n).$$

↖clearly nonsingular (as P is nonsingular)

(This is simply the aforementioned change from $x_1, \dots, x_{n-1}$ to $y_1, \dots, y_{n-1}$ along with the re-naming of x_n as y_n.) In (*) it transforms $q_{A_{n-1}}(x_1, \dots, x_{n-1})$, as before, into $y_1^2 + \dots + y_{n-1}^2$; and it will change $\sum_{i=1}^{n-1} \alpha_{in} x_i$ into some linear combination of $y_1, \ldots, y_{n-1}$, say $\sum_{i=1}^{n-1} \beta_i y_i$. Hence (*) becomes

$$q_A(x_1, \ldots, x_n) \equiv y_1^2 + \ldots + y_{n-1}^2 + \sum_{i=1}^{n-1} 2\beta_i y_i y_n + \alpha_{nn} y_n^2$$

$$\equiv \sum_{i=1}^{n-1} (y_i + \beta_i y_n)^2 + \gamma y_n^2,$$

where $\gamma = \alpha_{nn} - \sum_{i=1}^{n-1} \beta_i^2$,

$$\text{i.e. } q_A(x_1, \ldots, x_n) \equiv \hat{x}_1^2 + \ldots + \hat{x}_{n-1}^2 + \gamma \hat{x}_n^2, \qquad (\dagger)$$

where now the (clearly invertible) change of variables defined by

$$\hat{x}_i = y_i + \beta_i y_n \ (1 \leqslant i \leqslant n-1), \qquad \hat{x}_n = y_n$$

has been made.

This shows that, for a certain basis $\hat{L}$ of $\mathbb{R}_{n \times 1}$,

$$M(q_A; \hat{L}) = \text{diag}(1, 1, \ldots, 1, \gamma) \qquad (= D, \text{ say}).$$

So (cf. 75.2) $D = Q^T A Q$ for some nonsingular real $n \times n$ matrix Q. Clearly $\det D = \gamma$, and hence

$$\begin{aligned} \gamma &= \det(Q^T A Q) \\ &= (\det Q)^2 (\det A) \qquad \text{(by 32.6 and 31.1)} \\ &= (\det Q)^2 \Delta_n(A) \\ &> 0 \quad \text{(since } Q \text{ is a nonsingular real matrix and } \Delta_n(A) > 0). \end{aligned}$$

From (†) it is now apparent (cf. 79.1) that q_A is positive-definite; i.e. A is positive-definite.

This completes proof by induction of the desired result.

The corresponding theorem about negative-definiteness is:

80.4 Let A be a symmetric matrix in $\mathbb{R}_{n \times n}$. Then A is negative-definite if and only if the leading minors $\Delta_1(A), \Delta_2(A), \Delta_3(A), \Delta_4(A), \ldots$ of A are respectively negative, positive, negative, positive, ...

This can be deduced almost immediately from 80.3 by means of the facts that

(*a*) [fairly obviously] A is negative-definite $\Leftrightarrow$ $(-A)$ is positive-definite, and

(*b*) $\Delta_p(-A) = (-1)^p \Delta_p(A)$ (by 31.6).

One application of the theorems 80.3 and 80.4 to another part of mathematics arises in the discussion of the natures of stationary points of functions of several real variables. Here let us consider the simplest case of a function

$f(x,y)$ of two real variables x and y. We shall assume that, for (x,y) in a neighbourhood of the fixed point (a,b), $f(x,y)$ can be expanded as a Taylor series: i.e. for (h,k) sufficiently close to $(0,0)$,

$$f(a+h,b+k) = f(a,b)+\frac{1}{1!}(hf_x+kf_y)+\frac{1}{2!}(h^2f_{xx}+2hkf_{xy}+k^2f_{yy})+R(h,k),$$

where $R(h,k)$ involves terms of degree 3 and higher in h and k and where f_x, f_y, f_{xx}, f_{xy}, f_{yy} denote the partial derivatives $\dfrac{\partial f}{\partial x}, \dfrac{\partial f}{\partial y}, \dfrac{\partial^2 f}{\partial x^2}, \dfrac{\partial^2 f}{\partial x\,\partial y}, \dfrac{\partial^2 f}{\partial y^2}$, respectively, evaluated at (a,b). The above equation can be re-written

$$f(a+h,b+k) = f(a,b)+hf_x+kf_y+\tfrac{1}{2}q_A(h,k)+R(h,k),$$

where q_A is the quadratic form with matrix $A = \begin{bmatrix} f_{xx} & f_{xy} \\ f_{xy} & f_{yy} \end{bmatrix}$.

Suppose now that (a,b) is a stationary point of $f(x,y)$: that means that $f_x = f_y = 0$. Then, for (h,k) sufficiently close to $(0,0)$,

$$f(a+h,b+k) = f(a,b)+\eta,$$

where $\eta = \tfrac{1}{2}q_A(h,k)+R(h,k)$.

Consider what happens when the matrix A, and hence the quadratic form q_A, are positive-definite. It may be assumed that, when (h,k) is sufficiently close to, but not equal to, $(0,0)$, the positive quadratic term $\tfrac{1}{2}q_A(h,k)$ will dwarf $R(h,k)$ in magnitude, giving $\eta > 0$ and hence $f(a+h,b+k) > f(a,b)$. Therefore, when A is positive-definite, the stationary point at (a,b) is a local minimum. Similarly, one can see that when A is negative-definite, the stationary point at (a,b) is a local maximum.

Now the leading minors of A are f_{xx} and $\det A\,(= f_{xx}f_{yy}-f_{xy}^2)$. Hence the theorems 80.3 and 80.4 give us the following result.

80.5 With notation as in the above discussion,

(i) a sufficient condition for $f(x,y)$ to have a local minimum at (a,b) is the pair of inequalities

$$f_{xx} > 0 \quad \text{and} \quad f_{xx}f_{yy}-f_{xy}^2 > 0;$$

(ii) a sufficient condition for $f(x,y)$ to have a local maximum at (a,b) is the pair of inequalities

$$f_{xx} < 0 \quad \text{and} \quad f_{xx}f_{yy}-f_{xy}^2 > 0.$$

(There are, of course, other cases, which we shall not explore here: for example, it can be shown that if q_A is an indefinite quadratic form, then $f(x,y)$ has what is called a saddle point at (a,b).)

EXERCISES ON CHAPTER TEN

1. Given that $X = \text{col}(x_1, x_2, x_3)$ and $A = \begin{bmatrix} 3 & 2 & -1 \\ 2 & 0 & 4 \\ -1 & 4 & 5 \end{bmatrix}$, write out the full version of X^TAX as a quadratic polynomial in x_1, x_2, x_3.

2. Write down the unique 4×4 symmetric matrix A such that, for all values of x_1, x_2, x_3, x_4,

$$x_1^2+3x_3^2-x_4^2+4x_1x_2+6x_1x_3+x_1x_4-2x_2x_4+10x_3x_4$$

equals X^TAX, where X stands for $\text{col}(x_1, x_2, x_3, x_4)$.

3. Let B be an arbitrary matrix in $F_{n\times n}$. Show that there is a unique symmetric matrix A in $F_{n\times n}$ such that $X^TAX = X^TBX$ for all $X \in F_{n\times 1}$. (There is, therefore, no point in considering X^TBX for non-symmetric B: it would lead to nothing new and would merely spoil the uniqueness of the matrix of a quadratic form that can be claimed when we stick to symmetric matrices.)

4. Let $A = \begin{bmatrix} 3 & -1 \\ -1 & 2 \end{bmatrix}$, and let $\hat{L}$ be the basis $\left(\begin{bmatrix} 2 \\ 1 \end{bmatrix}, \begin{bmatrix} 3 \\ 2 \end{bmatrix}\right)$ of $\mathbb{R}_{2\times 1}$. Find $M(q_A; \hat{L})$ in two separate ways—(*a*) by making, in the general expression for $q_A(x_1, x_2)$, the change of component variables that corresponds to the change from the standard basis to the basis $\hat{L}$ (cf. the worked example preceding 75.1), (*b*) by using 75.2.

5. Each of the following formulae defines a quadratic form q. In each case diagonalize q by carrying out the rational reduction process (as described in §76), and then state the rank and signature of q.

(i) $q(x_1, x_2) \equiv x_1^2-6x_1x_2+14x_2^2.$

(ii) $q(x_1, x_2, x_3) \equiv x_1^2+3x_2^2+10x_3^2-4x_1x_2+6x_1x_3-10x_2x_3.$

(iii) $q(x_1, x_2, x_3) \equiv x_1^2-x_2^2+x_3^2+2x_1x_2-6x_1x_3+2x_2x_3.$

(iv) $q(x_1, x_2, x_3) \equiv x_1^2+4x_2^2+x_3^2-4x_1x_2+2x_1x_3-x_2x_3.$

(v) $q(x_1, x_2, x_3, x_4) \equiv x_1^2+3x_2^2-x_3^2+3x_4^2-2x_1x_2+4x_1x_4+4x_2x_3 -8x_2x_4+2x_3x_4.$

(vi) $q(x_1, x_2, x_3, x_4) \equiv x_1^2+4x_2^2+x_3^2+x_4^2+4x_1x_2-2x_1x_3+2x_1x_4 -2x_2x_3+6x_2x_4.$

(vii) $q(x_1, x_2, x_3, x_4) \equiv x_1x_2+2x_1x_3+4x_1x_4+6x_2x_3+8x_2x_4-16x_3x_4.$

(viii) $q(x_1, x_2, x_3, x_4) \equiv x_3^2+x_4^2+4x_1x_2+4x_2x_3+2x_3x_4.$

(ix) $q(x_1, x_2, x_3, \ldots, x_{10}) \equiv x_1x_2+x_2x_3+x_3x_4+x_4x_5+\ldots+x_8x_9 +x_9x_{10}-x_{10}x_1.$

6. In each of parts (i), (ii), (iii) of exercise 5, find a basis $\hat{L}$ of $\mathbb{R}_{n\times 1}$, where n is the order of q, such that $M(q; \hat{L})$ is diagonal.

7. What can be said about the rank and signature of a quadratic form q with the property that the general expression for $q(X)$ factorizes as the product of two linear combinations of $x_1, x_2, \ldots$ that are not constant multiples of each other?

8. Let q be the real quadratic form of order 3 defined by

$$q(x_1, x_2, x_3) \equiv x_1^2+3x_2^2+kx_3^2+2x_1x_2-2x_1x_3+6x_2x_3,$$

where k is a constant. (*a*) By first diagonalizing q, show that there exist values of x_1, x_2, x_3 that make $q(X)$ negative if and only if $k < 9$. (*b*) Suppose that $k < 9$. Find values of x_1, x_2, x_3 that make $q(X)$ negative (whatever the exact value of k).

9. Let q be the real quadratic form of order 3 defined by

$$q(x_1, x_2, x_3) \equiv x_1x_2 + \alpha x_1x_3 + \beta x_2x_3 \qquad (\alpha, \beta \text{ being constants}).$$

Find the rank and signature of q (*a*) when $\alpha\beta > 0$, (*b*) when $\alpha\beta < 0$, (*c*) when $\alpha\beta = 0$.

10. Show that the equation $\alpha x^2 + 2\beta xy + \gamma y^2 = 1$ defines an ellipse if $\alpha > 0$ and $\beta^2 < \alpha\gamma$. Show further that, when this condition is fulfilled, the area enclosed by the ellipse is $\dfrac{\pi}{\sqrt{(\alpha\gamma - \beta^2)}}$. (*Note:* the area enclosed by the standard ellipse $(x^2/a^2) + (y^2/b^2) = 1$ is πab.)

11. Show that the equation

$$3x^2 + 5y^2 + 4z^2 + 4xz + 4yz = 1$$

defines an ellipsoid, and find the lengths of its axes.

12. Let q be the real quadratic form (with diagonal matrix) defined by

$$q(x_1, x_2, x_3) \equiv kx_1^2 + k(k-1)x_2^2 + (k^2-1)x_3^2,$$

where k is a constant. For which values of k is q (*a*) positive-definite, (*b*) positive-semidefinite, (*c*) negative-definite, (*d*) negative-semidefinite, (*e*) indefinite?

13. (i) Show that the sum of two positive-definite real $n \times n$ matrices is also positive-definite. (ii) Is it true, for real $n \times n$ matrices A and B, that if A is positive-definite and B is positive-semidefinite, then $A + B$ is positive-definite?

14. Let A be a positive-definite real matrix. Show that A^{-1} (which certainly exists, by 79.4) is also positive-definite. (Try to find two different proofs, one using eigenvalues, the other not.)

15. Let A be an arbitrary $n \times n$ symmetric real matrix. Show that there exists a real number μ such that $\mu I + A$ is positive-definite.

16. Show that (i) for any real matrix A (not necessarily square), A^TA is a positive-semidefinite matrix (i.e. q_{A^TA} is a positive-semidefinite quadratic form), (ii) for any nonsingular real matrix A, A^TA is positive-definite.

17. Find a change of variable which diagonalizes both q_A and q_B, where $A = \begin{bmatrix} 1 & 2 \\ 2 & 5 \end{bmatrix}$, $B = \begin{bmatrix} 1 & 5 \\ 5 & 9 \end{bmatrix}$. (Here A is positive-definite.)

18. Let A and B be positive-definite matrices in $\mathbb{R}_{n \times n}$ such that $AB = BA$. By using the simultaneous diagonability theorem (79.5) with B replaced by AB (which is symmetric since $AB = BA$—cf. exercise 9(i) on chapter 2), show that the product AB is also positive-definite.

19.* Let A and B be real $n \times n$ positive-definite matrices, where $n \geqslant 2$. By using the simultaneous diagonability theorem, show that

$$\det(A + B) > \det A + \det B.$$

20.* Worked example 2 of §73 shows that every positive-definite real matrix has a positive-definite square root. Show that, in every case, a positive-definite real matrix has *just one* positive-definite square root. (*Hint:* use orthogonal diagonability of symmetric matrices and exercise 6 on chapter 2.)

21. Is it true that if the real symmetric $n \times n$ matrix A has rank n, then every leading minor of A must be nonzero?

22. Let q be the real quadratic form of order 3 defined by

$$q(x_1, x_2, x_3) \equiv \alpha(x_1^2 + 2x_2^2 + 3x_3^2) + 2\beta(x_1x_3 + 2x_2x_3),$$

where α and β are real constants. Use the leading minors theorem to prove that q is positive-definite if and only if $\alpha > |\beta|$.

23. Let α, β, γ be positive constants, and let q be the real quadratic form of order 3 defined by

$$q(x_1, x_2, x_3) \equiv \alpha x_1^2 + \beta x_2^2 + \gamma x_3^2 - (x_1 + x_2 + x_3)^2.$$

Show that q is positive-definite if and only if

$$\frac{1}{\alpha} + \frac{1}{\beta} + \frac{1}{\gamma} < 1.$$

24.* Let $A = [\alpha_{ik}]_{n \times n}$ be a positive-definite real matrix, so that A can be written as the partitioned matrix $\begin{bmatrix} A_{n-1} & Y \\ Y^T & \alpha_{nn} \end{bmatrix}$, where A_{n-1} is the leading $(n-1) \times (n-1)$ submatrix of A and Y is a column with $n-1$ entries. Let $P = \begin{bmatrix} I_{n-1} & -CY \\ O & 1 \end{bmatrix}$, where $C = A_{n-1}^{-1}$ (which, by 80.2 and exercise 14 above, is positive-definite). By considering the product AP, show that

$$\det A = [\alpha_{nn} - q_C(Y)] \times (\det A_{n-1}).$$

Deduce that $\det A \leqslant \alpha_{nn} \det A_{n-1}$. Deduce further the result that

$$\det A \leqslant \alpha_{11}\alpha_{22} \ldots \alpha_{nn}.$$

25. Let $f(x, y) = 2x^3 - 6xy + 3y^2$. Verify that $f(x, y)$ has a stationary point at $(1, 1)$, and find the nature of this stationary point.

APPENDIX

MAPPINGS

This appendix contains a very concise summary of what one needs to know about mappings in order to understand all the references to them in this textbook (especially from chapter 6 onwards). More details and explanations of all these matters will be found in the author's *An Introduction to Abstract Algebra* (Blackie, 1978).

Throughout this appendix S, T, U will denote non-empty sets.

1. A mapping from S to T means a rule (or idea) which associates with (or produces from) *each* element x of S a *single* element (depending on x) in the set T.

Let f denote such a mapping, and let $x \in S$. The single element of T that f associates with x is denoted by $f(x)$ and is called the **image** of x under f. If $f(x) = y$, we also say that "f maps x to y".

2. The notation

$$f : S \rightarrow T$$

is used to indicate that f is a mapping from S to T. When f is such a mapping, the set S is called the **domain** of f, and the set T is called the **codomain** of f.

3. To specify a particular mapping $f : S \rightarrow T$, one has to say what $f(x)$ is for each $x \in S$. This may be done by a general formula of the form

$$f(x) = \dots \qquad (x \in S),$$

the blank being filled by an expression for $f(x)$ in terms of x. Alternatively, the action of f may be indicated by the notation

$$x \mapsto \dots \qquad (x \in S),$$

the blank again being filled by an expression for $f(x)$ in terms of x.

4. Mappings f, g are equal to each other if and only if (1) f and g have the same domain (S, say), (2) f and g have the same codomain, *and* (3) $f(x) = g(x)$ for all $x \in S$.

5. Let $f: S \to T$ be a mapping and A a subset of S. Then $f(A)$ means the subset

$$\{f(x): x \in S\}$$

of T, i.e. the set of all things $f(x)$ can be as x varies through the set A. This subset of T is called the image of the set A under f.

An important particular case is $f(S)$, the image of the whole domain under f. This is called the **image set** of f, and may be denoted by $\operatorname{im} f$.

6. Let $f: S \to T$ be a mapping, and suppose that $U \subseteq S$. A mapping denoted by $f|_U$ and called the **restriction** of f to U as domain (or simply the restriction of f to U) is defined by

$$(f|_U)(x) = f(x) \qquad (x \in U).$$

Thus $f|_U$ has the same action as f, but acts on a smaller domain.)

7. If $f: S \to T$ is a mapping and $U \subseteq S$, then clearly $\operatorname{im}(f|_U) = f(U)$.

8. Let $f: S \to T$ and $g: T \to U$ be mappings (such that the domain of g coincides with the codomain of f). Then a mapping $g \circ f$, called the **composition** of f followed by g, is defined by

$$(g \circ f)(x) = g(f(x)) \qquad (x \in S).$$

9. If $f: S \to T$ and $g: T \to U$ are mappings and $A \subseteq S$, then

$$(g \circ f)(A) = g(f(A)).$$

10. Mapping composition is associative: i.e. if the mapping compositions $(h \circ g) \circ f$ and $h \circ (g \circ f)$ exist, they are equal.

11. A mapping $f: S \to T$ is described as **surjective** if and only if $\operatorname{im} f$ is the whole of T, i.e. if and only if, for every $y \in T$, there exists $x \in S$ such that $f(x) = y$.

12. A mapping $f: S \to T$ is **injective** if and only if different elements of S have, in all cases, different images under f, i.e. if and only if

$$f(a) = f(b) \Rightarrow a = b \qquad (a, b \in S).$$

13. A mapping f is **bijective** if and only if it is both surjective and injective.

14. The **identity mapping** of S to itself is the mapping $i_S: S \to S$ defined by $i_S(x) = x$ for every $x \in S$. Clearly, for every set S, i_S is a bijection (i.e. a bijective mapping).

15. Identity mappings behave like "ones" in composition: i.e. if $f: S \to T$ is any mapping, then $i_T \circ f = f \circ i_S = f$.

16. When $f: S \to T$ is bijective, a mapping $f^{-1}: T \to S$ (called the **inverse** of f) is defined by

$$f^{-1}(y) = \text{the (one and only) element of } S \text{ that } f \text{ maps to } y \quad (y \in T).$$

So, when $f: S \to T$ is bijective, we have:

$$f^{-1}(y) = x \Leftrightarrow f(x) = y \qquad (x \in S, y \in T).$$

17. When $f: S \to T$ is bijective,

$$f \circ f^{-1} = i_T \quad \text{and} \quad f^{-1} \circ f = i_S.$$

18. If $f: S \to T$ is a mapping and there is a mapping $g: T \to S$ such that $f \circ g = i_T$ and $g \circ f = i_S$, then f is bijective and $g = f^{-1}$. (This fact may be regarded as a converse of 17.)

ANSWERS TO EXERCISES

(Answers are given to numerical examples, excluding cases where there are several possible answers. Hints are given for other problems, excluding only the standard and straightforward.)

Chapter 1 (p. 24)

(In these solutions, **a** is used for $\mathbf{r}_A$, **b** for $\mathbf{r}_B$, etc.)

1. (*a*) $2\mathbf{x}$, (*b*) $-\frac{1}{2}\mathbf{x}$, (*c*) $\frac{1}{4}\mathbf{x}$ and $-\frac{1}{4}\mathbf{x}$. **2.** $BF/FE = 2$.

3. Let $\overrightarrow{BD}/\overrightarrow{DC}$, etc., be m/n. Prove $\frac{1}{3}(\mathbf{d}+\mathbf{e}+\mathbf{f}) = \frac{1}{3}(\mathbf{a}+\mathbf{b}+\mathbf{c})$.

4. $[M_1M_2] = [M_3M_4] = \frac{1}{2}(\mathbf{c}-\mathbf{a})$. $\mathbf{p} = \frac{1}{5}(\mathbf{a}+\mathbf{b}+\mathbf{c}+\mathbf{d})$, symmetrical in **a**, **b**, **c**, **d**.

5. (*a*) Parallelogram is rhombus if and only if its diagonals are perpendicular. (*b*) $(\mathbf{x}+\mathbf{y}).(\mathbf{x}-\mathbf{y}) = |\mathbf{x}|^2-|\mathbf{y}|^2$.

7. Find position vector of point on PQ whose position ratio with respect to P, Q is λ. This point is X when λ is chosen so that the position vector is a multiple of **b** (i.e. $\lambda = -\frac{1}{6}$); $PX/XQ = -\frac{1}{6}$; $\mathbf{r}_X = \frac{3}{5}\mathbf{b}$.

8. $AB/BC = 3$; D is $(-2, 3, -3)$. **9.** $(1, 1, 2)$.

10. $4\mathbf{i}-6\mathbf{j}+12\mathbf{k}$. **11.** $\frac{1}{4}\pi$ (or 45°).

12. Let $\mathbf{x} = \lambda\mathbf{i}+\mu\mathbf{j}+\nu\mathbf{k}$, and find that $\mathbf{x}.\mathbf{i} = \lambda$. Second part: $\mathbf{x}.\mathbf{i} = |\mathbf{x}||\mathbf{i}|\cos\alpha = |\mathbf{x}|\cos\alpha$, etc.

13. 7.

14. $\mathbf{a}.\mathbf{b}$, etc. are $\frac{1}{2}$. $[\overrightarrow{HA}].[\overrightarrow{HB}] = \frac{1}{2}(\mathbf{a}-\mathbf{b}-\mathbf{c}).\frac{1}{2}(\mathbf{b}-\mathbf{c}-\mathbf{a}) = \frac{1}{4}(\mathbf{c}^2-(\mathbf{a}-\mathbf{b})^2) = \frac{1}{4}(\mathbf{c}^2-\mathbf{a}^2-\mathbf{b}^2+2\mathbf{a}.\mathbf{b})$, etc.

15. Prove $\cos(\angle AOC) = \cos(\angle BOC) = (\mathbf{a}.\mathbf{b}+|\mathbf{a}||\mathbf{b}|)/|\mathbf{d}|$ where $\mathbf{d} = |\mathbf{b}|\mathbf{a}+|\mathbf{a}|\mathbf{b}$.

16. (i) In non-degenerate cases, construct parallelogram with diagonal representing **z** and sides parallel to **x**, **y**. (ii) Point described has position vector equal to $-(\alpha/(\beta+\gamma))\mathbf{a}$ and so lies on OA as well as on BC. $BD/DC = \gamma/\beta$, etc. $OA/AD = -(\beta+\gamma)$, etc.

Chapter 2 (p. 49)

(In these solutions α_{ik}, β_{ik} are used for (i, k)th entries of matrices A, B.)

1. $2, -1, 2, 0$, respectively.

2. $(2, 1)$th and $(2, 3)$th entries should be $4, -5$, respectively.

3. $A+B = (A+B)^2 = A+AB+BA+B$. Second part: $ABA = A(BA) = A(-AB) = -A^2B = -AB$, and similarly $ABA = (AB)A = -BA$.

4. $A = \frac{1}{2}((A+B)+(A-B)) = \frac{1}{2}((A+B)^2+(A-B)^2)$, etc. $B^2 = A-A^2 \Rightarrow B^2A = AB^2 = A^2-A^3$. Last part: take $x = 3, y = 1$.

5. $\operatorname{tr}(AB) = \operatorname{tr}(BA) = \sum_{i=1}^{n}\left(\sum_{j=1}^{n}\alpha_{ij}\beta_{ji}\right)$. $\operatorname{tr}(AB-BA) = \operatorname{tr}(AB)-\operatorname{tr}(BA) = 0$.

6. Let $C = \operatorname{diag}(\alpha_1, \ldots)$, $D = \operatorname{diag}(\delta_1, \ldots)$. Equality of (i, k)th entries of C^2A and $AD^2 \Rightarrow \gamma_i^2\alpha_{ik} = \delta_k^2\alpha_{ik} \Rightarrow \alpha_{ik}(\gamma_i-\delta_k)(\gamma_i+\delta_k) = 0 \Rightarrow \alpha_{ik}(\gamma_i-\delta_k) = 0$, since $\gamma_i+\delta_k > 0$; etc.

7. $\alpha_{ii} = \alpha_{kk}$ and all other entries of ith row, kth column of A zero.

8. Can be nicely done using the matrix units, with 17.6 and fact that $A = \sum\alpha_{ik}E_{ik}$. In (ii) and its generalization, note that nonzero entries retreat farther towards bottom left corner as successive powers are formed.

9. (i) $(AB)^T = AB \Leftrightarrow B^TA^T = AB$, etc. (ii) Show $(AB-BA)^T = -(AB-BA)$.

10. $(X^TSX)^T = X^TS^TX$, etc. And see remark (*c*) in §18.

11. Suppose A, B orthogonal. (i) $(AB)^TAB = I = AB(AB)^T$; (ii) $(A^{-1})^TA^{-1} = (A^T)^TA^{-1} = I = A^{-1}(A^{-1})^T$.

12. One way: use given facts to prove $ABA = -B^T$.

13. (i) Note that $(PBP^{-1})^m = PBP^{-1}PBP^{-1}PBP^{-1}\ldots = PB^mP^{-1}$.

(ii) $A^m = \begin{bmatrix} 3.2^m-2 & -2(2^m-1) \\ 3(2^m-1) & 3-2^{m+1} \end{bmatrix}$.

14. $A^{-1} = -(I+\frac{1}{4}A)$. One way for last part: observe $A+I = -\frac{1}{4}A^2$, and use 20.4, 20.2.

15. Pre- and postmultiply $(I+A)A = A(I+A)$ by $(I+A)^{-1}$.

16. $(\alpha I_n+\beta J)^{-1} = \dfrac{1}{\alpha}\left(I_n - \dfrac{\beta}{\alpha+n\beta}J\right)$. $\frac{1}{9}\begin{bmatrix} 7 & -2 & -2 & -2 \\ -2 & 7 & -2 & -2 \\ -2 & -2 & 7 & -2 \\ -2 & -2 & -2 & 7 \end{bmatrix}$.

17. (i) If A nonsingular, $A^{-1}AB = A^{-1}A$. (ii, iii) From $AB = A$, show AB^2, AB^3, etc. all equal A. (iv) Simplify ABA as $A(BA)$ and as $(AB)A$.

18. $P^TAP = B$ (symmetric) $\Rightarrow A = (P^T)^{-1}BP^{-1} \Rightarrow A^T = \ldots.$

19. If $A^r = O$ and A^{-1} exists, $O = (A^{-1})^{r-1}A^r = A$—contradiction.

20. $I = (A+B)(A^{-1}+B^{-1}) = 2I+C+C^{-1}$. $C+C^{-1} = -I \Rightarrow C^2 = -C-I$ and $C^3 = -C^2-C$, etc.

21. (i) True. (ii) False: consider $A = -B = I$. (iii) True. (iv) True. (v) False: consider $A = \begin{bmatrix} 1 & 1 \\ 1 & 1 \end{bmatrix}, B = \begin{bmatrix} 1 & 0 \\ 0 & -1 \end{bmatrix}$.

22. $A^3 = O \Rightarrow A^2 = tA$ and $O = A^3 = tA^2 \Rightarrow A^2 = O$ or $t = O$, etc.

23. From $MM^{-1} = I$, obtain $X = -AZ$ and $I = BX+CZ = B(-AZ)+CZ = (C-BA)Z$; consider also $M^{-1}M$. $Y = -VB$, $X = -AV$, $W = I+AVB$, where $V = (C-BA)^{-1}$.

24. $\begin{bmatrix} 1 & n \\ 0 & 1 \end{bmatrix}$; $\begin{bmatrix} 1 & n & 0 \\ 0 & 1 & 0 \\ 0 & 0 & 3^n \end{bmatrix}$.

Chapter 3 (p. 77)

2. (i) $\begin{bmatrix} 1 & 0 & 0 & 1 \\ 0 & 1 & 0 & -1 \\ 0 & 0 & 1 & 1 \\ 0 & 0 & 0 & 0 \end{bmatrix}$;

(ii) $\begin{bmatrix} 1 & 0 & 0 & -1 & 0 \\ 0 & 1 & 0 & -\frac{1}{2} & 0 \\ 0 & 0 & 1 & \frac{1}{2} & 0 \\ 0 & 0 & 0 & 0 & 1 \end{bmatrix}$.

Reduced echelon matrices obtained from matrices in 1 are:

(i) $\begin{bmatrix} 1 & 1 & 0 & 0 \\ 0 & 0 & 1 & 0 \\ 0 & 0 & 0 & 1 \end{bmatrix}$, (ii) $\begin{bmatrix} 0 & 1 & 0 & -3 \\ 0 & 0 & 1 & 1 \\ 0 & 0 & 0 & 0 \end{bmatrix}$,

(iii) $\begin{bmatrix} 1 & 0 & 0 & -\frac{3}{2} \\ 0 & 1 & 0 & -\frac{5}{2} \\ 0 & 0 & 1 & 2 \\ 0 & 0 & 0 & 0 \end{bmatrix}$, (iv) $\begin{bmatrix} 1 & 0 & 3 & 0 \\ 0 & 1 & 2 & 0 \\ 0 & 0 & 0 & 1 \\ 0 & 0 & 0 & 0 \end{bmatrix}$.

3. $Q_1 = \begin{bmatrix} 0 & 1 & 0 \\ 1 & 0 & 0 \\ 0 & 8 & 2 \end{bmatrix}$, $Q_2 = \begin{bmatrix} 0 & 1 & 0 \\ 1 & 0 & 0 \\ 4 & 0 & 2 \end{bmatrix}$.

4. (ii) is singular; the others have inverses as follows.

(i) $\frac{1}{2}\begin{bmatrix} 2 & -2 & -2 \\ -1 & 2 & 1 \\ 1 & 0 & 1 \end{bmatrix}$. (iii) $\frac{1}{3}\begin{bmatrix} 5 & -1 & -6 \\ -2 & 1 & 3 \\ -7 & 2 & 6 \end{bmatrix}$.

(iv) $\frac{1}{3}\begin{bmatrix} -1 & 1 & 2 \\ -7 & 4 & 8 \\ -4 & 1 & 5 \end{bmatrix}$. (v) $\frac{1}{2}\begin{bmatrix} -5 & -2 & 3 & -6 \\ -6 & -2 & 4 & -8 \\ 18 & 6 & -10 & 22 \\ 31 & 10 & -17 & 36 \end{bmatrix}$.

6. $ABAB = I \Rightarrow A$ nonsingular with $A^{-1} = BAB \Rightarrow (BAB)A = I$.
7. $(I-A)(I-B) = I \Rightarrow (I-B)(I-A) = I$.
8. Last part: $(I-AB)C = -A$; hence $C = -(I-AB)^{-1}A = -(I-CB)A$.

9. $\begin{bmatrix} 1 & 0 \\ 3 & 1 \end{bmatrix}\begin{bmatrix} 1 & 0 \\ 0 & -2 \end{bmatrix}\begin{bmatrix} 1 & 2 \\ 0 & 1 \end{bmatrix}$ is one answer.

10. Express Q as a product of elementary matrices and consider the corresponding e.r.o.s.

11. $k = 4$. $x_1 = 2-3\alpha+2\beta$, $x_2 = 1+\alpha-\beta$, $x_3 = \beta$, $x_4 = \alpha$.

12. (i) $x_1 = \frac{1}{2}(3-3\alpha-\beta-\gamma)$, $x_2 = \frac{1}{2}(1+\alpha-\beta+\gamma)$, $x_3 = \gamma$, $x_4 = \beta$, $x_5 = \alpha$.
(ii) $x_1 = 1+\alpha+\gamma$, $x_2 = 2+\beta+\gamma-2\delta$, $x_3 = \delta$, $x_4 = \gamma$, $x_5 = 3$, $x_6 = 4-2\alpha$, $x_7 = \beta$, $x_8 = \alpha$.

13. $x_1 = 1$, $x_2 = 1$, $x_3 = -1$.

14. Consider 26.2 and 28.3.

Chapter 4 (p. 92)

1. (i) 11, (ii) -65, (iii) 44, (iv) 15, (v) -92.
2. Second determinant $= (a-b)(b-c)(c-a)(a+b+c+abc)$.
3. $R_2 \to R_2 + R_3$ is useful.
4. If $A \in \mathbb{R}_{n\times n}$, $A = -A^T \Rightarrow \det A = \det(A^T) = \det(-A) = (-1)^n \det A$.
5. Second part: show $tI - B = P^{-1}(tI-A)P$.

6. (i) A orthogonal $\Rightarrow 1 = \det(A^T A) = (\det(A^T))(\det A) = (\det A)^2$. (ii) By 32.6 and 31.1, $\det(A^T(A+B)B^T) = -\det(A+B)$; but $A^T(A+B)B^T = (A+B)^T$; etc. (iii) $A = \operatorname{diag}(2, \frac{1}{2})$, $B = \operatorname{diag}(-\frac{1}{2}, 2)$.

8. $MN = \begin{bmatrix} A & O \\ O & I \end{bmatrix}\begin{bmatrix} I & O \\ O & B \end{bmatrix}$. Take determinants and use the multiplicative property. Note (e.g.) that $\begin{bmatrix} A & O \\ O & I \end{bmatrix}$ has determinant equal to $\det A$: consider repeated expansion by last row or column.

10. $(\det A)(\det(\operatorname{adj} A)) = \det(A(\operatorname{adj} A)) = (\det A)^n$ (by 33.3), etc. For second part, $\operatorname{adj}(\operatorname{adj} A) = \det(\operatorname{adj} A) \times (\operatorname{adj} A)^{-1}$; and 33.3 shows what $(\operatorname{adj} A)^{-1}$ is.

11. Work from $x_i = (1/\Delta) \times$ (ith entry of $(\operatorname{adj} A)K$).

Chapter 5 (p. 130)

2. Only (i) is a subspace.

4. (Pursuing hint) $\mathbf{x}+\mathbf{y} \in S \cup T$; so $\mathbf{x}+\mathbf{y} \in S$ or $\mathbf{x}+\mathbf{y} \in T$. If former, $(\mathbf{x}+\mathbf{y})-\mathbf{x} \in S$ (cf. 37.2); etc.

5. Use 38.3.

6. $\sin^4 x = \frac{3}{8} - \frac{1}{2}\cos 2x + \frac{1}{8}\cos 4x$ (for all $x \in \mathbb{R}$).

7. Clear when AX is rewritten in accordance with 21.4.

8. (Pursuing hint) Let m be the maximum of the degrees. Then every polynomial in W, being a linear combination of the fs, has degree $\leqslant m$—an obvious contradiction.

9. (i) $1(\mathbf{w}+\mathbf{x})+(-1)(\mathbf{x}+\mathbf{y})+1(\mathbf{y}+\mathbf{z})+(-1)(\mathbf{z}+\mathbf{w}) = \mathbf{0}$. (ii) Standard application of 39.2.

10. If $\alpha = -1$, then $\mathbf{u}+\mathbf{v}+\mathbf{w} = \mathbf{0}$. Prove $(\mathbf{u}, \mathbf{v}, \mathbf{w})$ L.I. in the remaining case by a standard application of 39.2.

11. (i) $(\mathbf{i}, \mathbf{j}, \mathbf{i}+\mathbf{j})$. (ii) $(\mathbf{i}, 2\mathbf{i}, \mathbf{j})$. (iii) $(\mathbf{0}, \mathbf{i}, \mathbf{j})$.

12. Suppose $\lambda\mathbf{y} + \sum_{j=2}^{n} \mu_j \mathbf{x}_j = \mathbf{0}$. Since $(\mathbf{x}_1, \ldots, \mathbf{x}_n)$ L.I., $\lambda\alpha_1 = 0 = \lambda\alpha_j + \mu_j$ for each j in the range 2 to n. Since $\alpha_1 \neq 0$, $\lambda = 0$. Etc.

13. Standard application of 39.2: premultiply by A^{-1} at appropriate stage.

14. $((-1,1,0,0), (-1,0,1,0), (-1,0,0,1))$ is a basis; dimension is 3.

15. (One answer) $(\operatorname{col}(-1,0,0,1), \operatorname{col}(1,2,3,0))$.

16. $\dim S = 3$; the sequence is L.D. One basis of $S \cap T$ is $((1,-1,0,0), (0,3,1,-4))$.

17. In the first part, $\dim(C(A)) = 2$. Second part: if A is a scalar multiple of I, $C(A)$ is the whole of $\mathbb{R}_{n\times n}$; otherwise $\operatorname{sp}(I, A)$ is a 2-dimensional subspace of $C(A)$.

18. No: the sequence spans a 3-dimensional subspace.

19. In $\mathbb{R}_{2\times 2}$, bases are, respectively, $(E_{11}, E_{22}, E_{12}+E_{21})$, $(E_{12}-E_{21})$, (E_{11}, E_{22}). In $\mathbb{R}_{n\times n}$, the dimensions are, respectively, $\frac{1}{2}n(n+1)$, $\frac{1}{2}n(n-1)$, n.

20. Suppose $\alpha_0 X + \alpha_1 AX + \alpha_2 A^2 X + \ldots + \alpha_{p-1}A^{p-1}X = O$. Premultiply this equation by A^{p-1} to deduce $\alpha_0 = 0$; then premultiply the equation by A^{p-2} to deduce $\alpha_1 = 0$; etc. $p \leqslant n$ follows from 40.8(ii).

21. 2 if $a = 0$, 3 if $a = -1$, 4 otherwise.

22. Follows from 40.2. **23.** $\operatorname{col}(2, 1, -4)$.

25. Apply 40.4 to the subspace $\operatorname{sp}(\mathbf{x}_1, \ldots, \mathbf{x}_n)$, whose dimension is $\leqslant n$.

26. For each positive integer n, the sequences which are zero after the nth entry form a subspace of dimension n. So, if F^∞ were f.d., its dimension would be $\geqslant n$ for every positive integer n.

27. $S \cap \text{sp}(\mathbf{x})$ is a subspace of $\text{sp}(\mathbf{x})$ and not the whole of it; hence $S \cap \text{sp}(\mathbf{x})$ has dimension < 1, by 41.4(iii).

28. $(\mathbf{x}_1, \mathbf{x}_2)$ is a basis of S. Express $\mathbf{y}_1, \mathbf{y}_2$ in terms of $\mathbf{x}_1, \mathbf{x}_2$, and hence prove $(\mathbf{y}_1, \mathbf{y}_2)$ L.I. $\dim U = 1$, and so $U \neq S$.

29. $S \cap (T+U)$ is a subspace containing $S \cap T$ and $S \cap U$: now use 42.1 (iii). Consider $S = \text{sp}(\mathbf{i})$, $T = \text{sp}(\mathbf{j})$, $U = \text{sp}(\mathbf{i}+\mathbf{j})$.

30. Use 41.5: for a start, $S \cap T = S + T$.

31. Use 42.3 to get $\dim T = \dim U$. This, with (3), gives $T = U$ by 41.5. Last part: same example as in 29.

32. $\dim(S \cap T) = 2$ or 3.

33. $S+T$ strictly contains S and so has strictly bigger dimension.

34. One method: note that $T = S + \text{sp}(\mathbf{x}_{n+1})$.

35. For each $A \in \mathbb{R}_{n \times n}$, the one and only way to express A as the sum of a matrix is S and one in T is $\frac{1}{2}(A+A^T)+\frac{1}{2}(A-A^T)$.

36. Use the fact that $(\mathbf{e}_1, \ldots, \mathbf{e}_n)$ spans V to prove $S+T$ is the whole of V. Use 43.2 and the fact that $(\mathbf{e}_1, \ldots, \mathbf{e}_n)$ is L.I. to prove the sum direct.

37. Let $(\mathbf{e}_1, \ldots, \mathbf{e}_s)$ be a basis of S; extend to a basis $(\mathbf{e}_1, \ldots, \mathbf{e}_n)$ of V; take T to be $\text{sp}(\mathbf{e}_{s+1}, \ldots, \mathbf{e}_n)$. For $\dim(T_1 \cap T_2) \geqslant \ldots$, use 42.3 and the fact that $\dim(T_i) = \dim V - \dim S$.

Chapter 6 (p. 161)

2. (i) $(2x_1+x_2,\ -x_1+3x_2,\ -4x_2+5x_3,\ 4x_1+7x_2+2x_3)$. (ii) $(3x_1-5x_2+3x_3,\ x_1-2x_3,\ -x_2+2x_3)$. (iii) Yes. (iv) No.

3. Apply 21.3.

4. $a(x_1, x_2) = (0, x_1)$.

5. im a has basis $((1,1,2))$; ker a has basis $((-1,0,1), (-2,1,0))$. im b has basis $((1,1,1),(0,1,-1))$; ker b has basis $((5,2,-3))$.

6. a is surjective because the arbitrary vector $\mathbf{y} \in V$ is the image under a of $b\mathbf{x}$. So a^{-1} exists, by 49.5, and thus $ba = a^{-1}(ab)a$, etc.

7. c is surjective but not injective; d is injective but not surjective; $cd = i_V$, which is nonsingular, while c is singular and $dc \neq i_V$.

8. The hint lets it be seen that $aV + \ker a$ is the whole of V. Proof of directness of the sum is a special case of the "$\ker a = \ker a^2 \Rightarrow aV \cap \ker a = \{\mathbf{0}\}$" step in worked example in §49.

9. $r(a) = r(a^2)$ gives $aV = a^2V$ (since, in any event, $a^2V \subseteq aV$). So, for arbitrary $\mathbf{x} \in V$, $a\mathbf{x} = a^2\mathbf{y}$ (for some $\mathbf{y}$); now use $a^3\mathbf{y} = a^2\mathbf{y}$.

10. In any event, $\ker c \subseteq \ker(bc)$; equality follows from $r(c) = r(bc)$, 49.2 and 41.5. To prove $cV \cap \ker b$ zero, start with arbitrary $\mathbf{x}$ in $cV \cap \ker b$: $\mathbf{x} = c\mathbf{y}$ for some $\mathbf{y}$ and $b\mathbf{x} = \mathbf{0} \Rightarrow \mathbf{y} \in \ker(bc) = \ker c$, etc. In the next part, use dimension to get $cV \oplus \ker b$ equal to whole of V. For the converse, 48.3 is useful.

11. Observe that if equality occurs at any point in the descending chain, it occurs for ever thereafter. So, since a^nV is strictly smaller than $a^{n-1}V$, the dimensions of $V, aV, a^2V, \ldots, a^nV$ must form a strictly descending sequence; the only possibility is $\dim(a^jV) = n-j$ $(j = 1, \ldots, n)$.

12. The columns of B form a L.D. sequence; and hence so do the columns of the first matrix to right of the vertical partition; hence so do all the columns of the first matrix (cf. 39.7).

13. (i) False. (ii) True. (iii) True.

14. $b = -1: x_1 = 8+5\alpha+2\beta, x_2 = 3+2\alpha+\beta, x_3 = \beta, x_4 = \alpha$.

15. No. In both parts, watch the tricky case $c = 1$.

16. Take $b = -a \neq O$ and $S \neq \{\mathbf{0}\}$.

17. (i) Use 52.7: $r(a^k) \geqslant kr(a)-(k-1)(\dim V)$. (ii) By matrix version, $r(A^4) \geqslant 4r(A) -3 \times 10$.

18. (i) $a = (a+b)+(-1)b$, and hence $r(a+b) \geqslant r(a)-r(b)$ by 52.7 and 52.4. (ii) $r(AB+CD) \geqslant r(AB)-r(CD)$, by part (i); $r(AB) \geqslant 5$ by 52.7M, $r(CD) \leqslant 3$ by 52.2M. (iii) Use matrix version of 17(i) to get $r(P+Q) \leqslant \frac{2}{3}n$: so $\frac{2}{3}n \geqslant r(P)-r(Q) = n-r(Q)$, P being nonsingular.

19. If they existed, 52.2M would give $r(A) \geqslant m$ and $r(A) \geqslant n$, contradicting 52.1M since $m \neq n$.

20. $s(a|_{bV}) = r(b)-r(ab)$ by 49.2; now use 49.3. Last part: $(bab)\mathbf{x} = b\mathbf{x}$ can be deduced from $(abab)\mathbf{x} = (ab)\mathbf{x}$ since $a|_{bV}$ is injective.

21. Key step is: $r(BA) \geqslant r(A)+r(B)-n \geqslant 2r(AB)-n$ (by 52.7M, 52.2M, respectively).

22. Apply 49.2 to $a|_W$. Second part: do as instructed and note that $\ker a \cap a^2V \subseteq \ker a \cap aV$ (since $a^2V \subseteq aV$). Last part: use middle part in conjunction with 17(i).

23. Because $a+b$ nonsingular, $r(a) = r(a(a+b)) = r((a+b)b) = r(b)$; and $\dim V = r(a+b) \leqslant r(a)+r(b)$ (52.8M).

24. Use 52.2 to get $r((cb)^2) = r(bc)$. In last paragraph, $r(cb) \geqslant n-1$ comes from 52.2 and $r((cb)^2) = r(bc) = n-1$. One way to get $r(cb) \leqslant n-1$ is via $r((cb)^2) \geqslant 2r(cb)-n$.

25. (i) is clear from examination of the detail of the proof of 52.8. (ii) By part (i) with a replaced by i_V+a, b by i_V-a, and $a+b$, therefore, by $2i_V$ (which has rank n), we obtain $(i_V+a)V \cap (i_V-a)V = \{\mathbf{0}\}$. Let $\mathbf{x}$ be arbitrary in V. Since $(i_V-a^2)\mathbf{x} = (i_V+a)[(i_V-a)\mathbf{x}]$, $(i_V-a^2)\mathbf{x} \in (i_V+a)V$. Similarly $(i_V-a^2)\mathbf{x} \in (i_V-a)V$. Hence $(i_V-a^2)\mathbf{x} \in \{\mathbf{0}\}$, i.e. $\mathbf{x}-a^2\mathbf{x} = \mathbf{0}$, i.e. $a^2\mathbf{x} = \mathbf{x}$.

26. In the descending chain $V \supseteq aV \supseteq a^2V \supseteq a^3V \supseteq \dots$ there cannot be more than n strict containments where $n = \dim V$. So $a^jV = a^{j+1}V$ for some positive integer j; and applying a repeatedly to both sides gives $a^{j+1}V = a^{j+2}V$, $a^{j+2}V = a^{j+3}V$, etc. (i) From 49.2, it follows that $\dim[\ker(a^j)] = \dim[\ker(a^{j+1})] = \dim[\ker(a^{j+2})] = \dots$. Hence, since $\ker(a^j) \subseteq \ker(a^{j+1}) \subseteq \ker(a^{j+2}) \subseteq \dots$, $\ker(a^j) = \ker(a^{j+1}) = \ker(a^{j+2}) = \dots$. (ii) Let $\mathbf{x} \in a^jV \cap \ker(a^j)$. Then $\mathbf{x} = a^j\mathbf{y}$ for some $\mathbf{y} \in V$, and $\mathbf{0} = a^j\mathbf{x} = a^{2j}\mathbf{y}$. Hence $\mathbf{y} \in \ker(a^{2j}) = \ker(a^j)$, so that $\mathbf{0} = a^j\mathbf{y} = \mathbf{x}$. It follows that the sum $a^jV+\ker(a^j)$ is a direct sum, and consideration of its dimension shows that it is the whole of V

Chapter 7 (p. 181)

1. (i) $2\mathbf{f}_1+5\mathbf{f}_2+8\mathbf{f}_3+11\mathbf{f}_4$. (ii) $\mathbf{f}_1+7\mathbf{f}_2+13\mathbf{f}_3+19\mathbf{f}_4$.

2. (i) Get columns of B by considering what $b\mathbf{i}$ and $b\mathbf{j}$ are. Clearly $b^2 = i$ (the identity transformation of E_2), and correspondingly $B^2 = I$ (cf. §56). (ii) For every $\mathbf{x} \in E_2$, $c\mathbf{x} = \frac{1}{2}(b+i)\mathbf{x}$ by (6.2): so $c = \frac{1}{2}(b+i)$; and correspondingly $C = \frac{1}{2}(B+I)$. Clearly $c^2 = c$, and so, correspondingly, $C^2 = C$. (iii) Consider (e.g.) BA, where A is the matrix of reflection in the x-axis.

3. $\begin{bmatrix} 1 & 3 & -1 \\ 2 & 0 & 1 \end{bmatrix}$.

4. (i) $\begin{bmatrix} 1 & -1 & 2 \\ 1 & 0 & -1 \end{bmatrix}$, (ii) $\begin{bmatrix} 1 & -\frac{1}{2} & \frac{1}{2} \\ 0 & -\frac{1}{2} & \frac{3}{2} \end{bmatrix}$,

(iii) $\begin{bmatrix} 0 & 1 & 3 \\ 1 & -1 & 0 \end{bmatrix}$, (iv) $\begin{bmatrix} \frac{1}{2} & 0 & \frac{3}{2} \\ -\frac{1}{2} & 1 & \frac{3}{2} \end{bmatrix}$.

5. $M(a;L)$ is $A = \begin{bmatrix} 1 & 1 \\ 0 & 1 \end{bmatrix}$. Easy to show that $A^n - I = n(A-I)$; correspondingly (cf. §56) $a^n - i = n(a-i)$. $a^n(x_1, x_2)$ is

$$((1-6n)x_1 + 9nx_2, -4nx_1 + (1+6n)x_2).$$

6. $BA = A^{-1}(AB)A$.

7. Take L to be $(\mathbf{e}_1, \ldots, \mathbf{e}_r, \mathbf{e}_{r+1}, \ldots, \mathbf{e}_n)$, where $(\mathbf{e}_1, \ldots, \mathbf{e}_r)$ is a basis of aV and $(\mathbf{e}_{r+1}, \ldots, \mathbf{e}_n)$ is a basis of ker a. Then, by 43.4(i), L is a basis of V. From detail of L, $M(a;L)(=A$, say) has the form $\begin{bmatrix} A_1 & O \\ O & O \end{bmatrix}$; and $r(A_1) = r(A) = r(a) = r$. Take b to be the linear transformation such that $M(b;L) = \begin{bmatrix} A_1^{-1} & O \\ O & O \end{bmatrix}$.

8. (i) Note that, in exercise 7, a idempotent $\Rightarrow A_1^2 = A_1 \Rightarrow A_1 = I$, ($A_1$ being nonsingular). (ii) Use the fact that $(1/\lambda)b$ is indempotent. (iii) Show $(c - \beta i_V)^2 = (\alpha - \beta)(c - \beta i_V)$, and apply part (ii). Last part: from their matrices $c - \alpha i_V$ and $c - \beta i_V$ have ranks equal to the numbers of βs and αs, respectively, in matrix of c.

9. $r(a) = 1$ from 52.7; so dim (ker a) $= 2$; sp$(\mathbf{e}_1)$ + ker a is V because strictly bigger than ker a; for directness of sum, cf. exercise 27 on chapter 5. Let $\mathbf{e}_2 = a\mathbf{e}_1$ (a nonzero vector in ker a); extend $(\mathbf{e}_2)$ to a basis $(\mathbf{e}_2, \mathbf{e}_3)$ of ker a; then $L = (\mathbf{e}_1, \mathbf{e}_2, \mathbf{e}_3)$ is a basis of V, and $M(a;L)$ is the matrix stated. Particular case: one answer is

$$((1,0,0),(1,2,1),(1,0,-1)).$$

10. By 59.3, $A = BK_rC$ (B, C nonsingular). Take $Q = C^{-1}K_r^TB^{-1}$. (Here $K_r = (K_r)_{m \times n}$, and $K_r^T = (K_r)_{n \times m}$.)

11. (ii) Prove θ linear and injective; surjectivity follows from considering dimensions.·

12. Suppose $a^r = O$ and $M(a;L)$ is a diagonal matrix D (for some basis L of V). Then $D^r = O$ (cf. §56); hence $D = O$, and so a must be O.

13. For arbitrary $\mathbf{x} \in V$, $b\mathbf{x} = a(b\mathbf{x}) \in aV$; hence $bV \subseteq aV$, and equality follows from $r(a) = r(b)$. Next, for arbitrary $\mathbf{x} \in V$, $a\mathbf{x} = b\mathbf{y}$ for some $\mathbf{y} \in V$; and so $a^2\mathbf{x} = ab\mathbf{y} = b\mathbf{y} = a\mathbf{x}$. There is basis L w.r.t. the idempotent A has matrix K_r, where $r = r(a) = r(b)$. From $ab = b$, $M(b;L)$ partitioned after its rth row has the form $\begin{bmatrix} B_1 \\ O \end{bmatrix}$, where the rows of B_1 form a L.I. sequence (since $r(b) = r$). There is a nonsingular matrix C whose top r rows form submatrix equal to B_1: take c to be the linear transformation whose matrix w.r.t. L is C.

14. First check that f is linear. f is surjective because $A + B$ is the whole of V. Further, for $\mathbf{x} \in A$, $\mathbf{y} \in B$, $f(\mathbf{x}, \mathbf{y}) = \mathbf{0} \Rightarrow \mathbf{x} + \mathbf{y} = \mathbf{0} \Rightarrow \mathbf{y} = -\mathbf{x} \in A \cap B \Rightarrow \mathbf{x} = \mathbf{y} = \mathbf{0} \Rightarrow (\mathbf{x}, \mathbf{y}) = \mathbf{0}_W$, and hence f is injective.

Chapter 8 (p. 204)

1. (i) $(t+3)(t-2)(t-3)$, (ii) $-3, 2, 3$, (iii) col$(7, 2, -10)$, col$(1, 1, 0)$, col$(2, 1, 1)$, respectively.

2. Only eigenvalue is 2; (col (0, 1, 1)).

3. -4 and 2; $(1, -1)$ and $(1, 1)$, respectively.

4. Only eigenvalue is 3; $((1, -1))$.

5. Take z-axis along axis of rotation; then consider matrix of the rotation w.r.t. $(\mathbf{i}, \mathbf{j}, \mathbf{k})$.

6. For arbitrary $\alpha \in \mathbb{C}$, $(1, \alpha, \alpha^2, \alpha^3, \alpha^4, \ldots)$ is an eigenvector of c corresponding to eigenvalue α; but $d\mathbf{x} = \alpha\mathbf{x} \Rightarrow \mathbf{x} = \mathbf{0}$.

7. Take $X = -A$, $Y = tI_n$.

8. $AX = \lambda X \Rightarrow A^2X = A(AX) = A(\lambda X) = \lambda(AX) = \lambda(\lambda X) = \lambda^2 X$; etc.

9. (i) λ an eigenvalue of $A \Rightarrow \lambda^s$ an eigenvalue of $I \Rightarrow \lambda^s = 1$. (ii) If $A^s = O$, then λ an eigenvalue of $A \Rightarrow \lambda^s$ an eigenvalue of $O \Rightarrow \lambda^s = 0 \Rightarrow \lambda = 0$.

10. If $\operatorname{tr}(A) = n$, then (by 64.3(i)) every eigenvalue is $+1$, so that $\chi_A(-1) \neq 0$; hence $\det(I + A) \neq 0$, and $A = I$ is deduced by premultiplying the equation $(A + I)(A - I) = O$ by $(A + I)^{-1}$.

11. $\det(tI_n - A^{-1}) = \det\left[(-tA^{-1})\left(\frac{1}{t}I_n - A\right)\right]$. Last part: note that the reciprocal of a root of unity equals its complex conjugate, and use 64.3(i).

12. (i) $((1, -1), (1, 1))$; (ii) $((i, 1), (1, i))$; in the case of b, geometric multiplicity of the eigenvalue 3 is only 1, less than its algebraic multiplicity.

13. $\begin{bmatrix} 7 & 1 & 2 \\ 2 & 1 & 1 \\ -10 & 0 & 1 \end{bmatrix}$.

14. Suppose matrix A has λ as sole eigenvalue and that A is diagonable. By 66.2, $P^{-1}AP = \lambda I$ for some nonsingular matrix P. Hence $A = P(\lambda I)P^{-1} = \lambda I$.

15. A_1 is diagonable. (For example) $P = \begin{bmatrix} 1 & 0 & 1 \\ 0 & 1 & -1 \\ 2 & 2 & -1 \end{bmatrix}$.

16. $\begin{bmatrix} 1 & 0 & 1 \\ -1 & 1 & 2 \\ 0 & -1 & -2 \end{bmatrix}$; (i) $\begin{bmatrix} 2 & 2 & 2 \\ 2 & 2 & 1 \\ -2 & -2 & -1 \end{bmatrix}$;

(ii) $\begin{bmatrix} 2^{2n} & 2^{2n} & 2^{2n} \\ 2^{2n+1}-2 & 2^{2n+1}-2 & 2^{2n+1}-3 \\ 2-2^{2n+1} & 2-2^{2n+1} & 3-2^{2n+1} \end{bmatrix}$.

17. Suppose A similar to the diagonal matrix D (and therefore A^2 similar to D^2). Then $r(A) = r(A^2) = n - m$ = number of nonzero entries on main diagonal of D. Last part: $\begin{bmatrix} 0 & 0 \\ 1 & 0 \end{bmatrix}$.

19. Note that $S^* = -S$; let λ be an eigenvalue of S, and follow the pattern of the proof of 68.1 to show that $\bar{\lambda} = -\lambda$. Because -1 is not an eigenvalue, $\chi_S(-1) \neq 0$; and hence $\det(I + S) \neq 0$. Last part: use 20.3 and, if necessary, the fact that $I + S$ and $I - S$ commute.

20. Use 64.3(ii) along with 68.2 and the following remark (*a*).

21. That $SX = 1X$ follows from the 3 identities such as $a = b\cos C + c\cos B$ (easily seen by drawing the altitude through A). The quadratic equation is: $t^2 + t$

$+2\cos A\cos B\cos C = 0$. The inequality follows because (by 68.1) this quadratic has real roots.

22. For first part, observe that $X^TA^TAX = (AX)^T(AX)$, and use 67.4(ii). $\ker(m_{A^TA}) \subseteq \ker(m_A)$ follows through noting that $A^TAX = O \Rightarrow X^TA^TAX = O \Rightarrow AX = O$; then $r(A^TA) = r(A)$ follows by 49.2. To deal with $r(AA^T)$, replace A by A^T(cf. 50.2). Last part: consider $r((BC)^T(BC))$ and use 52.2M.

23. The complex analogue is $r(A^*A) = r(A)$.

24. Where X is a corresponding real eigenvector, $\lambda(X^TX) = X^T(\lambda X) = X^T(I+A^TA)X = X^TX+(AX)^T(AX) \geqslant X^TX$; and $X^TX > O$, since $X \neq O$.

Chapter 9 (p. 224)

1. (*a*) 9, (*b*) 6.

2. (i) $\|\lambda\mathbf{x}+\mathbf{y}\|^2 = (\lambda\mathbf{x}+\mathbf{y}).(\lambda\mathbf{x}+\mathbf{y}) = a\lambda^2+b\lambda+c$. Since this quadratic in λ is never negative, $b^2-4ac \leqslant 0$—an inequality which simplifies to (*). (ii) For nonzero $\mathbf{x}$, $\mathbf{y}$, (*) shows that the number $t = \dfrac{\mathbf{x}.\mathbf{y}}{\|\mathbf{x}\|\,\|\mathbf{y}\|}$ lies in the interval $[-1, 1]$, so that the angle between $\mathbf{x}$ and $\mathbf{y}$ may be defined as $\cos^{-1}t$. (iii) $(\|\mathbf{x}\|+\|\mathbf{y}\|)^2-\|\mathbf{x}+\mathbf{y}\|^2 = (\|\mathbf{x}\|+\|\mathbf{y}\|)^2-(\mathbf{x}+\mathbf{y}).(\mathbf{x}+\mathbf{y}) = 2(\|\mathbf{x}\|\,\|\mathbf{y}\|-\mathbf{x}.\mathbf{y}) \geqslant 0$, by (*), since the real number $\mathbf{x}.\mathbf{y}$ is either equal to or less than its modulus. Hence $(\|\mathbf{x}+\mathbf{y}\|^2 \leqslant (\|\mathbf{x}\|+\|\mathbf{y}\|)^2$, and 70.2 now follows.

3. $\displaystyle\int_a^b f(x)g(x)\,dx \leqslant \left[\left(\int_a^b (f(x))^2\,dx\right)\left(\int_a^b (g(x))^2\,dx\right)\right]^{1/2}.$

4. One answer: $\left(\frac{1}{7}(2,3,6), \frac{1}{7\sqrt{5}}(15,-2,-4), \frac{1}{\sqrt{5}}(0,2,-1)\right)$.

5. One answer: $\left(\frac{1}{\sqrt{2}}(1,-1,0), \frac{1}{\sqrt{3}}(1,1,-1)\right)$.

6. One answer: $(1, 3^{1/2}(2x-1), 5^{1/2}(6x^2-6x+1))$.

7. One answer: $\dfrac{1}{\sqrt{110}}\begin{bmatrix}\sqrt{10} & 10 & 0\\ \sqrt{10} & -1 & 3\sqrt{11}\\ 3\sqrt{10} & -3 & -\sqrt{11}\end{bmatrix}$.

8. $\sqrt{n}$.

9. (The "$\Rightarrow$" half) Suppose $(\mathbf{e}_1,\ldots,\mathbf{e}_{n-1},\mathbf{x})$ is an orthonormal basis; and let $\mathbf{x} = \sum_{i=1}^{n}\lambda_i\mathbf{e}_i$. Prove $\lambda_1,\ldots,\lambda_{n-1}$ all zero, then $\lambda_n = \pm 1$.

10. Let $(\mathbf{e}_1,\ldots,\mathbf{e}_r)$ be an orthonormal basis of S; extend to an orthonormal basis $(\mathbf{e}_1,\ldots,\mathbf{e}_n)$ of V; show that $S^\perp = \operatorname{sp}(\mathbf{e}_{r+1},\ldots,\mathbf{e}_n)$.

11. (i) $\frac{1}{\sqrt{5}}\begin{bmatrix}1 & 2\\ -2 & 1\end{bmatrix}$; (ii) $\frac{1}{\sqrt{6}}\begin{bmatrix}\sqrt{3} & \sqrt{2} & 1\\ 0 & -\sqrt{2} & 2\\ -\sqrt{3} & \sqrt{2} & 1\end{bmatrix}$; (iii) $\frac{1}{3}\begin{bmatrix}-2 & 2 & 1\\ 1 & 2 & -2\\ 2 & 1 & 2\end{bmatrix}$.

12. If there were such a matrix, it would be similar to a real diagonal matrix D such that $D^3 = I$.

13. (i) Appeal to 63.3 or 64.3(ii). (ii) Take $Q = B^{-1}$, where B is the positive-definite symmetric matrix produced in the worked example. Introduce an orthogonal matrix R such that $R^T(Q^TSQ)R$ is diagonal; and take $P = QR$.

14. Let the eigenvalues of A be $\alpha_1,\ldots,\alpha_k,\beta_{k+1},\ldots,\beta_n$, where the αs are nonnegative and the βs are negative. For some orthogonal matrix P, $P^TAP = \operatorname{diag}(\alpha_1,\ldots,\alpha_k,$

$\beta_{k+1},\ldots,\beta_n) = D_1 - D_2$, where $D_1 = \text{diag}(\alpha_1,\ldots,\alpha_k,0,\ldots,0)$ and $D_2 = \text{diag}(0,\ldots,0,|\beta_{k+1}|,\ldots,|\beta_n|)$. Hence $A = P(D_1 - D_2)P^T = B - C$, where $B = PD_1P^T$ and $C = PD_2P^T$, which satisfy all the stated conditions. For the matrix A of 11(ii), an expression for A as a difference of the kind in question is

$$A = \tfrac{1}{6}\begin{bmatrix} 2 & 4 & 2 \\ 4 & 8 & 4 \\ 2 & 4 & 2 \end{bmatrix} - \tfrac{1}{6}\begin{bmatrix} 14 & -2 & -10 \\ -2 & 2 & -2 \\ -10 & -2 & 14 \end{bmatrix}.$$

15. M^TM is symmetric by 18.4. To prove M^TM positive-definite, introduce an arbitrary eigenvalue λ of M^TM and a corresponding eigenvector X, and prove $\lambda > 0$ by considering $X^T(M^TMX) = (MX)^T(MX) = \|MX\|^2 > 0$ (X being nonzero and M nonsingular). MB^{-1} is orthogonal because $(MB^{-1})^T(MB^{-1}) = B^{-1}M^TMB^{-1}$ (B being symmetric) $= B^{-1}(B^2)B^{-1} = I$. Finally, observe simply that $M = (MB^{-1})B$.

Chapter 10 (p. 254)

1. $3x_1^2 + 5x_3^2 + 4x_1x_2 - 2x_1x_3 + 8x_2x_3$.

2. $\begin{bmatrix} 1 & 2 & 3 & \tfrac{1}{2} \\ 2 & 0 & 0 & -1 \\ 3 & 0 & 3 & 5 \\ \tfrac{1}{2} & -1 & 5 & -1 \end{bmatrix}$.

3. For all $X \in F_{n\times 1}$, $X^TBX = X^TAX$ where A is the symmetric matrix $\tfrac{1}{2}(B + B^T)$. That there is only one possibility for A is clear from 74.2.

4. $M(q_A;\hat{L}) = \begin{bmatrix} 10 & 15 \\ 15 & 23 \end{bmatrix}$. Method (a): put $x_1 = 2\hat{x}_1 + 3\hat{x}_2$, $x_2 = \hat{x}_1 + 2\hat{x}_2$ in general expression for $q_A(x_1, x_2)$. Method (b): work out P^TAP where $P = \begin{bmatrix} 2 & 3 \\ 1 & 2 \end{bmatrix}$.

5. Ranks: (i) 2, (ii) 3, (iii) 2, (iv) 3, (v) 3, (vi) 4, (vii) 4, (viii) 3, (ix) 8. Signatures: (i) 2, (ii) 1, (iii) 0, (iv) 1, (v) 1, (vi) 0, (vii) -2, (viii) 1, (ix) 0. In (ix) the initial step of putting $x_1 = u+v$, $x_2 = u-v$ and completing the square twice to absorb all references to u and v gives $q(X) \equiv \hat{x}_1^2 - \hat{x}_2^2 + (x_3x_4 + x_4x_5 + \ldots + x_9x_{10} + x_{10}x_3) -$ at which point the resemblance of the bracketed residual expression to the original should be exploited to complete the diagonalization with minimal further effort.

6. In each case an answer is obtained by taking the columns of the coefficient matrix P of the equations giving the original variables $x_1, x_2, \ldots$ in terms of the variables $\hat{x}_1, \hat{x}_2, \ldots$ in a diagonal version. Notice that, in the standard diagonalization process, the equations that come immediately are those giving $\hat{x}_1, \hat{x}_2, \ldots$ in terms of the original variables: these equations require to be inverted to give us the matrix P. In each part there are many possible answers. The answers produced by the most obvious procedures are: (i) (col(1, 0), col(3, 1)); (ii) (col(1, 0, 0), col(2, 1, 0), col(1, 1, 1)); (iii) ((col(1, 0, 0), col(-1, 1, 0), col(1, 2, 1)).

7. Rank 2, signature 0. (An obvious change of variable gives the version y_1y_2; then put $y_1 = u+v$, $y_2 = u-v$.)

8. (a) The standard rational reduction procedure gives $q(X) \equiv \hat{x}_1^2 + 2\hat{x}_2^2 + (k-9)\hat{x}_3^2$, where $\hat{x}_1 = x_1 + x_2 - x_3$, $\hat{x}_2 = x_2 + 2x_3$, $\hat{x}_3 = x_3$. (b) If $k < 9$, $q(X) < 0$ when

$\hat{x}_1 = \hat{x}_2 = 0$, $\hat{x}_3 = 1$, i.e. $x_1 = 3$, $x_2 = -2$, $x_3 = 1$. (This is not the only possible answer.)

9. Ranks: (*a*) 3, (*b*) 3, (*c*) 2. Signatures: (*a*) -1, (*b*) 1, (*c*) 0.

10. Suppose $\alpha > 0$ and $\beta^2 < \alpha\gamma$, and let $A = \begin{bmatrix} \alpha & \beta \\ \beta & \gamma \end{bmatrix}$. It is easy to show (e.g. by completion of the square) that q_A has rank 2, signature 2. So (cf. invariance of rank and signature) orthogonal diagonalization ($\equiv$ change of coordinates) turns the given equation into $\lambda_1\hat{x}^2 + \lambda_2\hat{y}^2 = 1$ with λ_1, λ_2 (the eigenvalues of A) both positive. So equation defines ellipse. Area enclosed $= \pi/\sqrt{(\lambda_1\lambda_2)} = \pi/\sqrt{(\det A)}$.

11. Eigenvalues of matrix of left-hand side are 1, 4, 7 (all positive). So equation defines ellipsoid, with axes of lengths 2, 1, and $2/\sqrt{7}$.

12. (*a*) $k > 1$, (*b*) $k \geqslant 1$, (*c*) no values of k, (*d*) $k = 0$, (*e*) $k < 1$ but $\neq 0$.

13. (i) If A, B are positive-definite real $n \times n$ matrices, then $A+B$ is symmetric and, for all nonzero $X \in \mathbb{R}_{n\times 1}$, $X^T(A+B)X = X^TAX + X^TBX > 0$. (ii) Yes.

14. Note that A^{-1} is also symmetric. Method (*a*): if A has eigenvalues $\alpha_1, \ldots, \alpha_n$ (all positive), then A^{-1} has eigenvalues $1/\alpha_1, \ldots, 1/\alpha_n$ (cf. exercise 11 on chapter 8), and these too are all positive. Method (*b*): the change of variable $X = A\hat{X}$ transforms $X^TA^{-1}X$ into $\hat{X}^TA\hat{X}$.

15. Note that, for all $\mu \in \mathbb{R}$, $\mu I + A$ is symmetric. Further, if A has eigenvalues $\alpha_1, \ldots, \alpha_n$, then $\mu I + A$ has eigenvalues $\mu + \alpha_1, \ldots, \mu + \alpha_n$, which are all positive for sufficiently large μ.

16. (i) A^TA is symmetric; and, for every real column X of the relevant size, $X^T(A^TA)X = (AX)^T(AX) = \|AX\|^2 \geqslant 0$. (ii) Let A be nonsingular. By part (i) A^TA is positive-semidefinite. Further: $q_{A^TA}(X) = 0 \Rightarrow \|AX\| = 0 \Rightarrow AX = O \Rightarrow X = O$ (A being nonsingular).

17. Simply following the procedure considered in the proof of 79.5 (reducing $q_A(X)$ to $y_1^2 + y_2^2$, then orthogonally diagonalizing the resulting version of q_B) produces the answer $x_1 = c(\hat{x}_1 + 7\hat{x}_2)$, $x_2 = c(\hat{x}_1 - 3\hat{x}_2)$, where $c = 1/\sqrt{10}$. Omission of the c factors gives an alternative (simpler) answer.

18. By 79.5 there is a change of variable yielding $q_{AB}(X) \equiv \gamma_1\hat{x}_1^2 + \ldots + \gamma_n\hat{x}_n^2$, where $\gamma_1, \ldots, \gamma_n$ are the roots of $\det(tA - AB) = 0$, i.e. $(\det A)\det(tI - B) = 0$. Hence $\gamma_1, \ldots, \gamma_n$ are the eigenvalues of B, which are positive (B being positive-definite). Therefore q_{AB} is positive-definite.

19. By 79.5, for some nonsingular P, $P^TAP = I$ and $P^TBP = \operatorname{diag}(\gamma_1, \ldots, \gamma_n)$ for certain positive numbers $\gamma_1, \ldots, \gamma_n$; and $(\det P)^2 \det(A+B) = \det[P^T(A+B)P] = \det[\operatorname{diag}(1+\gamma_1, \ldots, 1+\gamma_n)] = (1+\gamma_1)(1+\gamma_2)\ldots(1+\gamma_n) > 1 + \gamma_1\gamma_2\ldots\gamma_n = \det(P^TAP) + \det(P^TBP) = (\det P)^2(\det A + \det B)$.

20. Suppose $B_1^2 = B_2^2 = A$, where A, B_1, B_2 are all positive-definite. (The problem is to prove that B_1, B_2 must then be equal.) By 73.1 there are orthogonal matrices, P, Q such that P^TB_1P and Q^TB_2Q are diagonal matrices—C and D, respectively, say—the entries on whose main diagonals are all positive (B_1, B_2 being positive-definite). $B_1^2 = B_2^2$ gives $(PCP^T)^2 = (QDQ^T)^2$, i.e. $PC^2P^T = QD^2Q^T$, i.e. $C^2P^TQ = P^TQD^2$. By exercise 6 on chapter 2, $CP^TQ = P^TQD$, i.e. $PCP^T = QDQ^T$, i.e. $B_1 = B_2$.

21. No. Consider, for example, $\begin{bmatrix} 0 & 1 \\ 1 & 1 \end{bmatrix}$.

22. The leading minors of the matrix of q are α, $2\alpha^2$, and $6\alpha(\alpha^2 - \beta^2)$, which are all positive precisely if $\alpha > |\beta|$. (Note that this condition implies $\alpha > 0$.)

23. The leading minors of the matrix of q can be written as

$$\alpha\left(1-\frac{1}{\alpha}\right), \quad \alpha\beta\left[1-\left(\frac{1}{\alpha}+\frac{1}{\beta}\right)\right], \quad \alpha\beta\gamma\left[1-\left(\frac{1}{\alpha}+\frac{1}{\beta}+\frac{1}{\gamma}\right)\right].$$

24. $AP = \begin{bmatrix} A_{n-1} & O \\ Y^T & \alpha_{nn}-q_C(Y) \end{bmatrix}$. Hence (via expansion by the final column) $\det(AP) = [\alpha_{nn}-q_C(Y)]\det A_{n-1}$. The first result follows since $\det(AP) = (\det A)(\det P) = (\det A)\times 1$. Next observe that $\alpha_{nn}-q_C(Y) \leqslant \alpha_{nn}$ since C is positive-definite, and $\det A_{n-1} > 0$ since A_{n-1} is positive-definite: hence $\det A \leqslant \alpha_{nn}\det A_{n-1}$. Now make repeated application of this to obtain the final result: e.g. $\det A_{n-1} \leqslant \alpha_{n-1,n-1}\det A_{n-2}$, etc.

25. There is a stationary point at $(1, 1)$ since both f_x and f_y vanish there. At this stationary point $f_{xx} = 12$, $f_{yy} = 6$, $f_{xy} = -6$, so that both f_{xx} and $f_{xx}f_{yy}-f_{xy}^2 > 0$: therefore the stationary point is a local minimum.

Index

Printed and bound by CPI Group (UK) Ltd, Croydon, CR0 4YY
17/10/2024
01775686-0010